AF443413

Encyclopaedia of Mathematical Sciences
Volume 130

Invariant Theory and Algebraic Transformation Groups

Subseries Editors:
R.V. Gamkrelidze   V.L. Popov

Springer
Berlin
Heidelberg
New York
Barcelona
Hong Kong
London
Milan
Paris
Tokyo

Harm Derksen Gregor Kemper

Computational Invariant Theory

 Springer

Harm Derksen
University of Michigan
Department of Mathematics
East Hall
525 East University
48109-1109 Ann Arbor, MI
USA
e-mail: hderksen@umich.edu

Gregor Kemper
University of Heidelberg
Institute for Scientific Computing
Im Neuenheimer Feld 368
69120 Heidelberg
Germany
e-mail: Gregor.Kemper@iwr.uni-heidelberg.de

Founding editor of the Encyclopedia of Mathematical Sciences:
R. V. Gamkrelidze

Mathematics Subject Classification (2000):
Primary: 13A50; secondary: 13H10, 13P10

ISSN 0938-0396
ISBN 3-540-43476-3 Springer-Verlag Berlin Heidelberg New York

Springer-Verlag Berlin Heidelberg New York
a member of BertelsmannSpringer Science+Business Media GmbH
http://www.springer.de
© Springer-Verlag Berlin Heidelberg 2002
Printed in Germany

Typeset by the authors using a Springer TEX macro package
Cover Design: E. Kirchner, Heidelberg, Germany
Printed on acid-free paper   SPIN: 10865151   46/3142 db   5 4 3 2 1 0

To Maureen, William, Claire

To Elisabeth, Martin, Stefan

Preface

Invariant theory is a subject with a long tradition and an astounding ability to rejuvenate itself whenever it reappears on the mathematical stage. Throughout the history of invariant theory, two features of it have always been at the center of attention: computation and applications. This book is about the computational aspects of invariant theory. We present algorithms for calculating the invariant ring of a group that is linearly reductive or finite, including the modular case. These algorithms form the central pillars around which the book is built. To prepare the ground for the algorithms, we present Gröbner basis methods and some general theory of invariants. Moreover, the algorithms and their behavior depend heavily on structural properties of the invariant ring to be computed. Large parts of the book are devoted to studying such properties. Finally, most of the applications of invariant theory depend on the ability to calculate invariant rings. The last chapter of this book provides a sample of applications inside and outside of mathematics.

Acknowledgments. Vladimir Popov and Bernd Sturmfels brought us together as a team of authors. In early 1999 Vladimir Popov asked us to write a contribution on algorithmic invariant theory for Springer's Encyclopaedia series. After we agreed to do that, it was an invitation by Bernd Sturmfels to spend two weeks together in Berkeley that really got us started on this book project. We thank Bernd for his strong encouragement and very helpful advice. During the stay at Berkeley, we started outlining the book, making decisions about notation, etc. After that, we worked separately and communicated by e-mail. Most of the work was done at MIT, Queen's University at Kingston, Ontario, Canada, the University of Heidelberg, and the University of Michigan at Ann Arbor. In early 2001 we spent another week together at Queen's University, where we finalized most of the book. Our thanks go to Eddy Campbell, Ian Hughes, and David Wehlau for inviting us to Queen's.

The book benefited greatly from numerous comments, suggestions, and corrections we received from a number of people who read a pre-circulated version. Among these people are Karin Gatermann, Steven Gilbert, Julia Hartmann, Gerhard Hiß, Jürgen Klüners, Hanspeter Kraft, Martin Lorenz, Kay Magaard, Gunter Malle, B. Heinrich Matzat, Vladimir Popov, Jim Shank, Bernd Sturmfels, Nicolas Thiéry, David Wehlau, and Jerzy Weyman.

We owe them many thanks for working through the manuscript and offering their expertise. The first author likes to thank the National Science Foundation for partial support under the grant 0102193. Last but not least, we are grateful to the anonymous referees for further valuable comments and to Ms. Ruth Allewelt and Dr. Martin Peters at Springer-Verlag for the swift and efficient handling of the manuscript.

Ann Arbor and Heidelberg,   *Harm Derksen*
March 2002   *Gregor Kemper*

Table of Contents

Introduction

"Like the Arabian phoenix rising out of the ashes, the theory of invariants, pronounced dead at the turn of the century, is once again at the forefront of mathematics. During its long eclipse, the language of modern algebra was developed, a sharp tool now at last being applied to the very purpose for which is was invented." (Kung and Rota [157])

A brief history. Invariant theory is a mathematical discipline with a long tradition, going back at least one hundred and fifty years. Sometimes its has blossomed, sometimes it has lain dormant. But through all phases of its existence, invariant theory has had a significant computational component. Indeed, the period of "Classical Invariant Theory", in the late 1800s, was championed by true masters of computation like Aronhold, Clebsch, Gordan, Cayley, Sylvester, and Cremona. This classical period culminated with two landmark papers by Hilbert. In the first [107], he showed that invariant rings of the classical groups are finitely generated. His non-constructive proof was harshly criticized by Gordan (see page 49 in this book). Hilbert replied in the second paper [108] by giving constructive methods for finding all invariants under the special and general linear group. Hilbert's papers closed the chapter of Classical Invariant Theory and sent this line of research into a nearly dormant state for some decades, but they also sparked the development of commutative algebra and algebraic geometry. Indeed, Hilbert's papers on invariant theory [107, 108] contain such fundamental results as the Nullstellensatz, the Basis Theorem, the rationality of what is now called the Hilbert series, and the Syzygy Theorem. The rise of algebraic geometry and commutative algebra had a strong influence on invariant theory—which never really went to sleep—as might be best documented by the books by Mumford et al. [169] (whose first edition was published in 1965) and Kraft [152].

The advent in the 1960s and 1970s of computational methods based on Gröbner bases[1] brought a decisive turn. These methods initiated the development of computational commutative algebra as a new field of research, and consequently they revived invariant theory. In fact, new algorithms and fast computers make many calculations now feasible that in the classical period

---

[1] It may be surprising that Gröbner bases themselves came much earlier. They appeared in an 1899 paper of Gordan [95], where he re-proved Hilbert's finiteness theorem for invariant rings.

were either simply impossible or carried a prohibitive cost. Furthermore, a heightened interest in modulo p questions led to a strong activity in modular invariant theory. An important role in boosting interest in computational invariant theory was also played by Sturmfels's book "Algorithms in Invariant Theory" [239]. Two other books (Benson [18] and Smith [225]) and numerous research articles on invariant theory have appeared recently, all evidence of a field in ferment.

Aims of this book. This book focuses on algorithmic methods in invariant theory. A central topic is the question how to find a generating set for the invariant ring. We deal with this question in the case of finite groups and linearly reductive groups. In the case of finite groups, we emphasize the modular case, in which the characteristic of the ground field divides the group order. In this case, many interesting theoretical questions in invariant theory of finite groups are still open, and new phenomena tend to occur. The scope of this book is not limited to the discussion of algorithms. A recurrent theme in invariant theory is the investigation of structural properties of invariant rings and their links with properties of the corresponding linear groups. In this book, we consider primarily the properties of invariant rings that are susceptible to algorithmic computation (such as the depth) or are of high relevance to the behavior and feasibility of algorithms (such as degree bounds). We often consider the geometric "incarnation" of invariants and examine, for example, the question of separating orbits by invariants. In addition, this book has a chapter on applications of invariant theory to several mathematical and non-mathematical fields. Although we are non-experts in most of the fields of application, we feel that it is important and hope it is worthwhile to include as much as we can from the applications side, since invariant theory, as much as it is a discipline of its own, has always been driven by what it was used for. Moreover, it is specifically the computational aspect of invariant theory that lends itself to applications particularly well.

Other books. Several books on invariant theory have appeared in the past twenty-five years, such as Springer [231], Kraft [152], Kraft et al. [153], Popov [193], Sturmfels [239], Benson [18], Popov and Vinberg [194], Smith [225], and Goodman and Wallach [93]. A new book by Neusel and Smith [181] has just arrived straight off the press. We hope that our book will serve as a useful addition to its predecessors. Our choice of material differs in several ways from that for previous books. In particular, of the books mentioned, Sturmfels's is the only one that strongly emphasizes algorithms and computation. Several points distinguish our book from Sturmfels [239]. First of all, this book is appearing nine years later, enabling us to include many new developments such as the first author's algorithm for computing invariant rings of linearly reductive groups and new results on degree bounds. Moreover, the modular case of invariant theory receives a fair amount of our attention in this book. Of the other books mentioned, only Benson [18], Smith [225], and

Neusel and Smith [181] have given this case a systematic treatment. On the other hand, Sturmfels's book [239] covers many aspects of Classical Invariant Theory and brings them together with modern algorithms. In contrast, our book touches only occasionally on Classical Invariant Theory. It is probably fair to say that most of the material covered in Chapters 3 and 4 (the core chapters of this book) has never appeared in a book before.

Readership. The intended readership of this book includes postgraduate students as well as researchers in geometry, computer algebra, and, of course, invariant theory. The methods used in this book come from different areas of algebra, such as algebraic geometry, (computational) commutative algebra, group and representation theory, Lie theory, and homological algebra. This diversity entails some unevenness in the knowledge that we assume on the readers' part. We have nevertheless tried to smooth out the bumps, so a good general knowledge of algebra should suffice to understand almost all of the text. The book contains many examples and explicit calculations that we hope are instructive. Generally, we aim to maximize the benefits of this book to readers. We hope that it, or at least parts of it, can also be used as a basis for seminars.

Proofs. When writing this book, we had to decide which proofs of particular statements to include or omit. Our primary consideration was whether a proof is, in our view, instructive. Of course, other factors also had some weight, such as the length of a proof, its novelty, its availability elsewhere in the literature, the importance of the result, and its relevance to computational matters. Some degree of arbitrariness is probably unavoidable in such decisions, but we do hope that our choices contribute to the readability of the book. When proofs are omitted, we give references.

Organization of the book. Most of the algorithms presented in this book rely in one way or another on Gröbner basis methods. Therefore we decided to devote the first chapter of this book to introducing Gröbner bases and methods in constructive ideal theory that are built on them. Since most of the material is also covered in several other books (see the references at the beginning of Chapter 1), we considered it justifiable and appropriate to give a concise presentation almost completely "unburdened" by proofs. The aim is to give the reader a quick overview of the relevant techniques. We cover most of the standard applications of Gröbner bases to ideal theory, such as the computation of elimination ideals, intersections, ideal quotients, dimension, syzygy modules and resolutions, radical ideals, and Hilbert series. In the section on radical calculation, we present a new algorithm that works in positive characteristic. Our treatment in Section 1.6 of de Jong's normalization algorithm goes beyond the material found in the standard texts. We believe that this algorithm has not previously appeared in a monograph. For this reason, we have decided to give full proofs in Section 1.6.

The second chapter gives a general introduction into invariant theory. The goal is to acquaint the reader with the basic objects and problems and, perhaps most important, to specify the notation. The presentation is enriched with many examples. In this chapter we aim to set the stage for later developments. In particular, Sections 2.4 through 2.6 are written with applications to Chapters 3 and 4 in mind. In Section 2.5.2, we present a proof of the Hochster-Roberts Theorem that is based on the concept of tight closure. Section 2.3.2 is devoted to separating invariants, a subject rarely or never mentioned in books on invariant theory. Here we go back to one of the original purposes for which invariant theory was invented and ask whether a subset of the invariant ring might have the same properties of separating group orbits as the full invariant ring, even if the subset may not generate the invariant ring. As it turns out, it is always possible to find a finite set with this property, even though the invariant ring itself may not be finitely generated (see Theorem 2.3.15). This result seems to be new.

Chapters 3 and 4 form the core of the book. In Chapter 3 we look at invariants of finite groups. Here the modular case, in which the characteristic of the ground field divides the group order, is included and indeed emphasized. The main goal of the chapter is to present algorithms for finding a finite set of generators of the invariant ring. As the reader will discover, these algorithms are much more cumbersome in the modular case. The importance of having algorithms for this case lies mainly in the fact that modular invariant theory is a field with many interesting problems that remain unsolved. Therefore it is crucial to be able explore the terrain by using computation. The main algorithms for computing generators and determining properties of invariant rings are presented in Sections 3.1 through 3.7. Many of the algorithms were developed by the second author. In Sections 3.10 and 3.11, we discuss methods applicable to special situations and ad hoc methods. A number of not strictly computational issues are addressed in Chapter 3, notably degree bounds. We present a recent proof found by Benson, Fleischmann, and Fogarty for the Noether bound that extends to the case of positive characteristic not dividing the group order, which was left open by Noether's original argument. In Section 3.9.3 we give a (very large) general degree bound for the modular case that depends only on the group order and the dimension of the representation. Such a bound has not appeared in the literature before. In Section 3.9.4 we revisit the topic of separating subalgebras and show that the Noether bound always holds for separating invariants even when it fails for generating invariants.

The fourth chapter is devoted to invariants of linearly reductive groups. We present a general algorithm for computing a finite set of generating invariants, which was found by the first author. This algorithm makes use of the Reynolds operator, which is studied systematically in Section 4.5. In Section 4.6 we discuss how the Hilbert series of the invariant ring can be calculated by using an integral similar to Molien's formula. As for finite groups, degree bounds are also an important issue in the case of reductive groups. In Section 4.7 we discuss an improvement of a degree bound given by Popov. An important special case of reductive groups are tori. In Section 4.3 we present a new algorithm for computing generating invariants of tori.

In Chapter 5 we embark on a tour of several applications of invariant theory. We start with applications to different areas in algebra. Here we discuss the computation of cohomology rings of finite groups, solving systems of algebraic equations with symmetries, the determination of Galois groups, and the construction of generic polynomials via a positive solution of Noether's problem. Then we move on to other mathematical disciplines. We address applications to graph theory, combinatorics, coding theory, and dynamical systems. Finally, we look at examples from computer vision and material science in which invariant theory can be a useful tool. This chapter is incomplete in (at least) three ways. First, the scope of fields where invariant theory is applied is much bigger than the selection that we present here. We aim to present applications that we consider to be typical and that represent a certain bandwidth. Second, we are non-experts in most of the fields addressed in this chapter. Therefore certain inaccuracies are unavoidable in our presentation, and many experts will probably find that we missed their favorite article on the subject. We apologize in advance and ask readers to bring such shortcomings to our attention. Third, we very intentionally limit ourselves to giving a short presentation of a few selected topics and examples for each field of application. We want to convey to the reader more a taste of the subject matter than a comprehensive treatment. So Chapter 5 is meant to operate a bit like a space probe originating from our home planet (algebra) and traveling outward through the solar system, visiting some planets and skipping others, and taking snapshots along the way.

Finally, the book has an appendix where we have compiled some standard facts about algebraic groups. The material of the appendix is not a prerequisite for every part of the book. In fact, the appendix is needed primarily for the second half of Chapter 4.

1 Constructive Ideal Theory

In this chapter we will provide the basic algorithmic tools which will be used in later chapters. More precisely, we introduce some algorithms of constructive ideal theory, almost all of which are based on Gröbner bases. As the reader will find out, these algorithms and thus Gröbner bases literally permeate this book. When Sturmfels' book [239] was published, not much introductory literature on Gröbner bases and their applications was available. In contrast, we now have the books by Becker and Weispfenning [15], Adams and Loustaunau [6], Cox et al. [48], Vasconcelos [250], Cox et al. [49], Kreuzer and Robbiano [155], and a chapter from Eisenbud [59]. This list of references could be continued further. We will draw heavily on these sources and restrict ourselves to giving a rather short overview of the part of the theory that we require. The algorithms introduced in Sections 1.1–1.3 of this chapter have efficient implementations in various computer algebra systems, such as Co-CoA [40], MACAULAY (2) [97], MAGMA [24], or SINGULAR [99], to name just a few, rather specialized ones. The normalization algorithm explained in Section 1.6 is implemented in MACAULAY and SINGULAR.

We will be looking at ideals $I \subseteq K[x_1, \ldots, x_n]$ in a polynomial ring over a field K. For polynomials $f_1, \ldots, f_k \in K[x_1, \ldots, x_n]$, the ideal generated by the f_i will be denoted by $(f_1, \ldots, f_k)K[x_1, \ldots, x_n]$ or by $(f_1, \ldots, f_k)$ if no misunderstanding can arise. The algorithms in this chapter will be mostly about questions in algebraic geometry, so let us introduce some basic notation. An **affine variety** is a subset X of the n-dimensional affine space $\mathbb{A}^n = \mathbb{A}^n(K) := K^n$ defined by a set $S \subseteq K[x_1, \ldots, x_n]$ of polynomials as

$$X = \mathcal{V}(S) := \{(\xi_1, \ldots, \xi_n) \in K^n \mid f(\xi_1, \ldots, \xi_n) = 0 \text{ for all } f \in S\}.$$

When we talk about varieties, we usually assume that K is algebraically closed. (Otherwise, we could work in the language of schemes.) The **Zariski topology** on $\mathbb{A}^n$ is defined by taking the affine varieties as closed sets. An affine variety (or any other subset of $\mathbb{A}^n$) inherits the Zariski topology from $\mathbb{A}^n$. A non-empty affine variety X is called **irreducible** if it is not the union of two non-empty, closed proper subsets. (In the literature varieties are often defined to be irreducible, but we do not make this assumption here.) The (Krull-) **dimension** of X is the maximal length k of a strictly increasing chain

$$X_0 \subsetneq X_1 \subsetneq \cdots \subsetneq X_k \subseteq X$$

of irreducible closed subsets.

For an affine variety $X = \mathcal{V}(S)$, let I be the radical ideal of the ideal in $K[x_1, \ldots, x_n]$ generated by S. Then $X = \mathcal{V}(I)$, and the quotient ring $K[X] := K[x_1, \ldots, x_n]/I$ is called the **coordinate ring**. X is irreducible if and only if $K[X]$ is an integral domain, and the dimension of X equals the Krull dimension of $K[X]$, i.e., the maximal length of a strictly increasing chain of prime ideals in $K[X]$. By Hilbert's Nullstellensatz, we can identify $K[X]$ with a subset of the ring K^X of functions from X into K. Elements from $K[X]$ are called **regular functions** on X. If X and Y are affine varieties, a **morphism** $\varphi\colon X \to Y$ is a mapping from X into Y such that the image of the induced mapping

$$\varphi^*\colon K[Y] \to K^X, \quad f \mapsto f \circ \varphi,$$

lies in $K[X]$.

1.1 Ideals and Gröbner Bases

In this section we introduce the basic machinery of monomial orderings and Gröbner bases.

1.1.1 Monomial Orderings

By a **monomial** in $K[x_1, \ldots, x_n]$ we understand an element of the form $x_1^{e_1} \cdots x_n^{e_n}$ with e_i non-negative integers. Let M be the set of all monomials. A **term** is an expression $c \cdot t$ with $0 \neq c \in K$ and $t \in M$. Thus every polynomial is a sum of terms.

Definition 1.1.1. *A **monomial ordering** is a total order ">" on M satisfying the following conditions:*

(i) $t > 1$ for all $t \in M \setminus \{1\}$,
(ii) $t_1 > t_2$ implies $st_1 > st_2$ for all $s, t_1, t_2 \in M$.

We also use a monomial ordering to compare terms. A non-zero polynomial $f \in K[x_1, \ldots, x_n]$ can be written uniquely as $f = ct + g$ such that $t \in M$, $c \in K \setminus \{0\}$, and every term of g is smaller (with respect to the order ">") than t. Then we write

$$\mathrm{LT}(f) = ct, \quad \mathrm{LM}(f) = t, \quad \textit{and} \quad \mathrm{LC}(f) = c$$

*for the **leading term**, **leading monomial**, and **leading coefficient** of f. For $f = 0$, all three values are defined to be zero.*

A monomial ordering is always a well-ordering. This follows from the fact that ideals in $K[x_1, \ldots, x_n]$ are finitely generated. We note that the usage of terminology is not uniform in the literature. Some authors (e.g. Becker and Weispfenning [15]) have monomials and terms interchanged, and some speak of initial or head terms, monomials and coefficients. Monomial orderings are often called term orders. When browsing through the literature one can find almost any combination of these pieces of terminology.

Example 1.1.2. We give a few examples of monomial orderings. Let $t = x_1^{e_1} \cdots x_n^{e_n}$ and $t' = x_1^{e_1'} \cdots x_n^{e_n'}$ be two distinct monomials.

(a) The lexicographic monomial ordering (with $x_1 > x_2 > \cdots > x_n$): t is considered greater than t' if $e_i > e_i'$ for the smallest i with $e_i \neq e_i'$. We sometimes write $t >_{\text{lex}} t'$ in this case. As an example, we have

$$\mathrm{LM}_{\text{lex}}(x_1 + x_2 x_4 + x_3^2) = x_1.$$

The lexicographic monomial ordering is useful for solving systems of algebraic equations.

(b) The graded lexicographic monomial ordering: $t >_{\text{glex}} t'$ if $\deg(t) > \deg(t')$, or if $\deg(t) = \deg(t')$ and $t >_{\text{lex}} t'$. Here $\deg(t)$ is the total degree $e_1 + \cdots + e_n$. For example,

$$\mathrm{LM}_{\text{glex}}(x_1 + x_2 x_4 + x_3^2) = x_2 x_4.$$

The graded lexicographic monomial ordering can be generalized by using a weighted degree $\deg(t) := w_1 e_1 + \cdots + w_n e_n$ with w_i fixed positive real numbers.

(c) The graded reverse lexicographic monomial ordering (grevlex-ordering for short): $t >_{\text{grevlex}} t'$ if $\deg(t) > \deg(t')$, or if $\deg(t) = \deg(t')$ and $e_i < e_i'$ for the *largest* i with $e_i \neq e_i'$. For example,

$$\mathrm{LM}_{\text{grevlex}}(x_1 + x_2 x_4 + x_3^2) = x_3^2.$$

The grevlex ordering is often very efficient for computations. It can also be generalized by using a weighted degree.

(d) Block orderings: Let $>_1$ be a monomial ordering on the monomials in $x_1, \ldots, x_r$, and $>_2$ a monomial ordering on the monomials in $x_{r+1}, \ldots, x_n$. Then the block ordering formed from $>_1$ and $>_2$ is defined as follows: $t > t'$ if $x_1^{e_1} \cdots x_r^{e_r} >_1 x_1^{e_1'} \cdots x_r^{e_r'}$, or if $x_1^{e_1} \cdots x_r^{e_r} = x_1^{e_1'} \cdots x_r^{e_r'}$ and $x_{r+1}^{e_{r+1}} \cdots x_n^{e_n} >_2 x_{r+1}^{e_{r+1}'} \cdots x_n^{e_n'}$. For example, the lexicographic monomial ordering is a block ordering. Block orderings are useful for the computation of elimination ideals (see Section 1.2). ◁

We say that a monomial ordering is **graded** if $\deg(t) > \deg(t')$ implies $t > t'$. So the orderings in (b) and (c) of the previous example are graded.

Given a monomial ordering, we write $x_i \gg x_j$ if $x_i > x_j^e$ for all non-negative integers e. For example, in the lexicographic monomial ordering we have $x_1 \gg x_2 \gg \cdots \gg x_n$. Moreover, if ">" is a block ordering with blocks $x_1, \ldots, x_r$ and $x_{r+1}, \ldots, x_n$, then $x_i \gg x_j$ for $i \leq r$ and $j > r$. If $x_i \gg x_j$ for all $j \in J$ for some $J \subset \{1, \ldots, n\}$, then x_i is greater than any monomial in the indeterminates x_j, $j \in J$. This follows directly from Definition 1.1.1.

1.1.2 Gröbner Bases

We fix a monomial ordering on $K[x_1, \ldots, x_n]$.

Definition 1.1.3. *Let $S \subseteq K[x_1, \ldots, x_n]$ be a set of polynomials. We write*

$$L(S) = (\mathrm{LM}(g) \mid g \in S)$$

for the ideal generated by the leading monomials from S. $L(S)$ is called the **leading ideal** *of S (by some authors also called the initial ideal).*

Let $I \subseteq K[x_1, \ldots, x_n]$ be an ideal. Then a finite subset $\mathcal{G} \subseteq I$ is called a **Gröbner basis** *of I (with respect to the chosen monomial ordering) if*

$$L(I) = L(\mathcal{G}).$$

It is clear that a Gröbner basis of I generates I as an ideal. Indeed, a (hypothetical) element $f \in I \setminus (\mathcal{G})$ with minimal leading monomial could be transformed into $g \in I \setminus (\mathcal{G})$ with smaller leading monomial by subtracting a multiple of an element from $\mathcal{G}$, which yields a contradiction. It is also clear that Gröbner bases always exist. Indeed, $\{\mathrm{LM}(f) \mid f \in I\}$ generates $L(I)$ by definition, hence by the Noether property a finite subset $\{\mathrm{LM}(f_1), \ldots, \mathrm{LM}(f_m)\}$ also generates $L(I)$, and so $\{f_1, \ldots, f_m\}$ is a Gröbner basis. This argument, however, is non-constructive. But we will see in Section 1.1.4 that there is in fact an algorithm for computing Gröbner bases.

The most obvious question about an ideal $I \subseteq K[x_1, \ldots, x_n]$ that can be decided with Gröbner bases is whether $I = K[x_1, \ldots, x_n]$. Indeed, this is the case if and only if $\mathcal{G}$ contains a (non-zero) constant polynomial.

1.1.3 Normal Forms

A central element in the construction and usage of Gröbner bases is the computation of so-called normal forms.

Definition 1.1.4. *Let $S \subseteq K[x_1, \ldots, x_n]$ be a set of polynomials.*

(a) *A polynomial $f \in K[x_1, \ldots, x_n]$ is said to be in* **normal form** *with respect to S if no term of f is divisible by the leading monomial of any $g \in S$.*

(b) *If f and $\tilde{f}$ are polynomials in $K[x_1, \ldots, x_n]$, then $\tilde{f}$ is said to be a* **normal form** *of f with respect to S if $\tilde{f}$ is in normal form with respect to S and $f - \tilde{f}$ lies in the ideal generated by S.*

The following algorithm, which mimics division with remainder in the univariate case, calculates a normal form with respect to a finite set S of polynomials.

Algorithm 1.1.5 (Normal form). Given a polynomial $f \in K[x_1, \ldots, x_n]$ and a finite subset $S = \{g_1, \ldots, g_s\} \subset K[x_1, \ldots, x_n]$, perform the following steps to obtain a normal form $\tilde{f}$ of f with respect to S, together with polynomials $h_1, \ldots, h_s \in K[x_1, \ldots, x_n]$ such that

$$f = \tilde{f} + \sum_{i=1}^{s} h_i g_i.$$

(1) Set $\tilde{f} := f$ and $h_i := 0$ for all i, and repeat the steps (2)–(4).
(2) If no term of $\tilde{f}$ is divisible by the leading monomial of any $g_i \in S$, return $\tilde{f}$ as a normal form of f, and return the h_i.
(3) Let ct be the maximal term of $\tilde{f}$ such that there exists $g_i \in S$ with $\mathrm{LM}(g_i)$ dividing t.
(4) Set

$$\tilde{f} := \tilde{f} - \frac{ct}{\mathrm{LT}(g_i)} g_i \quad \text{and} \quad h_i := h_i + \frac{ct}{\mathrm{LT}(g_i)}.$$

Of course the computation of the h_i can be omitted if only a normal form is desired. The termination of Algorithm 1.1.5 is guaranteed by the fact that the maximal monomials t of $\tilde{f}$ divisible by some $\mathrm{LM}(g_i)$ form a strictly decreasing sequence, but such a sequence is finite by the well-ordering property. The result of Algorithm 1.1.5 is in general not unique, since it depends on the choice of the g_i in step (3). However, if $\mathcal{G}$ is a Gröbner basis of an ideal I, then normal forms with respect to $\mathcal{G}$ are unique. In fact, if $\tilde{f}$ and $\hat{f}$ are two normal forms of f with respect to $\mathcal{G}$, then $\tilde{f} - \hat{f} \in I$, so $\mathrm{LM}(\tilde{f} - \hat{f})$ is divisible by some $\mathrm{LM}(g)$ with $g \in \mathcal{G}$. But if $\tilde{f} \neq \hat{f}$, then $\mathrm{LM}(\tilde{f} - \hat{f})$ must appear as a monomial in $\tilde{f}$ or $\hat{f}$, contradicting the fact that $\tilde{f}$ and $\hat{f}$ are in normal form. In the case of a Gröbner basis $\mathcal{G}$ we write $\tilde{f} =: \mathrm{NF}(f) = \mathrm{NF}_{\mathcal{G}}(f)$ for the normal form.

It should be mentioned that there is a variant of the normal form algorithm which stops when the leading term of $\tilde{f}$ is zero or not divisible by any $\mathrm{LM}(g)$, $g \in S$ ("top-reduction").

Using Algorithm 1.1.5, we obtain a membership test for ideals.

Algorithm 1.1.6 (Membership test in ideals). Let $I \subseteq K[x_1, \ldots, x_n]$ be an ideal, $\mathcal{G}$ a Gröbner basis of I, and $f \in K[x_1, \ldots, x_n]$ a polynomial. Then

$$f \in I \quad \Longleftrightarrow \quad \mathrm{NF}_{\mathcal{G}}(f) = 0.$$

One can also substitute $\mathrm{NF}_{\mathcal{G}}(f)$ by the result of top-reducing f.

Thus the map $\mathrm{NF}_{\mathcal{G}}: K[x_1, \ldots, x_n] \to K[x_1, \ldots, x_n]$ is K-linear with kernel I, and therefore provides a way to perform explicit calculations in the quotient ring $K[x_1, \ldots, x_n]/I$. In fact, this was the main objective for which Gröbner bases were invented.

A Gröbner basis $\mathcal{G}$ of an ideal I can be transformed into a **reduced Gröbner basis** by iteratively substituting an element from $\mathcal{G}$ by a normal form with respect to the other elements, until every element is in normal form. After deleting zero from the resulting set and making all leading coefficients equal to 1, the resulting monic reduced Gröbner basis is unique (i.e., it only depends on I and the chosen monomial ordering, see Becker and Weispfenning [15, Theorem 5.43]).

1.1.4 The Buchberger Algorithm

In order to present Buchberger's algorithm for the construction of Gröbner bases, we need to introduce s-polynomials. Let $f, g \in K[x_1, \ldots, x_n]$ be two non-zero polynomials, and set $t := \mathrm{lcm}(\mathrm{LM}(f), \mathrm{LM}(g))$ (the least common multiple). Then the **s-polynomial** of f and g is defined as

$$\mathrm{spol}(f, g) := \frac{\mathrm{LC}(g) \cdot t}{\mathrm{LM}(f)} f - \frac{\mathrm{LC}(f) \cdot t}{\mathrm{LM}(g)} g.$$

Note that the coefficients of t cancel in $\mathrm{spol}(f, g)$, and that $\mathrm{spol}(f, g) \in (f, g)$. The following lemma is the key step toward finding an algorithm for the construction of a Gröbner basis.

Lemma 1.1.7 (Buchberger [32]). *Let $\mathcal{G}$ be a basis (=generating set) of an ideal $I \subseteq K[x_1, \ldots, x_n]$. Then the following conditions are equivalent.*

(a) $\mathcal{G}$ is a Gröbner basis of I.
(b) If $f, g \in \mathcal{G}$, then $\mathrm{spol}(f, g)$ has 0 as a normal form with respect to $\mathcal{G}$.
(c) If $f, g \in \mathcal{G}$, then every normal form of $\mathrm{spol}(f, g)$ with respect to $\mathcal{G}$ is 0.

See Becker and Weispfenning [15, Theorem 5.48] for a proof. We can give Buchberger's algorithm in a rather coarse form now.

Algorithm 1.1.8 (Buchberger's algorithm). Given a finite basis S for an ideal $I \subseteq K[x_1, \ldots, x_n]$, construct a Gröbner basis (with respect to a given monomial ordering) by performing the following steps:

(1) Set $\mathcal{G} := S$ and repeat steps (2)–(4).
(2) For $f, g \in \mathcal{G}$ compute a normal form h of $\mathrm{spol}(f, g)$ with respect to $\mathcal{G}$.
(3) If $h \neq 0$, include h into $\mathcal{G}$.
(4) If h was found to be zero for all $f, g \in \mathcal{G}$, then $\mathcal{G}$ is the desired Gröbner basis.

This algorithm terminates after a finite number of steps since $L(S)$ strictly increases with every performance of steps (2)–(4).

Remark 1.1.9. The theoretical cost of Buchberger's algorithm is extremely high. In fact, no general upper bound for the running time is known. But Möller and Mora [168] proved an upper bound for the maximal degree of the Gröbner basis elements which depends doubly exponentially on the number of variables. They also proved that this doubly exponential behavior cannot be improved. What makes things even worse is the phenomenon of "intermediate expression swell", meaning that during the computation the number and size of polynomials can become much bigger than in the final result. It is known that the memory space required for the computation of Gröbner bases increases at most exponentially with the size of the input, and all problems with this behavior can be reduced to the problem of testing ideal membership; so the problem of computing Gröbner bases is "EXPSPACE-complete". We refer to von zur Gathen and Gerhard [79, Section 21.7] for a more detailed account of what is known about the complexity of Gröbner bases.

In spite of all this bad news, practical experience shows that the algorithm often terminates after a reasonable time (although this is usually not predictable in advance). Much depends on improvements of the algorithm given above, such as omitting some pairs f, g (by Buchberger's first and second criterion, see Becker and Weispfenning [15, Section 5.5]), by having a good strategy which pairs to treat first, and by choosing a suitable monomial ordering (if there is any freedom of choice). There are also algorithms which transform a Gröbner basis with respect to one monomial ordering into one with respect to another ordering (see Faugère et al. [66], Collart et al. [46]).
◁

There is a variant of Buchberger's algorithm which keeps track of how the polynomials in the Gröbner basis $\mathcal{G}$ arise as linear combinations of the polynomials in the original ideal basis S. This variant is called the extended Buchberger algorithm, and its output is an (ordered) Gröbner basis $\mathcal{G} = \{g_1, \ldots, g_r\}$ and an $r \times s$-matrix A with coefficients in $K[x_1, \ldots, x_n]$ such that

$$\begin{pmatrix} g_1 \\ \vdots \\ g_r \end{pmatrix} = A \cdot \begin{pmatrix} f_1 \\ \vdots \\ f_s \end{pmatrix},$$

where $S = \{f_1, \ldots, f_s\}$. On the other hand, it is straightforward to obtain an $s \times r$-matrix B such that $(f_1, \ldots, f_s)^{\mathrm{tr}} = B(g_1, \ldots, g_r)^{\mathrm{tr}}$ by applying the Normal Form Algorithm 1.1.5 to the f_i.

1.2 Elimination Ideals

Given an ideal $I \subseteq K[x_1, \ldots, x_n]$ and an integer $k \in \{1, \ldots, n\}$, the **elimination ideal** of I with respect to $x_k, \ldots, x_n$ is defined as the intersection $I \cap K[x_k, \ldots, x_n]$. It has the following geometric interpretation: If

$$\pi\colon \mathbb{A}^n \to \mathbb{A}^{n-k+1}, \ (\xi_1,\ldots,\xi_n) \mapsto (\xi_k,\ldots,\xi_n)$$

is the canonical projection, then for K algebraically closed we have

$$\overline{\pi\left(\mathcal{V}(I)\right)} = \mathcal{V}(I \cap K[x_k,\ldots,x_n]), \qquad (1.2.1)$$

where the left hand side is the Zariski-closure. (In scheme theoretic language, π is the intersection of a prime ideal in $K[x_1,\ldots,x_n]$ with $K[x_k,\ldots,x_n]$, and we do not need the hypothesis that K is algebraically closed.) An important feature of Gröbner bases is that they can be used to compute elimination ideals.

Algorithm 1.2.1 (Computing elimination ideals)**.** Given an ideal $I \subseteq K[x_1,\ldots,x_n]$ and an integer $k \in \{1,\ldots,n\}$, compute the elimination ideal $I \cap K[x_k,\ldots,x_n]$ as follows:

(1) Choose a monomial ordering such that $x_i \gg x_j$ for $i < k$ and $j \geq k$ (e.g., the lexicographic monomial ordering or a block ordering).
(2) Compute a Gröbner basis $\mathcal{G}$ of I with respect to this monomial ordering.
(3) $\mathcal{G} \cap K[x_k,\ldots,x_n]$ is a Gröbner basis of $I \cap K[x_1,\ldots,x_n]$.

It is elementary to see that this algorithm is correct (see Becker and Weispfenning [15, Proposition 6.15]). Equation (1.2.1) shows how elimination ideals can be used to solve a system of algebraic equations with a finite set of solutions.

We continue by presenting some applications of elimination ideals (and thus of Gröbner bases) which will be needed in the following chapters of this book.

1.2.1 Image Closure of Morphisms

Let X and Y be affine varieties and $f\colon X \to Y$ a morphism. (Again we assume that K is algebraically closed or use the language of schemes.) We want to compute the Zariski-closure of the image $f(X)$. Assume that X is embedded into $\mathbb{A}^n$ and Y into $\mathbb{A}^m$ for some n and m. Without loss of generality we can assume that $Y = \mathbb{A}^m$. If f is given by polynomials $(f_1,\ldots,f_m)$ with $f_i \in K[x_1,\ldots,x_n]$, and X is given by an ideal $I \subseteq K[x_1,\ldots,x_n]$, then the graph of f is given by the ideal

$$J := I \cdot K[x_1,\ldots,x_n,y_1,\ldots,y_m] + (f_1 - y_1,\ldots,f_m - y_m)$$

in $K[x_1,\ldots,x_n,y_1,\ldots,y_m]$. Thus by Equation (1.2.1), the closure of the image is

$$\overline{f(X)} = \mathcal{V}\left(J \cap K[y_1,\ldots,y_m]\right)$$

(see Vasconcelos [250, Proposition 2.1.3]), and can therefore be calculated by Algorithm 1.2.1.

1.2.2 Relations Between Polynomials

A further application of elimination ideals is the computation of relations between polynomials. More precisely, let $f_1, \ldots, f_m \in K[x_1, \ldots, x_n]$ be polynomials. We are interested in the kernel of the homomorphism

$$\Phi \colon K[t_1, \ldots, t_m] \to K[x_1, \ldots, x_n], \ t_i \mapsto f_i,$$

of K-algebras (where $t_1, \ldots, t_m$ are further indeterminates). The answer is as follows: Define the ideal

$$I := (f_1 - t_1, \ldots, f_m - t_m)$$

in $K[x_1, \ldots, x_n, t_1, \ldots, t_m]$. Then it is easy to show that

$$\ker(\Phi) = I \cap K[t_1, \ldots, t_m], \tag{1.2.2}$$

so the desired kernel is again an elimination ideal (see Eisenbud [59, Proposition 15.30]). Notice that generators for $\ker(\Phi)$ together with the f_i provide a presentation of the algebra generated by the f_i.

1.2.3 The Intersection of Ideals

The intersection of two ideals $I, J \subseteq K[x_1, \ldots, x_n]$ (which geometrically corresponds to the union of varieties) can be computed as follows: With a new indeterminate t, form the ideal L in $K[x_1, \ldots, x_n, t]$ generated by

$$I \cdot t + J \cdot (1 - t),$$

where the products are formed by multiplying each generator of I and J by t and $1 - t$, respectively. Then

$$I \cap J = L \cap K[x_1, \ldots, x_n] \tag{1.2.3}$$

(see Vasconcelos [250, Corollary 2.1.1]). A different method for computing the intersection of I and J involves the calculation of a syzygy module (see Vasconcelos [250, page 29]). We can apply any of these methods iteratively to obtain the intersection of a finite number of ideals, but there is also a direct method (involving further auxiliary indeterminates) given by Becker and Weispfenning [15, Corollary 6.20].

1.2.4 The Quotient of Ideals

Given two ideals $I, J \subseteq K[x_1, \ldots, x_n]$, it is often important to be able to calculate the **quotient ideal**

$$I : J := \{ g \in K[x_1, \ldots, x_n] \mid gf \in I \ \forall f \in J \}.$$

Sometimes $I : J$ is also referred to as the colon ideal. The quotient ideal has the following geometric interpretation: If I is a radical ideal and K is algebraically closed, then $I : J$ is precisely the ideal of all polynomials vanishing on $\mathcal{V}(I) \setminus \mathcal{V}(J)$. The quotient ideal is also of crucial importance for the computation of radical ideals (see Section 1.5) and primary decomposition.

If $J = (f)$ is a principal ideal, we sometimes write $I : f$ for the quotient ideal $I : (f)$. If $J = (f_1, \ldots, f_k)$, then clearly

$$I : J = \bigcap_{i=1}^{k} I : f_i,$$

which reduces the task to the case that J is a principal ideal. But clearly

$$I : f = (I \cap (f)) \cdot f^{-1} \tag{1.2.4}$$

(see Vasconcelos [250, Proposition 2.1.4(a)]), which can be computed by Equation (1.2.3). Thus quotient ideals can be obtained by using any algorithm for the intersection of ideals.

For an ideal $I \subseteq K[x_1, \ldots, x_n]$ and a polynomial $f \in K[x_1, \ldots, x_n]$ we can also consider the ideal

$$I : f^\infty := \bigcup_{i \in \mathbb{N}} I : f^i,$$

which is sometimes referred to as the saturation ideal of I with respect to f. The saturation ideal can be calculated by successively computing the quotient ideals $J_i := I : f^i = J_{i-1} : f$. This gives an ascending chain of ideals, thus eventually we get $J_{k+1} = J_k$, so $I : f^\infty = J_k$. But there is a more efficient algorithm, based on the following proposition.

Proposition 1.2.2. *Let $I \subseteq K[x_1, \ldots, x_n]$ be an ideal and $f \in K[x_1, \ldots, x_n]$ a polynomial. Introduce an additional indeterminate t and form the ideal J in $K[x_1, \ldots, x_n, t]$ generated by I and $tf - 1$. Then*

$$I : f^\infty = J \cap K[x_1, \ldots, x_n].$$

A proof can be found in Becker and Weispfenning [15, Proposition 6.37].

1.2.5 The Krull Dimension

We define the dimension of an ideal $I \subseteq K[x_1, \ldots, x_n]$ to be the Krull dimension of the quotient $K[x_1, \ldots, x_n]/I$. There is a method which computes the dimension by using elimination ideals (Becker and Weispfenning [15, Section 6.3]). However, this technique involves a large number of Gröbner basis computations and is therefore not very efficient. A better algorithm (also given in the book of Becker and Weispfenning [15]) is based on the following lemma, which follows from Cox et al. [48, Proposition 4 of Chapter 9, §3].

Lemma 1.2.3. *If ">" is a graded monomial ordering, then the dimensions of I and of the leading ideal $L(I)$ coincide.*

To prove this lemma, one uses the fact that the normal form provides an isomorphism of K-vector spaces (not of algebras) between $K[x_1, \ldots, x_n]/I$ and $K[x_1, \ldots, x_n]/L(I)$. Lemma 1.2.3 reduces our problem to the computation of the dimension of $L(I)$, which is a monomial ideal. But the variety defined by a monomial ideal is a finite union of so-called coordinate subspaces, i.e., varieties of the form $\mathcal{V}(\mathcal{M})$ with $\mathcal{M} \subseteq \{x_1, \ldots, x_n\}$. Clearly such a variety is contained in the zero set of the monomial ideal J if and only if every generator of J involves at least one variable x_i lying in $\mathcal{M}$. We obtain the following algorithm (see Cox et al. [48, Proposition 3 of Chapter 9, §1]).

Algorithm 1.2.4 (Dimension of an ideal).
Given an ideal $I \subseteq K[x_1, \ldots, x_n]$, calculate the dimension of I by performing the following steps:

(1) Compute a Gröbner basis $\mathcal{G}$ of I with respect to a graded monomial ordering.
(2) If $\mathcal{G}$ contains a non-zero constant, then $I = K[x_1, \ldots, x_n]$, and the dimension is (by convention) -1.
(3) Otherwise, find a subset $\mathcal{M} \subseteq \{x_1, \ldots, x_n\}$ of minimal cardinality such that for every non-zero $g \in \mathcal{G}$ the leading monomial $\mathrm{LM}(g)$ involves at least one variable from $\mathcal{M}$.
(4) The dimension of I is $n - |\mathcal{M}|$.

Step (3) of the above algorithm is purely combinatorial and therefore usually much faster than the Gröbner basis computation. An optimized version of this step can be found in Becker and Weispfenning [15, Algorithm 9.6].

The set $\mathcal{M} \subseteq \{x_1, \ldots, x_n\}$ occurring in Algorithm 1.2.4 has an interesting interpretation. In fact, let $\mathcal{M}' := \{x_1, \ldots, x_n\} \setminus \mathcal{M}$ be the complement of $\mathcal{M}$. Then for every non-zero $g \in \mathcal{G}$ the leading monomial $\mathrm{LM}(g)$ involves at least one variable *not* in $\mathcal{M}'$. This implies that every non-zero polynomial in $L(I)$ involves a variable not in $\mathcal{M}'$, so $L(I) \cap K[\mathcal{M}'] = \{0\}$. From this it follows that

$$I \cap K[\mathcal{M}'] = \{0\}. \tag{1.2.5}$$

Indeed, if $f \in I \cap K[\mathcal{M}']$ were non-zero, then $\mathrm{LM}(f)$ would lie in $L(I) \cap K[\mathcal{M}']$. Subsets $\mathcal{M}' \subseteq \{x_1, \ldots, x_n\}$ which satisfy (1.2.5) are called independent modulo I (see Becker and Weispfenning [15, Definition 6.46]). Consider the rational function field $L := K(\mathcal{M}')$ in the variables lying in $\mathcal{M}'$, and let $L[\mathcal{M}]$ be the polynomial ring over L in the variables from $\mathcal{M}$. Then (1.2.5) is equivalent to the condition that the ideal $IL[\mathcal{M}]$ generated by I in $L[\mathcal{M}]$ is not equal to $L[\mathcal{M}]$. Since we have $|\mathcal{M}'| = \dim(I)$, it follows that $\mathcal{M}'$ is *maximally* independent modulo I. (Indeed, if there existed a strict superset $\mathcal{N} \supsetneq \mathcal{M}$ of variables which is independent modulo I, the $\mathcal{N}$ would also be independent modulo some minimal prime P containing I. But this would imply that the transcendence degree of $K[x_1, \ldots, x_n]/P$ is at least $|\mathcal{N}|$, hence by Eisenbud [59,

Section 8.2, Theorem A] we would get $\dim(I) \geq \dim(P) \geq |\mathcal{N}| > |\mathcal{M}'|$.)
The maximality of $\mathcal{M}'$ means that no non-empty subset of $\mathcal{M}$ is independent
modulo $IL[\mathcal{M}]$. By Algorithm 1.2.4, the dimension of $IL[\mathcal{M}]$ must therefore
be zero. Thus we have shown:

Proposition 1.2.5. *Let $I \subsetneq K[x_1, \ldots, x_n]$ be an ideal and $\mathcal{M} \subseteq \{x_1, \ldots, x_n\}$
as in Algorithm 1.2.4. Set $\mathcal{M}' := \{x_1, \ldots, x_n\} \backslash \mathcal{M}$, and take the rational function field $L := K(\mathcal{M}')$ in the variables lying in $\mathcal{M}'$, and the polynomial ring
$L[\mathcal{M}]$. Then the ideal $J := IL[\mathcal{M}]$ generated by I in $L[\mathcal{M}]$ is not equal to
$L[\mathcal{M}]$, and $\dim(J) = 0$.*

1.3 Syzygy Modules

In this section we write $R := K[x_1, \ldots, x_n]$ for the polynomial ring and
R^k for a free R-module of rank k. The standard basis vectors of R^k are
denoted by $e_1, \ldots, e_k$. Given polynomials $f_1, \ldots, f_k \in R$, we ask for the set
of all $(h_1, \ldots, h_k) \in R^k$ such that $h_1 f_1 + \cdots + h_k f_k = 0$. This set is a
submodule of R^k, called the **syzygy module** of $f_1, \ldots, f_k$ and denoted by
$\mathrm{Syz}(f_1, \ldots, f_k)$. More generally, we ask for the kernel of an R-homomorphism
$\varphi \colon R^k \to R^l$ between two free R-modules. If $f_i := \varphi(e_i) \in R^l$, then the kernel
of φ consists of all $(h_1, \ldots, h_k) \in R^k$ with $h_1 f_1 + \cdots + h_k f_k = 0$. Again
$\mathrm{Syz}(f_1, \ldots, f_k) := \ker(\varphi)$ is called the syzygy module of the f_i.

1.3.1 Computing Syzygies

In order to explain an algorithm which computes syzygy modules, we have to
give a brief introduction into Gröbner bases of submodules of R^k. A **monomial** in R^k is an expression of the form te_i with t a monomial in R. The notion
of a monomial ordering is given as in Definition 1.1.1, with condition (i) replaced by $te_i > e_i$ for all i and $1 \neq t$ a monomial in R, and demanding (ii) for
monomials $t_1, t_2 \in R^k$ and $s \in R$. Given a monomial ordering, we can now
define the leading submodule $L(M)$ of a submodule $M \subseteq R^k$ and the concept
of a Gröbner basis of M as in Definition 1.1.3. Normal forms are calculated by
Algorithm 1.1.5, with the extra specification that te_i is said to be divisible by
$t'e_j$ if $i = j$ and t divides t', so the quotients are always elements in R. Moreover, the s-polynomial of f and $g \in R^k$ with $\mathrm{LM}(f) = te_i$ and $\mathrm{LM}(g) = t'e_j$
is defined to be zero if $i \neq j$. With these provisions, Buchberger's algorithm
can be formulated as in Algorithm 1.1.8.

Suppose that $\mathcal{G} = \{g_1, \ldots, g_k\}$ is a Gröbner basis of a submodule $M \subseteq
R^l$. Then for $g_i, g_j \in \mathcal{G}$ we have that $\mathrm{NF}_{\mathcal{G}}(\mathrm{spol}(g_i, g_j)) = 0$, so there exist
$h_1, \ldots, h_k \in R$ with

$$\mathrm{spol}(g_i, g_j) = h_1 g_1 + \cdots + h_k g_k, \tag{1.3.1}$$

and the h_i can be computed by the Normal Form Algorithm 1.1.5. Since $\mathrm{spol}(g_i, g_j)$ is an R-linear combination of g_i and g_j, Equation (1.3.1) yields a syzygy $r_{i,j} \in \mathrm{Syz}(g_1, \ldots, g_k)$. Of course $r_{i,j} = 0$ if the leading monomials of g_i and of g_j lie in different components of R^l.

The following monomial ordering "$>_{\mathcal{G}}$" on R^k, which depends on $\mathcal{G}$, was introduced by Schreyer [210]: te_i is considered bigger than $t'e_j$ if $t\,\mathrm{LM}(g_i) > t'\,\mathrm{LM}(g_j)$ (with "$>$" the given ordering on R^l), or if $t\,\mathrm{LM}(g_i) = t'\,\mathrm{LM}(g_j)$ and $i < j$. It is easy to see that "$>_{\mathcal{G}}$" satisfies the properties of a monomial ordering.

Theorem 1.3.1 (Schreyer [210]). *Let $\mathcal{G} = \{g_1, \ldots, g_k\}$ be a Gröbner basis with respect to an arbitrary monomial ordering "$>$" of a submodule $M \subseteq R^l$. Then, with the above notation, the $r_{i,j}$ $(1 \leq i < j \leq k)$ form a Gröbner basis of $\mathrm{Syz}(g_1, \ldots, g_k)$ with respect to the monomial ordering "$>_{\mathcal{G}}$".*

This settles the case of syzygies for Gröbner bases. Now assume that $f_1, \ldots, f_k \in R^l$ are arbitrary. Using the extended Buchberger algorithm (see at the end of Section 1.1), we can calculate a Gröbner basis $\{g_1, \ldots, g_{k'}\}$ of the submodule generated by $f_1, \ldots, f_k$, along with representations of the g_i as R-linear combinations of the f_j. Using the Normal Form Algorithm 1.1.5, we can also express the f_j in terms of the g_i. The choice of the f_j and g_i is equivalent to giving homomorphisms $R^k \to R^l$ and $R^{k'} \to R^l$, and expressing the f_j in terms of the g_i and vice versa is equivalent to giving homomorphisms φ and ψ such that the diagram

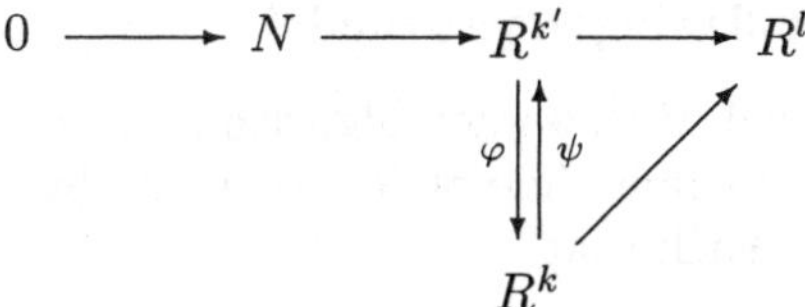

commutes (both along φ and ψ), where $N := \mathrm{Syz}(g_1, \ldots, g_{k'})$ can be computed by Theorem 1.3.1. The following lemma tells us how to compute $\mathrm{Syz}(f_1, \ldots, f_k) = \ker(R^k \to R^l)$.

Lemma 1.3.2. *Let A be a commutative ring and*

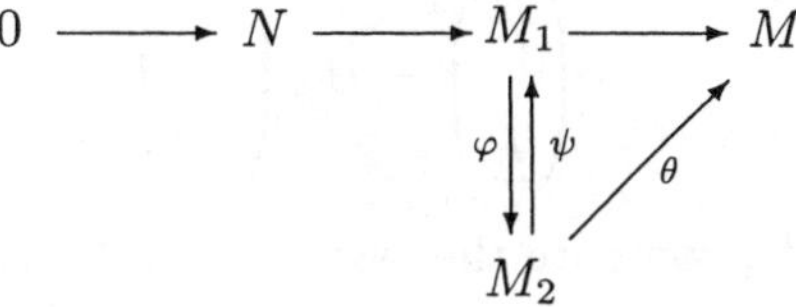

a commutative diagram (both along φ and ψ) of A-modules, with the upper row exact. Then we have an exact sequence

$$0 \longrightarrow (\mathrm{id} - \psi \circ \varphi)(M_1) \longrightarrow N \oplus (\mathrm{id} - \varphi \circ \psi)(M_2) \longrightarrow M_2 \xrightarrow{\theta} M$$

with maps

$$(\mathrm{id} - \psi \circ \varphi)(M_1) \to N \oplus (\mathrm{id} - \varphi \circ \psi)(M_2), \ m \mapsto (m, -\varphi(m)), \quad and$$
$$N \oplus (\mathrm{id} - \varphi \circ \psi)(M_2) \to M_2, \qquad\qquad (n, m) \mapsto \varphi(n) + m.$$

In particular,

$$\ker(\theta) = \varphi(N) + (\mathrm{id} - \varphi \circ \psi)(M_2).$$

Proof. It follows by a simple diagram chase that $(\mathrm{id} - \psi \circ \varphi)(M_1) \subseteq N$, so the first map is indeed into $N \oplus (\mathrm{id} - \varphi \circ \psi)(M_2)$. We show the exactness at M_2. Again by a diagram chase $\theta(\varphi(n) + m) = 0$ for $n \in N$ and $m \in (\mathrm{id} - \varphi \circ \psi)(M_2)$. Conversely, for $m \in \ker(\theta)$ we have

$$m = \varphi(\psi(m)) + (\mathrm{id} - \varphi \circ \psi)(m)$$

with $\psi(m) \in N$. To show the exactness at $N \oplus (\mathrm{id} - \varphi \circ \psi)(M_2)$, take $(n, m_2 - \varphi(\psi(m_2))) \in N \oplus (\mathrm{id} - \varphi \circ \psi)(M_2)$ with $\varphi(n) + m_2 - \varphi(\psi(m_2)) = 0$. Then

$$n = (\mathrm{id} - \psi \circ \varphi)(n - \psi(m_2)) \in (\mathrm{id} - \psi \circ \varphi)(M_1),$$

and $(n, -\varphi(n)) = (n, m_2 - \varphi(\psi(m_2)))$. This completes the proof. $\qquad\square$

In summary, we obtain the following algorithm.

Algorithm 1.3.3 (Calculation of a syzygy module). Given elements $f_1, \ldots, f_k \in R^l$, perform the following steps to find the syzygy module $\mathrm{Syz}(f_1, \ldots, f_k)$:

(1) Using the extended Buchberger algorithm, calculate a Gröbner basis $\{g_1, \ldots, g_{k'}\}$ of the submodule of R^l generated by the f_i together with a matrix $A \in R^{k' \times k}$ such that

$$\begin{pmatrix} g_1 \\ \vdots \\ g_{k'} \end{pmatrix} = A \cdot \begin{pmatrix} f_1 \\ \vdots \\ f_k \end{pmatrix}.$$

(2) Using the Normal Form Algorithm 1.1.5, compute a matrix $B \in R^{k \times k'}$ with

$$\begin{pmatrix} f_1 \\ \vdots \\ f_k \end{pmatrix} = B \cdot \begin{pmatrix} g_1 \\ \vdots \\ g_{k'} \end{pmatrix}.$$

(3) For $1 \leq i < j \leq k'$, compute the syzygies $r_{i,j} \in \mathrm{Syz}(g_1, \ldots, g_{k'})$ given by Equation (1.3.1).

(4) $\mathrm{Syz}(f_1, \ldots, f_k)$ is generated by the $r_{i,j} \cdot A$ and the rows of $I_k - BA$.

1.3.2 Free Resolutions

For a submodule $M \subseteq R^l$ (with $R = K[x_1, \ldots, x_n]$ as before) with generating set $f_1, \ldots, f_k$, we can compute generators for $N := \mathrm{Syz}(f_1, \ldots, f_k) \subseteq R^k$ by using Algorithm 1.3.3. Continuing by computing the syzygies of these generators and so on, we obtain a free resolution of M, i.e., an exact sequence

$$0 \longrightarrow F_r \longrightarrow F_{r-1} \longrightarrow \cdots \longrightarrow F_2 \longrightarrow F_1 \longrightarrow F_0 \longrightarrow M \longrightarrow 0 \qquad (1.3.2)$$

with the F_i free R-modules. Hilbert's syzygy theorem (see Eisenbud [59, Corollary 19.8] or the "original" reference Hilbert [107]) guarantees that there exists a free resolution of finite length (bounded by n, in fact), as given above. Free resolutions are of great interest because they contain a lot of information about the structure of M. Theorem 1.3.1 provides the following method for calculating a free resolution with only a single Gröbner basis computation.

Algorithm 1.3.4 (Schreyer's algorithm). Let $M \subseteq R^l$ be a submodule given by a generating set. Obtain a free resolution of M as follows:

(1) Compute a Gröbner basis $\mathcal{G} = \{g_1, \ldots, g_k\}$ of M with respect to an arbitrary monomial ordering ">". Set $i := 0$ and repeat steps (2)–(4).
(2) Set $F_i := R^k$ and obtain the map $F_i \to F_{i-1}$ (with $F_{-1} := M$) from (1.3.2) by $(h_1, \ldots, h_k) \mapsto h_1 g_1 + \cdots + h_k g_k$.
(3) Compute the relations $r_{i,j}$ from Equation (1.3.1). By Theorem 1.3.1, the $r_{i,j}$ form a Gröbner basis with respect to ">$_\mathcal{G}$" of the kernel of the map defined in (2).
(4) If all $r_{i,j}$ are zero, the resolution is complete. Otherwise, let $\mathcal{G} \subseteq R^k$ be the set of the non-zero $r_{i,j}$ and set $i := i + 1$.

The termination of Algorithm 1.3.4 after at most n iterations is guaranteed by (the proof of) Theorem 2.1 in Chapter 6 of Cox et al. [49] (which provides a new, constructive proof of Hilbert's syzygy theorem).

Now suppose that the polynomial ring R is made into a graded algebra by defining the degrees $\deg(x_i)$ of the indeterminates to be positive integers. Then the free module R^l can be made into a graded R-module by defining the $\deg(e_i)$ to be integers. Moreover, suppose that M is a graded submodule, i.e., generated by homogeneous elements. Then we want to find a graded free resolution , i.e., one that consists of graded free modules F_i with all mappings degree-preserving. Applying Buchberger's algorithm to a homogeneous generating set of M yields a homogeneous Gröbner basis, too, and by inspection of the way in which the syzygies $r_{i,j}$ are formed from Equation (1.3.1), we see that the resolution obtained by Algorithm 1.3.4 is indeed graded (with the proper choice of the degrees of the free generators, i.e., each generator gets the same degree as the relation to which it is mapped).

In the case that R^l is graded and M is a graded submodule, we are also interested in obtaining a **minimal free resolution** of M, i.e., a free resolution such that the free generators of each F_i are mapped to a minimal generating set of the image of F_i. Such a resolution is unique up to isomorphism of complexes (see Eisenbud [59, Theorem 20,2]), and in particular its length is unique. This length is called the **homological dimension** of M, written as $\mathrm{hdim}(M)$, and is an important structural invariant of M. A graded resolution (1.3.2) calculated by Algorithm 1.3.4 is usually not minimal, so how can it be transformed into a minimal resolution, preferably without computing any further Gröbner bases? As a first step, we can use linear algebra to select a minimal subset of the free generators of F_0 whose image in M generates M. Thus we obtain a free submodule $F_0' \subseteq F_0$ and a commutative diagram

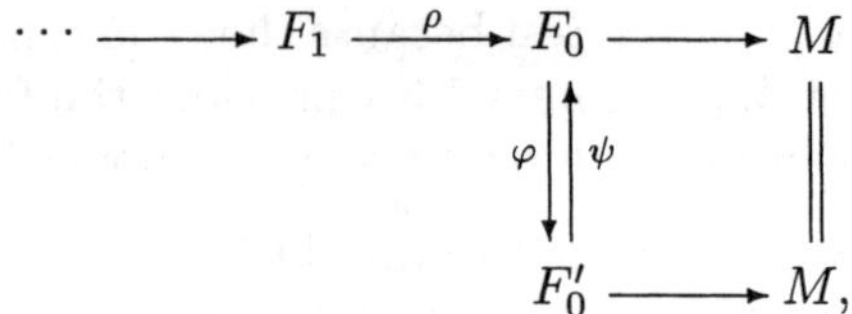

where $\varphi \circ \psi = \mathrm{id}$. Lemma 1.3.2 yields an exact sequence

$$0 \longrightarrow (\mathrm{id} - \psi \circ \varphi)(F_0) \longrightarrow \mathrm{im}(\rho) \overset{\varphi}{\longrightarrow} F_0' \longrightarrow M. \qquad (1.3.3)$$

Observe that $(\mathrm{id} - \psi \circ \varphi)$ maps a free generator e_i from F_0 either to zero (if it is also a generator of F_0') or to a non-zero element of $(\mathrm{id} - \psi \circ \varphi)(F_0)$ corresponding to the representation of the image of e_i in M in terms of the images of those e_j contained in F_0'. These non-zero elements are linearly independent, hence $(\mathrm{id} - \psi \circ \varphi)(F_0)$ is a free module. We can use linear algebra to compute preimages under ρ of the free generators of $(\mathrm{id} - \psi \circ \varphi)(F_0)$ in F_1. This yields a free submodule $\hat{F}_1 \subseteq F_1$ such that $\rho(\hat{F}_1) = (\mathrm{id} - \psi \circ \varphi)(F_0)$ and the restriction of ρ to $\hat{F}_1$ is injective. Now it is easy to see that (1.3.3) and (1.3.2) lead to the exact sequence

$$0 \longrightarrow F_r \longrightarrow F_{r-1} \longrightarrow \cdots \longrightarrow F_3 \longrightarrow F_2 \oplus \hat{F}_1 \longrightarrow F_1 \overset{\varphi \circ \rho}{\longrightarrow} F_0' \longrightarrow M \longrightarrow 0.$$

Thus we have managed to replace (1.3.2) by a free resolution with the first free module minimal. Iterating this process, we obtain the desired minimal free resolution of M. Notice that the only computationally significant steps are the selection of minimal generators for M and the computation of preimages of $e_i - \psi(\varphi(e_i))$ for some free generators e_i of F_0. Both of these are accomplished by linear algebra. Thus a minimal resolution of M can be computed by just one Gröbner basis computation and linear algebra.

1.4 Hilbert Series

In this section, we prove some results about Hilbert series of rings, and how we can use ideal theory to compute them.

Definition 1.4.1. *For a graded vector space $V = \bigoplus_{d=k}^{\infty} V_d$ with V_d finite dimensional for all d we define the **Hilbert series** of V as the formal Laurent series*

$$H(V,t) := \sum_{d=k}^{\infty} \dim(V_d) t^d.$$

In the literature, Hilbert series are sometimes called Poincaré series. In our applications, V will always be a graded algebra or a graded module.

Example 1.4.2. Let us compute the Hilbert series of $K[x_1, \ldots, x_n]$. There are $\binom{n+d-1}{n-1}$ monomials of degree d, therefore the Hilbert series is

$$H(K[x_1, \ldots, x_n], t) = \sum_{d=0}^{\infty} \binom{n+d-1}{n-1} t^d.$$

This is exactly the power series expansion of $(1-t)^{-n}$. ◁

Remark 1.4.3. If V and W are two graded vector spaces, then the tensor product $V \otimes W$ also has a natural grading, namely

$$(V \otimes W)_d = \bigoplus_{d_1+d_2=d} V_{d_1} \otimes W_{d_2}.$$

It is obvious from this formula that $H(V \otimes W, t) = H(V,t)H(W,t)$. Suppose that $R = K[x_1, \ldots, x_n]$ and x_i has degree $d_i > 0$. Then we have $R = K[x_1] \otimes K[x_2] \otimes \cdots \otimes K[x_n]$ as graded algebras and $H(K[x_i], t) = (1 - t^{d_i})^{-1}$. It follows that

$$H(R,t) = \frac{1}{(1 - t^{d_1}) \cdots (1 - t^{d_n})} \tag{1.4.1}$$

◁

Remark 1.4.4. If

$$0 \to V^{(1)} \to V^{(2)} \to \cdots \to V^{(r)} \to 0 \tag{1.4.2}$$

is an exact sequence of graded vector spaces (all maps respect degree) with $V_d^{(i)}$ finite dimensional for all i and d, then

$$\sum_{i=1}^{r} (-1)^i H(V^{(i)}, t) = 0.$$

This is clear because the degree d part of (1.4.2) is exact for all d. ◁

Proposition 1.4.5 (Hilbert). *If $R = \bigoplus_{d=0}^{\infty} R_d$ is a finitely generated graded algebra over a field $K = R_0$, then $H(R,t)$ is the power series of a rational function. The radius of convergence of this power series is at least 1. Moreover, if $M = \bigoplus_{d=k}^{\infty} M_d$ is a finitely generated graded R-module, then $H(M,t)$ is the Laurent series of a rational function (which may have a pole at 0).*

Proof. Let $A = K[x_1, x_2, \ldots, x_n]$ be the polynomial ring, graded in such a way that $\deg(x_i) = d_i > 0$. Then $H(A, t)$ is a rational function by (1.4.1), and the radius of convergence of the power series is 1 if $n > 0$, and ∞ if $n = 0$. For any integer e, we define the A-module $A(e)$ by $A(e) = \bigoplus_{d=-e}^{\infty} A(e)_d$ with $A(e)_d := A_{e+d}$. It is clear that $H(A(e), t) = t^{-e} H(A, t)$ is again a rational function. A module is free if it is isomorphic to a direct sum $\bigoplus_i A(e_i)$. The Hilbert series of a finitely generated free module M is a rational function. If M is a finitely generated A-module, then by Hilbert's syzygy theorem (see Eisenbud [59, Theorem 1.13]), there exists a resolution

$$0 \to F^{(r)} \to F^{(r-1)} \to \cdots \to F^{(1)} \to F^{(0)} \to M \to 0, \qquad (1.4.3)$$

where $F^{(i)}$ is a finitely generated free A-module for all i, and the sequence is exact. It follows from Remark 1.4.4 that

$$H(M, t) = \sum_{i=0}^{r} (-1)^i H(F^{(i)}, t), \qquad (1.4.4)$$

so $H(M, t)$ is a rational function. If M is non-negatively graded, then the same is true for all F_i, so the radius of convergence of $H(M, t)$ is at least 1.

Let R be an arbitrary finitely generated graded algebra over $K = R_0$. Then for some n and some $d_1, \ldots, d_n > 0$, there exists a homogeneous ideal $I \subseteq A$ such that $A/I \cong R$. Hence R is a finitely generated, non-negatively graded A-module, and the claim follows. Moreover, any finitely generated graded R-module M is also a finitely generated graded A-module. $\square$

The above proof gives an easy way to compute the Hilbert series of a graded module M over a graded polynomial ring $R = K[x_1, \ldots, x_n]$, if we have a graded free resolution (1.4.3) of M. Indeed, we only have to combine (1.4.4) and (1.4.1). A graded free resolution can be calculated by Algorithm 1.3.4, which involves the computation of a Gröbner basis of M. Given a Gröbner basis of M, there is also a more direct way to find the Hilbert series, which will be discussed in Section 1.4.1.

The Hilbert series encodes geometric information as the following lemma shows.

Lemma 1.4.6. *Let $R = \bigoplus_{d \geq 0} R_d$ be a graded algebra, finitely generated over the field $R_0 = K$. Then $r := \dim(R)$ is equal to the pole order of $H(R, t)$ at $t = 1$.*

Proof. The proof requires the concept of homogeneous systems of parameters. For the definition and the proof of existence, we refer forward to Section 2.4.2. Let $f_1, \ldots, f_r$ be a homogeneous system of parameters for R, and set $A := K[f_1, \ldots, f_r]$. It follows from (1.4.1) that $H(A, t)$ has pole order r. In fact, $\lim_{t \nearrow 1} (1-t)^r H(A, t) = \prod_{i=1}^{r} d_i^{-1}$, where $\lim_{t \nearrow 1}$ denotes the limit from below (see Example 1.4.8 below). There exists an A-free resolution

$$0 \to F^{(r)} \to F^{(r-1)} \to \cdots \to F^{(0)} \to R \to 0.$$

Using (1.4.4) we conclude that $H(R,t)$ has pole order $\leq r$ because the same holds for all $H(F^{(i)},t)$. Note that $H(R,t) \geq H(A,t)$ for $0 < t < 1$ since $A \subseteq R$. If the pole order of $H(R,t)$ were strictly smaller than r, then

$$0 = \lim_{t \nearrow 1}(1-t)^r H(R,t) \geq \lim_{t \nearrow 1}(1-t)^r H(A,t) = \prod_{i=1}^{r} d_i^{-1} > 0.$$

This contradiction shows that $H(R,t)$ has in fact pole order r. $\qquad\square$

Definition 1.4.7. *Let $R = \bigoplus_d R_d$ be a graded algebra, finitely generated over $R_0 = K$. Then the* **degree of R** *is defined as*

$$\deg(R) = \lim_{t \nearrow 1}(1-t)^r H(R,t)$$

where $r := \dim(R)$ is the Krull dimension of R and $\lim_{t \nearrow 1}$ means the limit from below.

Up to a sign, the degree of R is the first coefficient of the Laurent series expansion of $H(R,t)$ at $t = 1$.

Example 1.4.8. If $A = K[x_1, \ldots, x_n]$ with $\deg(x_i) = d_i$, then

$$\deg(A) = \lim_{t \nearrow 1} \frac{(1-t)^n}{\prod_{i=1}^{n}(1-t^{d_i})} = \lim_{t \nearrow 1} \frac{1}{\prod_{i=1}^{n}(1+t+\cdots+t^{d_i-1})} = \frac{1}{\prod_{i=1}^{n} d_i},$$

so we have $\deg(A) = (\prod_{i=1}^{n} d_i)^{-1}$. $\qquad\triangleleft$

If $A = K[x_1, \ldots, x_n]$ (all x_i of degree 1) and $I \subset A$ is a homogeneous ideal, then I corresponds to a projective variety $Y \subset \mathbb{P}^{n-1}$. Then the degree of A/I is the same as the degree of Y as a projective variety (see Hartshorne [102, page 52]).

1.4.1 Computation of Hilbert Series

Again, let $R = K[x_1, \ldots, x_n]$ be a polynomial ring, graded by $\deg(x_i) = d_i$, and suppose that $I \subseteq R$ is a homogeneous ideal. We want to compute $H(R/I,t)$, or equivalently $H(I,t) = H(R,t) - H(R/I,t)$. We choose a monomial ordering ">" on R and use the Buchberger Algorithm 1.1.8 to compute a Gröbner basis $\mathcal{G} = \{g_1, \ldots, g_r\}$ of I with respect to ">". The leading monomials $\mathrm{LM}(g_1), \ldots, \mathrm{LM}(g_r)$ generate the leading ideal $L(I)$. If $m_1, \ldots, m_l \in L(I)$ are distinct monomials which span $L(I)_d$, then we can find homogeneous $f_1, \ldots, f_l \in I_d$ such that $\mathrm{LM}(f_i) = m_i$. It is clear that $f_1, \ldots, f_l$ is a basis of I_d. It follows that

$$\dim(L(I)_d) = \dim(I_d).$$

We conclude $H(L(I),t) = H(I,t)$, so we have reduced the problem to computing the Hilbert series of a monomial ideal.

So suppose that $I = (m_1, \ldots, m_l) \subseteq R$ is a monomial ideal. We will show how to compute $H(I, t)$ using recursion with respect to l. Let $J = (m_1, \ldots, m_{l-1})$, then we have an isomorphism

$$J/(J \cap (m_l)) \cong I/(m_l)$$

of graded R-modules. Notice that

$$J \cap (m_l) = (\mathrm{lcm}(m_1, m_l), \mathrm{lcm}(m_2, m_l), \ldots, \mathrm{lcm}(m_{l-1}, m_l)),$$

where lcm means least common multiple. By recursion we have $H(J, t)$ and $H(J \cap (m_l), t)$, and $H((m_l), t) = t^{\deg(m_l)} \prod_{i=1}^{n} (1 - t^{d_i})^{-1}$. So we can compute $H(I, t)$ as

$$H(I, t) = H((m_l), t) + H(J, t) - H(J \cap (m_l), t). \tag{1.4.5}$$

See Bayer and Stillman [13] for more details. A slightly different approach was taken in Bigatti et al. [21].

Example 1.4.9. Let us compute the Hilbert series of the ideal $I = (xz - y^2, xw - yz, yw - z^2) \subset A := K[x, y, z, w]$, where all indeterminates have degree 1. Note that $H(A, t) = (1 - t)^{-4}$ and $H((f), t) = t^d (1 - t)^{-4}$ if f is a homogeneous polynomial of degree d. We choose the lexicographic ordering ">" with $x > y > z > w$. Then $\mathcal{G} = \{xz - y^2, xw - yz, yw - z^2\}$ is a Gröbner basis of I. It follows that the initial ideal $L(I)$ is generated by xz, xw, yw. Observe that $(xz, xw) \cap (yw) = (xyzw, xyw) = (xyw)$. By (1.4.5) we get

$$H(L(I), t) = H((xz, xw, yw), t) = H((yw), t) + H((xz, xw), t) - H((xyw), t). \tag{1.4.6}$$

We know that $H((yw), t) = t^2/(1 - t)^4$ and $H((xyw), t) = t^3/(1 - t)^4$. We only need to find $H((xz, xw), t)$. Repeating the above process and making use of $(xz) \cap (xw) = (xzw)$, we obtain (again by 1.4.5)

$$H((xz, xw), t) = H((xw), t) + H((xz), t) - H((xzw), t) = \frac{2t^2 - t^3}{(1 - t)^4}. \tag{1.4.7}$$

Substituting (1.4.7) in (1.4.6) gives

$$H(I, t) = H(L(I), t) = \frac{t^2}{(1 - t)^4} + \frac{2t^2 - t^3}{(1 - t)^4} - \frac{t^3}{(1 - t)^4} = \frac{3t^2 - 2t^3}{(1 - t)^4},$$

and finally

$$H(A/I, t) = H(A, t) - H(I, t) = \frac{1}{(1 - t)^4} - \frac{3t^2 - 2t^3}{(1 - t)^4} = \frac{1 + 2t}{(1 - t)^2}.$$

The pole order of $H(A/I, t)$ at $t = 1$ is 2, so $\dim(A/I) = 2$. If we take $\lim_{t \nearrow 1} (1 - t)^2 H(A/I, t)$, we get $\deg(A/I) = 3$. The ideal I defines a curve of degree 3 in $\mathbb{P}^3$ (the twisted cubic curve). $\triangleleft$

1.5 The Radical Ideal

The computation of the radical ideal $\sqrt{I}$ of an ideal $I \subseteq K[x_1, \ldots, x_n]$ is one of the basic tasks of constructive ideal theory. For the purposes of this book, radical computation is important since it is used in de Jong's normalization algorithm, which we present in Section 1.6. An important point for us is that we want an algorithm which works in any characteristic. As we will see, radical computation is a quite cumbersome task. Almost all methods that were proposed approach the problem by reducing it to the zero-dimensional case (see, for example, Gianni et al. [83], Krick and Logar [156], Alonso et al. [10], Becker and Weispfenning [15]). To the best of our knowledge, the only exception is a "direct" method given by Eisenbud et al. [60]. However, the limitation of this algorithm is that it requires the ground field K to be of characteristic 0, or that $K[x_1, \ldots, x_n]/I$ is generated by elements whose index of nilpotency is less than $\mathrm{char}(K)$ (see Theorem 2.7 in [60]). In our presentation, we will adhere to the strategy of reducing to the zero-dimensional case. We first explain how this reduction works, and then address the problem of zero-dimensional radical computation. Concerning the latter problem, we present a new variant of the "traditional" algorithm, which was given by Kemper [139] and works in positive characteristic.

1.5.1 Reduction to Dimension Zero

The material in this subsection is largely drawn from Becker and Weispfenning [15, Section 8.7]. Given an ideal $I \subsetneq K[x_1, \ldots, x_n]$, we may apply Algorithm 1.2.4 to find the dimension of I and a subset $\mathcal{M} \subseteq \{x_1, \ldots, x_n\}$ such that the complement $\mathcal{M}' := \{x_1, \ldots, x_n\} \setminus \mathcal{M}$ is independent modulo I, and $|\mathcal{M}'| = \dim(I)$. Changing the ordering of the variables, we may assume that $\mathcal{M} = \{x_1, \ldots, x_r\}$ and $\mathcal{M}' = \{x_{r+1}, \ldots, x_n\}$. By Proposition 1.2.5, the ideal $J := IK(x_{r+1}, \ldots, x_n)[x_1, \ldots, x_r]$ is zero-dimensional. The main idea in the reduction step is to calculate $\sqrt{J}$ first. In order to work out the radical of I from this, one first has to be able to form the intersection of $\sqrt{J}$ with $K[x_1, \ldots, x_n]$. An algorithm for this purpose is given by the following lemma.

Lemma 1.5.1 (Becker and Weispfenning [15, Lemma 8.91]). *Let* $L = K(x_{r+1}, \ldots, x_n)$ *be a rational function field and* $J \subseteq L[x_1, \ldots, x_r]$ *an ideal in a polynomial ring over* L. *Furthermore, let* $\mathcal{G}$ *be a Gröbner basis of* J *with respect to any monomial ordering such that* $\mathcal{G} \subset K[x_1, \ldots, x_n]$. *Set*

$$f := \mathrm{lcm}\{\mathrm{LC}(g) \mid g \in \mathcal{G}\},$$

where the least common multiple is taken in $K[x_{r+1}, \ldots, x_n]$, *and let* I *be the ideal in* $K[x_1, \ldots, x_n]$ *generated by* $\mathcal{G}$. *Then*

$$J \cap K[x_1, \ldots, x_n] = I : f^\infty.$$

In the above lemma, the condition $\mathcal{G} \subset K[x_1, \ldots, x_n]$ can always be achieved by multiplying each element from the Gröbner basis by the least common multiple of the denominators of its coefficients. The saturation $I : f^\infty$ can be calculated by means of Proposition 1.2.2. Thus we are able to compute the intersection $J \cap K[x_1, \ldots, x_n]$, which is sometimes called the contraction ideal of J.

If $I \subseteq K[x_1, \ldots, x_n]$ is an ideal, we can form the ideal J in $K(x_{r+1}, \ldots, x_n)[x_1, \ldots, x_r]$ generated by I and then calculate the contraction ideal of J. However, this is not enough for our purposes, since we also need to be able to express I as the intersection of the contraction ideal of J with another ideal. This is achieved by the following lemma.

Lemma 1.5.2 (Becker and Weispfenning [15, Propos. 8.94, Lemma 8.95]). *Let $I \subseteq K[x_1, \ldots, x_n]$ be an ideal. Choose monomial orders "$>_1$" and "$>_2$" on $K[x_1, \ldots, x_r]$ and $K[x_{r+1}, \ldots, x_n]$, respectively, and let "$>$" be the block ordering obtained from "$>_1$" and "$>_2$" (see Example 1.1.2(d)). Furthermore, let $\mathcal{G}$ be a Gröbner basis of I with respect to "$>$" and form*

$$f := \operatorname{lcm}\{\mathrm{LC}_{>_1}(g) \mid g \in \mathcal{G}\},$$

where $\mathrm{LC}_{>_1}(g)$ is formed by considering g as a polynomial in $K(x_{r+1}, \ldots, x_n)[x_1, \ldots, x_r]$ and taking the leading coefficient with respect to "$>_1$". Then the contraction ideal of $J := IK(x_{r+1}, \ldots, x_n)[x_1, \ldots, x_r]$ is

$$J \cap K[x_1, \ldots, x_n] = I : f^\infty.$$

Moreover, if $I : f^\infty = I : f^k$ for some $k \in \mathbb{N}$, then

$$I = \left(I + (f^k)\right) \cap (I : f^\infty).$$

We have now provided all ingredients which allow to reduce the problem of radical computation to the zero-dimensional case.

Algorithm 1.5.3 (Higher dimensional radical computation). Let $I \subseteq K[x_1, \ldots, x_n]$ be an ideal. Perform the following steps to obtain the radical ideal $\sqrt{I}$.

(1) Use Algorithm 1.2.4 to compute the dimension d of I. If $d = -1$, then $I = K[x_1, \ldots, x_n] = \sqrt{I}$, and we are done. Otherwise, let $\mathcal{M} \subseteq \{x_1, \ldots, x_n\}$ be the subset produced by Algorithm 1.2.4. Renumber the variables such that $\mathcal{M} = \{x_1, \ldots, x_r\}$.

(2) Use Lemma 1.5.2 to find $f \in K[x_{r+1}, \ldots, x_n]$ such that

$$I = \left(I + (f^k)\right) \cap (IL[x_1, \ldots, x_r] \cap K[x_1, \ldots, x_n]) \tag{1.5.1}$$

for some $k \in \mathbb{N}$, where $L := K(x_{r+1}, \ldots, x_n)$.

(3) Compute $J := \sqrt{IL[x_1, \ldots, x_r]}$. (Note that $IL[x_1, \ldots, x_r]$ is zero-dimensional by Proposition 1.2.5.)

(4) Use Lemma 1.5.1 to compute

$$J^c := J \cap K[x_1, \ldots, x_n].$$

(5) Apply this algorithm recursively to compute $\sqrt{I + (f)}$. Then

$$\sqrt{I} = \sqrt{I + (f)} \cap J^c, \tag{1.5.2}$$

which can be computed by Equation (1.2.3).

In order to convince ourselves that Algorithm 1.5.3 works correctly, we must show that (1.5.2) holds, and that the recursion will terminate. Indeed, (1.5.1) yields

$$\sqrt{I} = \sqrt{I + (f^k)} \cap \sqrt{IL[x_1, \ldots, x_r] \cap K[x_1, \ldots, x_n]} =$$
$$= \sqrt{I + (f)} \cap (J \cap K[x_1, \ldots, x_n]),$$

which is (1.5.2). Moreover, $I \cap K[x_{r+1}, \ldots, x_n] = \{0\}$ by Equation (1.2.5). Therefore $f \notin I$, so $I + (f)$ is a strictly larger ideal than I. Hence the recursion terminates since $K[x_1, \ldots, x_n]$ is Noetherian.

1.5.2 Zero-dimensional Radicals

Algorithm 1.5.3 reduces the computation of a radical ideal to the zero-dimensional case, but at the expense of having to compute over a larger field L. This field L is a rational function field over the original ground field K, so if K is a finite field, for example, then in general L is no longer perfect.

Let K be a field and $f \in K[x]$ a non-zero polynomial with coefficients in K. We call f **separable** if f has no multiple roots in a splitting field $L \geq K$. This is equivalent with $\gcd(f, f') = 1$ (see Becker and Weispfenning [15, Proposition 7.33]). If

$$f = c \cdot \prod_{i=1}^{m} (x - \alpha_i)^{e_i}$$

with $c \in K \setminus \{0\}$ and $\alpha_i \in L$ pairwise distinct roots of f, we write

$$\mathrm{sep}(f) := c \cdot \prod_{i=1}^{m} (x - \alpha_i) \in L[x]$$

for the **separable part** of f. If $\mathrm{char}(K) = 0$, then we have

$$\mathrm{sep}(f) = \frac{f}{\gcd(f, f')},$$

where the greatest common divisor is taken to be monic. Note that the computation of the gcd can be performed by the Euclidean algorithm (see Geddes et al. [80, Section 2.4]). Thus in characteristic 0 the separable part is very easy to get, and it coincides with the squarefree part. We will consider the case of positive characteristic below. The algorithm for zero-dimensional radical computation is based on the following result.

Proposition 1.5.4 (Seidenberg [214, Lemma 92]). *Let* $I \subseteq K[x_1, \ldots, x_n]$ *be an ideal in a polynomial ring over a field* K. *If* $I \cap K[x_i]$ *contains a separable polynomial for each* $i = 1, \ldots, n$, *then* $I = \sqrt{I}$.

A proof can also be found in Becker and Weispfenning [15, Lemma 8.13]. If I is zero-dimensional, then $I \cap K[x_i] \neq \{0\}$ for every i, since there exists no variables which are independent modulo I (see after Algorithm 1.2.4). Non-zero polynomials in $I \cap K[x_i]$ can most easily found by the following algorithm, which goes back to Faugère et al. [66].

Algorithm 1.5.5 (Finding univariate polynomials). Given an ideal $I \subseteq K[x_1, \ldots, x_n]$ and an index $i \in \{1, \ldots, n\}$ such that $I \cap K[x_i] \neq \{0\}$, find a non-zero polynomial $f \in I \cap K[x_i]$ as follows:

(1) Compute a Gröbner basis $\mathcal{G}$ of I with respect to an arbitrary monomial ordering.
(2) For $d = 0, 1, 2, \ldots$ perform steps (3)–(4).
(3) Compute the normal form $\mathrm{NF}_{\mathcal{G}}(x_i^d)$.
(4) Test whether the sequence $\mathrm{NF}_{\mathcal{G}}(x_i^0), \ldots, \mathrm{NF}_{\mathcal{G}}(x_i^d)$ is linearly independent over K. If it is, continue the loop for the next d. Otherwise, go to step 5.
(5) If

$$\sum_{j=0}^{d} \alpha_j \, \mathrm{NF}_{\mathcal{G}}(x_i^j) = 0$$

is a K-linear relation found in step (4), then $f := \sum_{j=0}^{d} \alpha_j x_i^j \in I$ is the desired polynomial.

It is clear that the f from Algorithm 1.5.5 lies in I, since $\mathrm{NF}_{\mathcal{G}}(f) = 0$ by the linearity of the normal form. We can now present the algorithm for zero-dimensional radical computation in characteristic 0.

Algorithm 1.5.6 (Zero-dimensional radical in characteristic 0). Given a zero-dimensional ideal $I \subseteq K[x_1, \ldots, x_n]$ with $\mathrm{char}(K) = 0$, perform the following steps:

(1) For $i = 1, \ldots, n$, use Algorithm 1.5.5 to obtain a non-zero $f_i \in I \cap K[x_i]$.
(2) For each i, compute $g_i := \mathrm{sep}(f_i) = f_i / \gcd(f_i, f_i')$, where the derivative is with respect to x_i.
(3) Set $\sqrt{I} := I + (g_1, \ldots, g_n)$.

The correctness of the above algorithm follows from Proposition 1.5.4. Now we come to the case of positive characteristic. Our presentation is largely drawn from Kemper [139]. The following example shows that applying Algorithm 1.5.6 may produce false results in this case.

Example 1.5.7 (Becker and Weispfenning [15, Example 8.16]). Let $K = \mathbb{F}_p(t)$ be the rational function field over $\mathbb{F}_p$ and consider the ideal

$$I = (x^p - t, y^p - t) \subseteq K[x, y].$$

$x^p - t$ and $y^p - t$ are both squarefree (in fact, irreducible), but $(x - y)^p = x^p - y^p \in I$. Hence $x - y \in \sqrt{I}$, so I is not a radical ideal.   ◁

The trouble is that in positive characteristic a univariate may be squarefree but not separable, and the separable part is only defined over a larger field. We have the following algorithm, which works over a rational function field over a finite field.

Algorithm 1.5.8 (Separable part). Given a non-zero polynomial $f \in k(t_1, \ldots, t_m)[x]$ with coefficients in a rational function field over a perfect field k of characteristic $p > 0$, compute the separable part of f as a polynomial in $k(\sqrt[q]{t_1}, \ldots, \sqrt[q]{t_m})[x]$ with q a power of p.

(1) Set $h := \gcd(f, f')$.
(2) Set $g_1 := f/h$.
(3) Set $\tilde{h} := \gcd(h, h')$.
(4) If $\tilde{h} = h$, go to (6).
(5) Set $h := \tilde{h}$ and go to (3).
(6) If $h = 1$ then return g_1.
(7) Write $h = u(x^p)$ with $u \in k(t_1, \ldots, t_m)[x]$. *(This is possible since $h' = 0$.)*
(8) Form $v \in k(\sqrt[p]{t_1}, \ldots, \sqrt[p]{t_m})[x]$ from u by replacing every t_i occurring in u with $\sqrt[p]{t_i}$ and every $a \in k$ in u with $\sqrt[p]{a} \in k$. *(Thus $v^p = h$.)*
(9) Compute $g_2 := \mathrm{sep}(v)$ by a recursive call.
(10) Compute $g_3 := \mathrm{sep}(g_1 g_2)$ by a recursive call and return g_3.

The proof of correctness of Algorithm 1.5.8 is straightforward and can be found in Kemper [139] or Kreuzer and Robbiano [155, Proposition 3.7.12] (the latter contains essentially the same algorithm). The following algorithm uses the separable part, as computed by Algorithm 1.5.8, to obtain the radical of a zero-dimensional ideal in positive characteristic.

Algorithm 1.5.9 (Zero-dimensional radical in characteristic p). Given a zero-dimensional ideal $I \subseteq K[x_1, \ldots, x_n]$ in a polynomial ring over the rational function field $K = k(t_1, \ldots, t_m)$ with k a perfect field of characteristic $p > 0$, obtain $\sqrt{I}$ as follows:

(1) For $i = 1, \ldots, n$, use Algorithm 1.5.5 to obtain a non-zero $f_i \in I \cap K[x_i]$.
(2) For each i, compute $\mathrm{sep}(f_i) \in k(\sqrt[p^{r_i}]{t_1}, \ldots, \sqrt[p^{r_i}]{t_m})[x_i]$ by using Algorithm 1.5.8.
(3) For each i, write $\mathrm{sep}(f_i) = g_i(\sqrt[q]{t_1}, \ldots, \sqrt[q]{t_m}, x_i)$, where $q := p^r$, $r := \max\{r_1, \ldots, r_n\}$, and $g_i \in K[y_1, \ldots, y_m, x_i]$ with new indeterminates $y_1, \ldots, y_m$.

(4) Form the ideal

$$J := IK[y_1, \ldots, y_m, x_1, \ldots, x_n] + (g_1, \ldots, g_n) + (y_1^q - t_1, \ldots, y_m^q - t_m)$$
$$\subseteq K[y_1, \ldots, y_m, x_1, \ldots, x_n].$$

(5) Calculate the elimination ideal

$$\tilde{J} := J \cap K[x_1, \ldots, x_n]$$

by using Algorithm 1.2.1 and return $\sqrt{I} = \tilde{J}$.

Again, it is straightforward to see that Algorithm 1.5.9 is correct. In fact, the g_i together with I generate the radical ideal over the larger field $k(\sqrt[q]{t_1}, \ldots, \sqrt[q]{t_m})$, and then in step (5) this radical is intersected with the original polynomial ring $k(t_1, \ldots, t_m)[x_1, \ldots, x_n]$. A formal proof is given in Kemper [139].

Remark 1.5.10. Although we formulated Algorithm 1.5.9 only for ground fields that are rational function fields over a perfect field, it can be made to work over any field of positive characteristic which is finitely generated over a perfect field. For details, we refer to Kemper [139].   ◁

Example 1.5.11. It is interesting to see how Algorithm 1.5.9 handles Example 1.5.7. So consider the ideal $I = (x_1^p - t, x_2^p - t) \subseteq \mathbb{F}_p(t)[x_1, x_2]$. We have

$$\mathrm{sep}(x_i^p - t) = x_i - \sqrt[p]{t},$$

so in step (4) of Algorithm 1.5.9 we obtain the ideal

$$J = (x_1 - y, x_2 - y, y^p - t) \subseteq \mathbb{F}_p(t)[x_1, x_2, y].$$

We choose the lexicographic monomial ordering with $y > x_1 > x_2$ on $\mathbb{F}_p(t)[x_1, x_2, y]$. By replacing $x_1 - y$ and $y^p - t$ by their normal forms with respect to $x_2 - y$, we obtain the new basis

$$\mathcal{G} = \{x_1 - x_2, x_2 - y, x_2^p - t\}.$$

$\mathcal{G}$ is a Gröbner basis since the polynomials in $\mathcal{G}$ have pairwise coprime leading monomials. Hence step (5) of Algorithm 1.5.9 yields

$$\sqrt{I} = J \cap \mathbb{F}_p(t)[x_1, x_2] = (x_1 - x_2, x_2^p - t),$$

which is the correct result.   ◁

1.6 Normalization

Let R be an integral domain and $\tilde{R}$ the integral closure of R in its field of fractions. We call $\tilde{R}$ the **normalization** of R. If $\tilde{R} = R$, we say that R is

normal. One reason why normalization is interesting in invariant theory, is that every invariant ring $K[x_1, \ldots, x_n]^G$ is normal (see Proposition 2.3.11). The usefulness of normalization is further underlined by Theorem 3.9.15. In this section we will describe a new (or at least newly re-discovered) algorithm by de Jong [121] for computing the normalization of an integral domain that is finitely generated as an algebra over a field (i.e., an "affine domain"). In fact, the algorithm is based on a theorem by Grauert and Remmert (see the references in [121]). Let R be a Noetherian integral domain. If $I \subseteq R$ is a non-zero ideal, choose $0 \neq f \in I$ and consider the mapping

$$\Psi\colon \operatorname{Hom}_R(I, R) \to \operatorname{Quot}(R), \quad \varphi \mapsto \frac{\varphi(f)}{f},$$

where $\operatorname{Quot}(R)$ denotes the field of fractions of R.

Lemma 1.6.1. *The map Ψ is independent of the choice of f, and it is a monomorphism of R-modules. Moreover, the restriction of Ψ to $\operatorname{End}_R(I)$ is a homomorphism of R-algebras, and*

$$R \subseteq \Psi\left(\operatorname{End}_R(I)\right) \subseteq \tilde{R}.$$

Proof. For $0 \neq g \in I$ we have

$$\frac{\varphi(g)}{g} = \frac{f\varphi(g)}{fg} = \frac{\varphi(fg)}{fg} = \frac{g\varphi(f)}{fg} = \frac{\varphi(f)}{f}.$$

This implies the independence of f and the injectivity of Ψ. It is clear that Ψ is a homomorphism of R-modules, and of R-algebras if restricted to $\operatorname{End}_R(I)$. The image of $\operatorname{End}_R(I)$ is contained in $\tilde{R}$ since $\operatorname{End}_R(I)$ is finitely generated as an R-module, and R is naturally embedded into $\operatorname{End}_R(I)$. $\square$

Lemma 1.6.2. *Let $I \subseteq R$ be a non-zero radical ideal. Then*

$$\operatorname{End}_R(I) = \Psi^{-1}(\tilde{R}).$$

Proof. The inclusion "$\subseteq$" follows from Lemma 1.6.1. For the reverse inclusion, take $\varphi \in \operatorname{Hom}_R(I, R)$ such that $h := \Psi(\varphi) \in \tilde{R}$. Then $hI \subseteq R$, and we have to show that $hI \subseteq I$. There exists an equation

$$h^k = a_0 + a_1 h + \cdots + a_{k-1} h^{k-1}$$

with $a_i \in R$. Hence for $f \in I$ we have

$$(hf)^k = a_0 f^k + a_1(hf)f^{k-1} + \cdots + a_{k-1}(hf)^{k-1}f \in I,$$

hence $hf \in I$ by the hypothesis. This proves the lemma. $\square$

By Lemma 1.6.1, normality of R implies $\Psi(\mathrm{End}_R(I)) = R$ for all non-zero ideals I. We can now give conditions on I under which the converse holds. We write $X := \mathrm{Spec}(R)$ and

$$X_{\mathrm{nn}} := \{x \in X \mid R_x \text{ is not normal}\}$$

for the non-normal locus. For an ideal $I \subseteq R$ we write $\mathcal{V}_X(I) := \{x \in X \mid I \subseteq x\}$. (The inclusion makes sense since the $x \in X$ are prime ideals in R.)

Theorem 1.6.3. *With the notation introduced above, let $I \subseteq R$ be a non-zero radical ideal such that $X_{\mathrm{nn}} \subseteq \mathcal{V}_X(I)$. Then the equivalence*

$$R \text{ is normal} \quad \Longleftrightarrow \quad \Psi(\mathrm{End}_R(I)) = R$$

holds.

Proof. The implication "$\Longrightarrow$" follows from Lemma 1.6.1. For the converse, assume that $\Psi(\mathrm{End}_R(I)) = R$ and take $h \in \tilde{R}$. With $J := \{f \in R \mid fh \in R\}$ we have

$$P(h) := \{x \in X \mid h \notin R_x\} = \mathcal{V}_X(J).$$

On the other hand, $P(h) \subseteq X_{\mathrm{nn}}$ by definition of $P(h)$. By hypothesis, $P(h) \subseteq \mathcal{V}_X(I)$, and therefore $I = \sqrt{I} \subseteq \sqrt{J}$. Thus there exists a non-negative integer d with $I^d \subseteq J$, hence $hI^d \subseteq R$ by definition of J. Let d be minimal with this property and assume, by way of contradiction, that $d > 0$. Then there exists an element $a \in I^{d-1}$ with $ha \notin R$. We have $ha \in \tilde{R}$ and $haI \subseteq hI^d \subseteq R$, hence $ha \in \Psi(\mathrm{Hom}_R(I, R)) \cap \tilde{R}$. Lemma 1.6.2 yields $haI \subseteq I$ and therefore, by the hypothesis that $\Psi(\mathrm{End}_R(I)) = R$, ha lies in R, a contradiction. Hence $d = 0$ after all, so $h \in R$. Since h was an arbitrary element from $\tilde{R}$, this completes the proof. $\square$

An apparent difficulty about Theorem 1.6.3 is that it seems to be hard to get one's hands on $\mathrm{End}_R(I)$. But this turns out to be surprisingly easy. In fact, multiplication by a non-zero $f \in I$ gives an isomorphism

$$\Psi(\mathrm{End}_R(I)) \to (f \cdot I) : I$$

(see Greuel and Pfister [98, Remark 3.1]). Thus we only need to compute a quotient ideal to obtain $\mathrm{End}_R(I)$. We summarize the results and translate them to the situation where $R = K[x_1, \ldots, x_n]/I$ is an affine domain.

Theorem 1.6.4. *Let $I \subset K[x_1, \ldots, x_n]$ be a prime ideal, $J \subseteq K[x_1, \ldots, x_n]$ an ideal containing I, and $f \in J \setminus I$. Then with $R := K[x_1, \ldots, x_n]/I$ we have:*

(a) For every $g \in (f \cdot J + I) : J$, the quotient $(g + I)/(f + I)$ lies in the normalization $\tilde{R}$.

(b) Assume moreover that J is a radical ideal such that the non-normal locus of $X := \mathrm{Spec}(R)$ is contained in $\mathcal{V}_X(J)$. Then R is normal if and only if

$$(f) + I = (f \cdot J + I) : J.$$

Remark 1.6.5. If K is a perfect field, an ideal J satisfying the conditions of Theorem 1.6.4(b) can be found as follows. Let $f_1, \ldots, f_m$ be generators of I and let $\mathfrak{J} = (\partial f_i / \partial x_j)_{i,j}$ be the Jacobian matrix. Then by Eisenbud [59, Theorem 16.19], the singular locus of $X = \mathrm{Spec}(R)$ is

$$X_{\mathrm{sing}} = \{P \in X \mid \mathfrak{J} \text{ reduced modulo } P \text{ has rank} < n - \dim(R)\}.$$

But the non-normal locus X_{nn} is contained in X_{sing} (see Eisenbud [59, Theorem 19.19]). If $J_0 \subseteq K[x_1, \ldots, x_n]$ is the ideal generated by I and all $(h \times h)$-minors of $\mathfrak{J}$, where $h := n - \dim(R)$, it follows that

$$X_{\mathrm{nn}} \subseteq X_{\mathrm{sing}} = \mathcal{V}_X(J_0).$$

In particular, $J_0 \supsetneq I$. But then for any ideal J with $I \subseteq J \subseteq \sqrt{J_0}$ we have

$$X_{\mathrm{nn}} \subseteq \mathcal{V}_X(J).$$

Therefore J can be chosen by taking an $(h \times h)$-minor f of $\mathfrak{J}$ which is not contained in I, and setting $J := \sqrt{I + (f)}$. ◁

We can use Theorem 1.6.4 to calculate the normalization by iteratively adding new generators $(g + I)/(f + I)$ to R until the condition in (b) is satisfied. But for each new iteration we need a presentation for the updated algebra R. De Jong [121] proposed a method for getting a presentation of $\mathrm{End}_R(I)$ consisting of relations of degree one and two. We take a somewhat different approach, given by the following lemma which reduces the task to the computation of an elimination ideal.

Lemma 1.6.6. *Let $K \subseteq L$ be a field extension and $\varphi \colon K[x_1, \ldots, x_n] \to L$ a homomorphism of K-algebras with kernel I. Furthermore, let $a = \varphi(g)/\varphi(f) \in L$ and consider the homomorphism*

$$\Phi \colon K[x_1, \ldots, x_n, y] \to L, \quad x_i \mapsto \varphi(x_i), \quad y \mapsto a,$$

where y is an indeterminate. With an additional indeterminate t, set

$$J := I \cdot K[x_1, \ldots, x_n, y, t] + (fy - g, ft - 1).$$

Then

$$\ker(\Phi) = J \cap K[x_1, \ldots, x_n, y].$$

Proof. J lies in the kernel of the homomorphism $K[x_1, \ldots, x_n, y, t] \to L$ with $x_i \mapsto \varphi(x_i)$, $y \mapsto a$, $t \mapsto 1/\varphi(f)$. Hence $J \cap K[x_1, \ldots, x_n, y] \subseteq \ker(\Phi)$. To prove the converse, we first remark that $fh \in J$ for $h \in K[x_1, \ldots, x_n, y, t]$ implies $h \in J$, since

$$h = tfh - h(tf - 1).$$

Furthermore, $f^i y^i - g^i \in J$ for any non-negative integer i, since

$$f^{i+1} y^{i+1} - g^{i+1} = (f^i y^i - g^i) fy + g^i (fy - g).$$

It follows that $f^d \left(y^i - (g/f)^i \right) \in J$ for $d \geq i$. Let $h \in \ker(\Phi)$ and set $d := \deg_y(h)$. Write $h(g/f)$ for the result of substituting y by g/f in h. Then $\varphi \left(f^d h(g/f) \right) = 0$, so $f^d h(g/f) \in I$. By the preceding argument we also have $f^d \left(h - h(g/f) \right) \in J$, hence $f^d h \in J$. But this implies $h \in J$, completing the proof. $\qquad\qquad\square$

We can now give the ensuing algorithm, whose termination is guaranteed by the fact that it generates a strictly ascending sequence of R-modules between R and $\tilde{R}$, with $\tilde{R}$ being Noetherian.

Algorithm 1.6.7 (de Jong's algorithm). Given a prime ideal $I \subseteq K[x_1, \ldots, x_n]$ with K a perfect field, perform the following steps to obtain the normalization $\tilde{R}$ of $R := K[x_1, \ldots, x_n]/I$, given by a presentation $\tilde{R} \cong K[x_1, \ldots, x_{n+m}]/\tilde{I}$ (and the embedding $R \subseteq \tilde{R}$ given by $x_i + I \mapsto x_i + \tilde{I}$):

(1) Set $m := 0$ and $\tilde{I} := I$.
(2) Compute the Jacobian matrix $\mathfrak{J} := (\partial f_i / \partial x_j)_{i,j}$, where $\tilde{I} = (f_1, \ldots, f_k)$.
(3) With $l := n + m - \dim(R)$, compute the ideal generated by $\tilde{I}$ and the $(l \times l)$-minors of $\mathfrak{J}$. Call this ideal J_{sing}. Choose an ideal J_0 such that $\tilde{I} \subsetneq J_0 \subseteq J_{\mathrm{sing}}$ and an element $f \in J_0 \setminus \tilde{I}$.
(4) Compute $J := \sqrt{J_0}$ and the quotient ideal $(f \cdot J + \tilde{I}) : J$.
(5) If $(f \cdot J + \tilde{I}) : J \subseteq \tilde{I} + (f)$ (test this by computing normal forms of the generators of the left hand side with respect to a Gröbner basis of the right hand side), we are done.
(6) Otherwise, choose $g \in \left((f \cdot J + \tilde{I}) : J \right) \setminus \left(\tilde{I} + (f) \right)$.
(7) Set $m := m + 1$ and form the ideal

$$J' := \tilde{I} \cdot K[x_1, \ldots, x_{n+m}, t] + (f x_{n+m} - g, ft - 1)$$

in $K[x_1, \ldots, x_{n+m}, t]$ with x_{n+m} and t new indeterminates.
(8) Compute $\tilde{I} := J' \cap K[x_1, \ldots, x_{n+m}]$ and go to step (2).

Remark. It is in step (4) that Algorithm 1.6.7 requires radical computation. This is the reason why the ability to calculate radical ideals is important in this book. $\qquad\qquad\triangleleft$

We conclude the section with an example.

Example 1.6.8. We can use Algorithm 1.6.7 to de-singularize curves. As an example, consider the curve $\mathcal{C}$ in $\mathbb{C}^2$ given by the ideal

$$I = (x^6 + y^6 - xy),$$

which has genus 9 and a double point at the origin. A Gröbner basis of $J_{\text{sing}} = (x^6 + y^6 - xy, 6x^5 - y, 6y^5 - x)$ is $\{x, y\}$. Therefore we can choose $f = x$ and $J = J_0 = J_{\text{sing}}$. We obtain

$$(f \cdot J + \tilde{I}) : J = (x, y^5) \quad \text{and} \quad \tilde{I} + (f) = (x, y^6).$$

Thus we can choose $g := y^5$ to obtain a new element $a := (g + I)/(f + I)$ in $\tilde{R}$. By step (8) of Algorithm 1.6.7 we calculate the kernel $\tilde{I}$ of the map

$$\mathbb{C}[x, y, z] \to \tilde{R}, \ x \mapsto x + I, \ y \mapsto y + I, \ z \mapsto a$$

and obtain

$$\tilde{I} = (y^5 - xz, x^5 + yz - y, x^4y^4 + z^2 - z).$$

The last equation confirms the integrality of a over R. Going into the next iteration of Algorithm 1.6.7 yields no new elements in $\tilde{R}$, hence $\tilde{R} = \mathbb{C}[x, y, z]/\tilde{I}$. $\tilde{I}$ defines a curve $\tilde{\mathcal{C}}$ in $\mathbb{C}^3$ which maps onto $\mathcal{C}$ by projecting on the first two coordinates. With the exception of the origin, every point of $\mathcal{C}$ has a fiber consisting of a single point, and the fiber of the origin consists of the points $(0, 0, 0)$ and $(0, 0, 1)$. ◁

2 Invariant Theory

For convenience, we will assume throughout this chapter that our base field K is algebraically closed, unless stated otherwise. Many results in this chapter remain true for arbitrary fields K as long as all relevant morphisms and varieties are defined over K.

We need to introduce some basic notation concerning algebraic groups. A **linear algebraic group** is an affine variety G which has a structure as a group such that the multiplication $\mu\colon G \times G \to G$ and the inversion $\eta\colon G \to G$ are morphisms of affine varieties. Examples of linear algebraic groups will be given below. We say that a linear algebraic group G **acts regularly** on an affine variety X if an action of G on X is given by a morphism $G \times X \to X$. A representation of G is a finite dimensional K-vector space V together with a group homomorphism $G \to \mathrm{GL}(V)$. A representation V is called **rational** if G acts regularly on V (considered as an affine n-space). A basic result is that every linear algebraic group has a faithful rational representation. For more information on linear algebraic groups we refer the reader to the Appendix of this book or to the books of Borel [23], Humphreys [118], and Springer [233].

2.1 Invariant Rings

Suppose that G is a linear algebraic group acting regularly on an affine variety X. If $f \in K[X]$ and $\sigma \in G$, then we define $\sigma \cdot f \in K[X]$ by

$$(\sigma \cdot f)(x) := f(\sigma^{-1} \cdot x) \quad \text{for all} \quad x \in X.$$

This defines an action of G on the coordinate ring of X. If $f \in K[X]$ and $\sigma \cdot f = f$ for all $\sigma \in G$, then f is called an **invariant**. In general, we are interested in the set

$$K[X]^G := \{ f \in K[X] \mid \sigma \cdot f = f \text{ for all } \sigma \in G \}$$

of all G-invariants. The set $K[X]^G$ is a subring of $K[X]$ and we call it the **invariant ring** of G. In this book we will focus on the case where $X = V$ is a representation of G. Then $K[V]$ is isomorphic to the polynomial ring $K[x_1, \ldots, x_n]$, where n is the dimension of V as a K-vector space. The polynomial ring $K[V] = \bigoplus_{d=0}^{\infty} K[V]_d$ is graded with respect to the total degree.

The G-action on $K[V]$ preserves degree and $K[V]^G \subseteq K[V]$ inherits the grading. This grading of $K[V]$ and $K[V]^G$ is very useful and it makes polynomial invariants of representations easier to deal with than regular invariant functions of affine varieties.

A fundamental problem in invariant theory is to find generators of the invariant ring $K[V]^G$. So a basic question is:

Can one always find finitely many generators $f_1, \ldots, f_r$ such that $K[V]^G = K[f_1, \ldots, f_r]$?

It is clear that the invariant ring $K[V]^G$ is in fact the intersection of the invariant field $K(x_1, \ldots, x_n)^G$ with the polynomial ring $K[x_1, \ldots, x_n]$. Hilbert asked in his fourteenth problem (see Hilbert [109]) the more general question whether the intersection of any subfield L of $K(x_1, \ldots, x_n)$ with $K[x_1, \ldots, x_n]$ gives a finitely generated ring. Both questions have a negative answer, since there exists a counter-example due to Nagata (see Nagata [171]). In many cases however, the invariant ring is finitely generated.

Related to finding generators of the invariant ring is the problem of finding degree bounds. For a graded ring $R = \bigoplus_{d=0}^{\infty} R_d$ we define

$$\beta(R) := \min\{k \mid R \text{ is generated by } \bigoplus_{d=0}^{k} R_d\}. \tag{2.1.1}$$

The problem is to find a good upper bound D such that $\beta(K[V]^G) \leq D$ (see Derksen and Kraft [55] for an overview). Once such an upper bound is known, there is an easy but inefficient method for finding generators of the invariant ring: Using linear algebra, we can find $K[V]_d^G$ using for all $d \leq D$ as follows. If $\sigma_1, \ldots, \sigma_r \in G$ generate a subgroup which is Zariski dense in G, then $K[V]_d^G$ is the kernel of the linear map $K[V]_d \to K[V]^r$ defined by

$$f \mapsto (\sigma_1 \cdot f - f, \sigma_2 \cdot f - f, \cdots, \sigma_r \cdot f - f).$$

Alternatively, if G is connected, $K[V]_d^G$ can be computed using the Lie algebra action.

There are also more sophisticated algorithms for computing invariants as we will see later on. Gröbner bases are used for most algorithms related to invariant rings. Most importantly, they will be used to compute generators of invariant rings in the next chapters. Moreover, Gröbner bases can be employed to find the relations between the generators $f_1, \ldots, f_r$ of an invariant ring $K[V]^G = K[f_1, \ldots, f_r]$ (see Section 1.2.2) and to express an arbitrary invariant as a polynomial in the generators.

Example 2.1.1. Suppose the symmetric group S_n act on $V = K^n$ by

$$\sigma \cdot (x_1, x_2, \ldots, x_n) = (x_{\sigma(1)}, x_{\sigma(2)}, \ldots, x_{\sigma(n)}), \qquad \sigma \in S_n.$$

Let us write

$$\varphi(t) := (t - x_1)(t - x_2) \cdots (t - x_n) = t^n - f_1 t^{n-1} + f_2 t^{n-2} - \cdots + (-1)^n f_n$$

with $f_1, \ldots, f_n \in K[x_1, \ldots, x_n]$ the so-called **elementary symmetric polynomials**. Formulas for $f_1, \ldots, f_n$ are given by:

$$
\begin{aligned}
f_1 &= x_1 + x_2 + \cdots + x_n \\
f_2 &= x_1 x_2 + x_1 x_3 + \cdots + x_1 x_n + x_2 x_3 + \cdots + x_{n-1} x_n \\
&\vdots \\
f_r &= \textstyle\sum_{i_1 < i_2 < \cdots < i_r} x_{i_1} x_{i_2} \cdots x_{i_r} \\
&\vdots \\
f_n &= x_1 x_2 \cdots x_n.
\end{aligned}
$$

The invariant ring of S_n in this representation is generated by the algebraically independent invariants $f_1, \ldots, f_n$ (see Theorem 3.10.1).   ◁

Example 2.1.2. Suppose that $\mathrm{char}(K) = 0$. The invariant rings of the 2-dimensional special linear group SL_2 over K were studied intensively in the nineteenth century. Gordan proved that invariant rings for SL_2 are always finitely generated (cf. Gordan [94]). Let V_d be the vector space

$$\{a_0 x^d + a_1 x^{d-1} y + \cdots + a_d y^d \mid a_0, a_1, \ldots, a_d \in K\}$$

of homogeneous polynomials of degree d in x and y. Such polynomials are often referred to as **binary forms**. The coordinate ring $K[V_d]$ can be identified with $K[a_0, a_1, \ldots, a_d]$. We can define an action of SL_2 on V_d by

$$\sigma \cdot g(x, y) := g(\alpha x + \gamma y, \beta x + \delta y), \qquad \sigma = \begin{pmatrix} \alpha & \beta \\ \gamma & \delta \end{pmatrix} \in \mathrm{SL}_2.$$

There are many ways of constructing invariants for binary forms. One important invariant of a binary form is the discriminant. Suppose that $g = a_0 x^d + a_1 x^{d-1} y + \cdots + a_d y^d$ and $h = b_0 x^e + b_1 x^{e-1} y + \cdots + b_e y^e$ are binary forms of degree d and e respectively. We define the **resultant** of g and h as the $(d + e) \times (d + e)$-determinant

$$
\mathrm{Res}(g, h) = \begin{vmatrix}
a_0 & a_1 & \cdots & a_d & & & & \\
& a_0 & a_1 & \cdots & a_d & & & \\
& & \ddots & & & \ddots & & \\
& & & a_0 & a_1 & \cdots & a_d & \\
b_0 & b_1 & \cdots & b_e & & & & \\
& b_0 & b_1 & \cdots & b_e & & & \\
& & \ddots & & & \ddots & & \\
& & & b_0 & b_1 & \cdots & b_e &
\end{vmatrix}.
$$

Since K is algebraically closed, we can write g and h as a product of linear functions

$$g = \prod_{i=1}^{d}(p_i x + q_i y), \qquad h = \prod_{i=1}^{e}(r_i x + s_i y).$$

Then we find

$$\mathrm{Res}(g, h) = \prod_{i=1}^{d}\prod_{j=1}^{e}(p_i s_j - q_i r_j). \tag{2.1.2}$$

From (2.1.2), it is easy to check that $\mathrm{Res}(g, h)$ is an invariant function in $K[V_d \oplus V_e]$ which vanishes if and only if g and h have a common zero in $\mathbb{P}^1$. The **discriminant** of g is defined as

$$\Delta(g) = \frac{(-1)^{d(d-1)/2}}{a_0} \, \mathrm{Res}\left(g, \frac{\partial}{\partial x}g\right).$$

Another formula for $\Delta(g)$ is given by

$$\prod_{i<j}(p_i q_j - q_i p_j)^2. \tag{2.1.3}$$

It is clear from (2.1.3) that the discriminant is an invariant in $K[V_d]$ and it vanishes exactly when g has a multiple zero.

In the literature, discriminants and resultants are usually defined for polynomials in one variable x (see Lang [158, §10]). If g and h are inhomogeneous polynomials, then we just define $\mathrm{Res}(g, h) = \mathrm{Res}(\widehat{g}, \widehat{h})$ and $\Delta(g) = \Delta(\widehat{g})$ where $\widehat{g}$ and $\widehat{h}$ are the homogenized polynomials. For the proofs of all formulas (for inhomogeneous polynomials) we refer to Lang [158, Chapter V, § 10].

For any d, let $g_d = a_0 x^d + a_1 x^{d-1} y + \cdots + a_d y^d$ be the general binary form of degree d. For $d = 2$ it turns out that $K[V_2]^{\mathrm{SL}_2} = K[\Delta(g_2)]$, where $\Delta(g_2) = a_1^2 - 4a_0 a_2$ is the well-known discriminant of a quadratic polynomial $g_2 = a_0 x^2 + a_1 xy + a_2 y^2$. Similarly, for $d = 3$ we get $K[V_3]^{\mathrm{SL}_2} = K[\Delta(g_3)]$, where now

$$\Delta(g_3) = a_1^2 a_2^2 - 4a_0 a_2^3 - 4a_1^3 a_3 - 27a_0^2 a_3^2 + 18a_0 a_1 a_2 a_3.$$

For binary forms of degree 4, one finds $K[V_4]^{\mathrm{SL}_2} = K[f_2, f_3]$, where

$$f_2 = a_0 a_4 - \tfrac{1}{4}a_1 a_3 + \tfrac{1}{12}a_2^2 \quad \text{and} \quad f_3 = \begin{vmatrix} a_0 & a_1/4 & a_2/6 \\ a_1/4 & a_2/6 & a_3/4 \\ a_2/6 & a_3/4 & a_4 \end{vmatrix}.$$

The discriminant $\Delta(g_4)$ can be expressed in f_2 and f_3, namely $\Delta(g_4) = 2^8(f_2^3 - 27f_3^2)$. For $d = 5, 6, 8$, the invariant rings are also explicitly known (see Shioda [222] and Springer [231]).   ◁

Example 2.1.3. Let $V = K^n$. The group $\mathrm{GL}(V)$ acts on $\mathrm{End}(V)$ by conjugation:

$$\sigma \cdot A := \sigma A \sigma^{-1}, \qquad \sigma \in \mathrm{GL}(V), \ A \in \mathrm{End}(V).$$

The characteristic polynomial of $A \in \mathrm{End}(V)$ is given by

$$\chi(t) := \det(tI - A) = t^n - g_1 t^{n-1} + g_2 t^{n-2} - \cdots + (-1)^n g_n.$$

We view $g_1, \ldots, g_n$ as functions of A. The coefficients $g_1, g_2, \ldots, g_n \in K[\mathrm{End}(V)]$ are clearly invariant under the action of $\mathrm{GL}(V)$. Let us show that $K[\mathrm{End}(V)]^G = K[g_1, \ldots, g_n]$. Consider the set of diagonal matrices

$$\mathfrak{t} := \left\{ \begin{pmatrix} x_1 & & & \\ & x_2 & & \\ & & \ddots & \\ & & & x_n \end{pmatrix} \mid x_1, x_2, \ldots, x_n \in K \right\}.$$

The group S_n can be viewed as the subgroup of GL_n of permutation matrices. The set $\mathfrak{t}$ is stable under the action of S_n. The restriction of $\chi(t)$ to $\mathfrak{t}$ is S_n-invariant, in fact, it is equal to $\varphi(t)$ as defined in Example 2.1.1. Restricting g_i to $\mathfrak{t}$ yields the elementary symmetric polynomial f_i. It follows that $g_1, \ldots, g_n$ are algebraically independent. If $h \in K[\mathrm{End}(V)]^{\mathrm{GL}(V)}$, then the restriction of h to $\mathfrak{t}$ is S_n-invariant. We can find a polynomial ψ such that the restriction of h to $\mathfrak{t}$ is equal to $\psi(f_1, \ldots, f_n)$. Let U be the set of matrices which have distinct eigenvalues. Every matrix with distinct eigenvalues can be conjugated into $\mathfrak{t}$, so $U \subseteq G \cdot \mathfrak{t}$. The set U is Zariski dense, because it is the complement of the Zariski closed set defined by $\Delta(\chi) = 0$. It follows that $h = \psi(g_1, \ldots, g_n)$ because $h - \psi(g_1, \ldots, g_n)$ vanishes on $G \cdot \mathfrak{t} \supset U$. The trick in this example (reducing the computation of $K[V]^G$ to the computation of $K[W]^H$ with $W \subseteq V$ and $H \subseteq G$) works in a more general setting (see Popov and Vinberg [194]). $\triangleleft$

Example 2.1.4. This is the counter-example of Nagata to Hilbert's fourteenth problem. Take $K = \mathbb{C}$ and complex numbers $a_{i,j}$ algebraically independent over $\mathbb{Q}$ where $i = 1, 2, 3$ and $j = 1, 2, \ldots, 16$. Let $G \subset \mathrm{GL}_{32}$ be the group of all block diagonal matrices

$$\begin{pmatrix} A_1 & & & \\ & A_2 & & \\ & & \ddots & \\ & & & A_{16} \end{pmatrix},$$

where

$$A_j = \begin{pmatrix} c_j & c_j b_j \\ 0 & c_j \end{pmatrix}$$

for $j = 1, 2, \ldots, 16$. Here the c_j and b_j are arbitrary complex numbers such that $c_1 c_2 \cdots c_{16} = 1$ and $\sum_{j=1}^{16} a_{i,j} b_j = 0$ for $i = 1, 2, 3$. Then $K[x_1, \ldots, x_{32}]^G$ is not finitely generated (see Nagata [171]). $\triangleleft$

Example 2.1.5. Let $\mathbb{G}_a = K$ be the additive group. It acts on K^2 by

$$\sigma \cdot (x, y) \mapsto (x, y + \sigma x).$$

Consider the non-reduced subscheme X of K^2 defined by $x^2 = 0$. $\mathbb{G}_a$ also acts on X and its coordinate ring $R = K[x, y]/(x^2)$. It is easy to see that the invariant ring $R^{\mathbb{G}_a}$ is generated by (the images of) xy^n for all n. For every n, xy^n is not in the $R^{\mathbb{G}_a}$-ideal generated by $x, xy, \ldots, xy^{n-1}$. This shows that $R^{\mathbb{G}_a}$ is not Noetherian and not finitely generated. Although this example is very simple, it does not quite fit into the general setting, since usually we consider actions on affine varieties which are by definition reduced.   ◁

Example 2.1.6. Let K be an algebraically closed field of characteristic 0. Roberts found a (nonlinear) action of the additive group $\mathbb{G}_a$ on K^7 such that the invariant ring is not finitely generated (see Roberts [204]). Recently, Daigle and Freudenburg found the following counterexample in dimension 5. Consider the action of $\mathbb{G}_a$ on K^5 defined by

$$\sigma \cdot (a, b, x, y, z) =$$
$$(a, b, x + \sigma a^2, y + \sigma(ax + b) + \tfrac{1}{2}\sigma^2 a^3, z + \sigma y + \tfrac{1}{2}\sigma^2(ax + b) + \tfrac{1}{6}\sigma^3 a^3).$$

Then $K[a, b, x, y, z]^{\mathbb{G}_a}$ is not finitely generated (see Daigle and Freudenburg [52]). For $n \le 3$ it is known for any rational action of $\mathbb{G}_a$ on K^n that the invariant ring is finitely generated (see Zariski [266]).   ◁

Remark 2.1.7. Suppose that K has characteristic 0 and that V is a representation of the additive group $\mathbb{G}_a$. Weitzenböck proved that $K[V]^{\mathbb{G}_a}$ is finitely generated (see Weitzenböck [257] and Seshadri [216]). In positive characteristic this is still an open question.

Remark 2.1.8. We can generalize the notion of invariants. An element $f \in K[X]$ is called a **semi-invariant** or **relative invariant** if there exists a map $\chi : G \to K^*$ such that $\sigma \cdot f = \chi(\sigma)f$ for all $\sigma \in G$. Because the action of G is regular, χ is necessarily a morphism of algebraic groups. In this case χ is called the **weight** of f.

2.2 Reductive Groups

As we have remarked in the previous section, the invariant ring $K[V]^G$ is not always finitely generated. A sufficient condition for the invariant ring $K[V]^G$ to be finitely generated is that G is a reductive group. There are different notions of reductivity, namely, linearly reductive, geometrically reductive and group theoretically reductive (also referred to as just reductive). In characteristic zero all notions coincide. In positive characteristic geometric reductivity and reductivity are still the same, but linear reductivity is stronger. Typical examples of reductive groups are GL_n, all semi-simple groups including SL_n,

O_n and Sp_n, finite groups and tori. In positive characteristic the only linearly reductive groups are finite groups whose order is not divisible by the prime characteristic, tori, and extensions of tori by finite groups whose order are not divisible by the prime characteristic.

If G is a geometrically reductive group, then the invariant ring is finitely generated. In this book, we will only show this for linearly reductive groups.

2.2.1 Linearly Reductive Groups

Let us first give the definition.

Definition 2.2.1. *A linear algebraic group G is called* **linearly reductive** *if for every rational representation V and every $v \in V^G \setminus \{0\}$, there exists a linear invariant function $f \in (V^*)^G$ such that $f(v) \neq 0$.*

See Example 2.2.18 for an example of a linear algebraic group which is not linearly reductive. Linearly reductive groups have a "nice" representation theory. As we will see in Theorem 2.2.5, every representation is fully reducible, i.e., is a direct sum of irreducible representations. Another useful property of linearly reductive groups is that there is a notion of "averaging", the so-called Reynolds operator.

Definition 2.2.2. *Suppose that X is an affine G-variety where G is a linear algebraic group. A* **Reynolds operator** *is a G-invariant projection, i.e., a linear map $\mathcal{R} : K[X] \to K[X]^G$ such that*

(a) $\mathcal{R}(f) = f$ for all $f \in K[X]^G$;
(b) $\mathcal{R}$ is G-invariant, i.e., $\mathcal{R}(\sigma \cdot f) = \mathcal{R}(f)$ for all $f \in K[X]$ and all $\sigma \in G$.

For finite groups, the Reynolds operator is just averaging.

Example 2.2.3. Suppose that G is a finite group such that $\mathrm{char}(K)$ does not divide the group order $|G|$. If X is an affine variety on which G acts, then a Reynolds operator is defined by

$$\mathcal{R}(f) = \frac{\sum_{\sigma \in G} \sigma \cdot f}{|G|}.$$

It is easy to see that the conditions of Definition 2.2.2 are satisfied.   ◁

For infinite groups, averaging over the whole group does not really make sense. Over $\mathbb{C}$ for example, the group $\mathbb{C}^*$ is not compact. However, it does contain the compact subgroup S^1 of all complex numbers of norm 1. If $f \in \mathbb{C}[X]$ is any regular function, then we can average it over S^1 (using a Haar measure $d\mu$) and the result $\int_{S^1} \sigma \cdot f \, d\mu$ will be S^1-invariant. The function $\int_{S^1} \sigma \cdot f \, d\mu$ lies in the finite-dimensional subspace of $K[X]$ spanned by all $\sigma \cdot f$ with $\sigma \in \mathbb{C}^*$, in particular $\int_{S^1} \sigma \cdot f \, d\mu$ is also a regular function on X. It is $\mathbb{C}^*$-invariant because S^1 is Zariski-dense in $\mathbb{C}^*$. This shows that $\mathcal{R}(f) = \int_{S^1} \sigma \cdot f \, d\mu$ defines a Reynolds operator. This construction of a Reynolds operator can be generalized to other groups. Over $\mathbb{C}$, a linearly reductive group G always has a maximal compact subgroup C which is Zariski-dense in G (see Anhang II of Kraft [152] or the references there).

In this book we will use an algebraic construction of the Reynolds operator (see Theorem 2.2.5), and we do not need to restrict ourselves to the complex numbers.

Example 2.2.4. Let $\mathbb{G}_m := K^*$ be the multiplicative group. The coordinate ring of $\mathbb{G}_m$ is isomorphic to $K[t, t^{-1}]$. If X is an affine variety with a regular $\mathbb{G}_m$-action, then the action $\mu : \mathbb{G}_m \times X \to X$ induces a ring homomorphism

$$\mu^* : K[X] \to K[\mathbb{G}_m] \otimes K[X] \cong K[X][t, t^{-1}]$$

with the property

$$\mu^*(f)(\sigma, x) = f(\sigma \cdot x).$$

For $f \in K[X]$ write $\mu^*(f) = \sum_i f_i t^i$ (finite sum) with $f_i \in K[X]$ for all i. Define $\mathcal{R}(f) = f_0$. In case the ground field K is $\mathbb{C}$, this really means we are averaging $f(\sigma \cdot x)$ over all $\sigma \in S^1$. We will check the properties of Definition 2.2.2. First of all, we have

$$\sum_i f_i(\tau^{-1} \cdot x)\sigma^i = \mu^*(f)(\sigma, \tau^{-1} \cdot x) =$$

$$f(\sigma\tau^{-1} \cdot x) = \mu^*(f)(\sigma\tau^{-1}, x) = \sum_i f_i(x)\tau^{-i}\sigma^i$$

for all $\tau, \sigma \in \mathbb{G}_m$ and $x \in X$. It follows that $(\tau \cdot f_i)(x) = f_i(\tau^{-1} \cdot x) = \tau^{-i} f_i(x)$ for all i, and in particular $f_0 \in K[X]^{\mathbb{G}_m}$. If $f \in K[X]^{\mathbb{G}_m}$, we get $\mu^*(f)(\sigma, x) = f(\sigma \cdot x) = f(x)$, so $\mathcal{R}(f) = f$. Finally, we have

$$\mu^*(\tau \cdot f)(\sigma, x) = (\tau \cdot f)(\sigma \cdot x) = f(\sigma\tau^{-1} \cdot x) = \sum_i f_i(x)\tau^{-i}\sigma^i,$$

so $\mathcal{R}(\tau \cdot f) = \mathcal{R}(f)$. $\triangleleft$

There are many other properties which characterize linearly reductive groups. In other books one might find different definitions. One important property is the fact that every representation is a direct sum of irreducible ones. Another characterization of linearly reductive groups is that there always exists a Reynolds operator. The next theorem shows that all these notions are equivalent.

Theorem 2.2.5. *The following properties are equivalent:*

(a) G is linearly reductive;

(b) for every rational representation V there exists a unique subrepresentation $W \subseteq V$ such that $V = V^G \oplus W$, and we have $(W^*)^G = 0$;

(c) for every affine G-variety X there exists a unique Reynolds operator $\mathcal{R} : K[X] \to K[X]^G$;

(d) for every rational representation V and subrepresentation $W \subseteq V$ there exists a subrepresentation $Z \subseteq V$ such that $V = W \oplus Z$;

(e) for every rational representation V there exist irreducible subrepresentations $V_1, V_2, \ldots, V_r \subseteq V$ such that $V = V_1 \oplus V_2 \oplus \cdots \oplus V_r$.

Proof. (a)$\Rightarrow$(b) From (a) and Definition 2.2.1 it follows that $V^G \subseteq V$ and $(V^*)^G \subseteq V^*$ are dual to each other with respect to the canonical pairing $V \times V^* \to K$. Take $W = ((V^*)^G)^\perp$, then $V = V^G \oplus W$. If $V = V^G \oplus Z$ is another such decomposition, then G acts trivially on $Z^\perp \subseteq V^*$, because $Z^\perp$ is dual to V^G. So $Z^\perp = (V^*)^G$ and $W = Z$. The representation W^* is isomorphic to the complement of $(V^*)^G$ in V^*. Hence $(W^*)^G$ must be trivial.

(b)$\Rightarrow$(c) For any finite dimensional and G-stable $V \subseteq K[X]$ we have a decomposition $V = V^G \oplus W$ as in (b). If $\mathcal{R}$ is any Reynolds operator $K[X] \twoheadrightarrow K[X]^G$, then the restriction of $\mathcal{R}$ to V^G must be the identity and the restriction of $\mathcal{R}$ to W must be 0 because otherwise we would have a non-zero element in $(W^*)^G$. Uniqueness follows from this. Now we will prove existence. Define $\mathcal{R}_V : V = V^G \oplus W \twoheadrightarrow V^G$ as the projection onto V^G along W. If V' is another G-stable, finite dimensional subspace with $V \subseteq V'$, then the restriction of $\mathcal{R}_{V'}$ to W is 0 because otherwise there would exist a non-zero element in $(W^*)^G$. Since $f - \mathcal{R}_V(f) \in W$ for $f \in V$, we obtain $0 = \mathcal{R}_{V'}(f - \mathcal{R}_V(f)) = \mathcal{R}_{V'}(f) - \mathcal{R}_V(f)$. So the restriction of $\mathcal{R}_{V'}$ to V is $\mathcal{R}_V$. For an arbitrary $f \in K[X]$, we define $\mathcal{R}(f) = \mathcal{R}_V(f)$, where $V \subseteq K[X]$ is a finite dimensional G-stable subspace containing f as in Lemma A.1.8. This is well-defined, because if V' is another subspace with these properties, then $\mathcal{R}_V(f) = \mathcal{R}_{V+V'}(f) = \mathcal{R}_{V'}(f)$. The properties in Definition 2.2.2 are easily checked.

(c)$\Rightarrow$(a) We have $V \subseteq K[V^*]$ as the set of linear functions on V^*. Let $\mathcal{R}_V$ be the restriction of the Reynolds operator $\mathcal{R} : K[V^*] \twoheadrightarrow K[V^*]^G$ to V. Suppose that $v \in V^G \setminus \{0\}$. Choose some projection $p : K[V^*]^G \twoheadrightarrow K$ such that $p(v) \neq 0$. Let f be the composition $p \circ \mathcal{R}_V : V \to K$. Clearly $f \in (V^*)^G$ and $f(v) = p(\mathcal{R}_V(v)) = p(v) \neq 0$.

(a)$\Rightarrow$(d) First assume that W is irreducible. The dual space of $\mathrm{Hom}(W, V)$ is $\mathrm{Hom}(V, W)$. In fact, the pairing is

$$(A, B) \in \mathrm{Hom}(W, V) \times \mathrm{Hom}(V, W) \mapsto \mathrm{Tr}(AB) \in K,$$

where $\mathrm{Tr}(AB)$ is the trace of $AB \in \mathrm{End}(V)$. Let $A \in \mathrm{Hom}(W, V)^G$ be the inclusion. By (a) there exists a $B \in \mathrm{Hom}(V, W)^G$ such that $\mathrm{Tr}(BA) \neq 0$. Because W is irreducible and $BA \in \mathrm{End}(W, W)^G$, BA must be a non-zero multiple of the identity by Schur's Lemma (see Fulton and Harris [74, Schur's Lemma 1.7], the proof there works for arbitrary groups G and arbitrary finite dimensional irreducible representations). We can write $V = W \oplus Z$ where Z is the kernel of B. If W is not irreducible, then take an irreducible subrepresentation $W' \subseteq W$. Then $V = W' \oplus Z'$ for some G-stable complement Z' and $W = W' \oplus W \cap Z'$. By induction hypothesis $W \cap Z'$ has a G-stable complement Z in Z' and we get $V = W \oplus Z$.

(d)$\Rightarrow$(e),(e)$\Rightarrow$(a) are left to the reader. $\square$

Remark 2.2.6. It is worth noting the construction of the Reynolds operator in the previous proof. Suppose that $f \in K[X]$. Choose a finite dimensional G-stable vector space V containing f. This always can be done, for example take for V the vector space spanned by all $\sigma \cdot f$ with $\sigma \in G$ (see Lemma A.1.8). Now $V^G \subset V$ has a unique G-stable complement W, and $\mathcal{R}(f)$ is the projection of f onto V^G. From the proof of Theorem 2.2.5 it follows that this is well-defined. $\triangleleft$

Corollary 2.2.7. *If G is linearly reductive and X an affine G-variety, then the Reynolds operator $K[X] \twoheadrightarrow K[X]^G$ has the following properties:*

(a) If $W \subseteq K[X]$ is a G-stable subspace, then $\mathcal{R}(W) = W^G$;
(b) $\mathcal{R}$ is a $K[X]^G$-module homomorphism, i.e., $\mathcal{R}(fh) = f\mathcal{R}(h)$ if $f \in K[X]^G$ and $h \in K[X]$.

Proof. (a) This is clear from Remark 2.2.6.

(b) Choose a G-stable finite dimensional subspace $V \subseteq K[X]$ with $h \in V$. Then $V = V^G \oplus W$ for some G-stable complement W and $\mathcal{R}(h)$ is the projection of h onto V^G. Notice that $(fW)^G = 0$ because otherwise $(W^*)^G \neq 0$ which contradicts Theorem 2.2.5(b). We have $fV = fV^G \oplus fW$ and $fV^G = (fV)^G$. It follows that $\mathcal{R}(fh)$ is the projection of fh onto fV^G which is $f\mathcal{R}(h)$. $\square$

Corollary 2.2.8. *If G is linearly reductive, V, W are rational representations, and $A : V \twoheadrightarrow W$ is a surjective G-equivariant linear map, then $A(V^G) = W^G$.*

Proof. Let Z be the kernel of A, then there exists a G-stable complement W' such that $V \cong Z \oplus W'$. The restriction of A to W' gives an isomorphism $W' \cong W$ of representations of G, in particular $(W')^G$ maps onto W^G. $\square$

This is a very useful fact. The following corollary shows that for linearly reductive groups, the computation of invariant rings $K[X]^G$ can always be reduced to the case of a rational representation.

Corollary 2.2.9. *Suppose that X is an affine G-variety. There exists a G-equivariant embedding $i : X \hookrightarrow V$ where V is a rational representation of G. The surjective G-equivariant ring homomorphism $i^* : K[V] \twoheadrightarrow K[X]$ has the property that $i^*(K[V]^G) = K[X]^G$.*

Proof. The first statement will be proved in A.1.9. For any finite dimensional G-stable subspace $W \subseteq K[X]$ there exists a G-stable finite dimensional subspace $Z \subseteq K[V]$ such that i^* maps Z onto W. We have $i^*(Z^G) = W^G$ by Corollary 2.2.8. It follows that $i^*(K[V]^G) = K[X]^G$, since $K[X]^G$ is the union of such W. $\square$

Now we come to the most important theorem in invariant theory, namely Hilbert's finiteness theorem. It had already been proven by Gordan that $K[V]^{\mathrm{SL}_2}$ is finitely generated (see Gordan [94]). The methods there do not generalize to other groups. It was in 1890 that Hilbert [107] surprised everyone by giving a proof which works for all GL_n, SL_n, in fact, all classical groups. We will see that the proof actually works whenever there is a Reynolds operator, i.e., whenever we are dealing with a linearly reductive group. The methods Hilbert used were abstract, non-constructive, and completely new in those days. The proof was criticized for not being constructive and it led Gordan, the "king of invariant theory" at that time, to make his famous exclamation "Das ist Theologie und nicht Mathematik."[1] Hilbert wrote a paper in 1893 (see Hilbert [108]), in which he gave a constructive proof. The abstract methods of both papers contain many important basics of modern commutative algebra, for example Hilbert's Basissatz, the Nullstellensatz, the Noether normalization lemma, and Hilbert's syzygy theorem.

We give the first short and elegant, non-constructive proof of Hilbert's finiteness theorem.

Theorem 2.2.10. *If G is a linearly reductive group and V is a rational representation, then $K[V]^G$ is finitely generated over K.*

Proof. Let I be the ideal in $K[V]$ generated by all homogeneous invariants of degree > 0. Because $K[V]$ is Noetherian, there exist finitely many homogeneous invariants $f_1, \ldots, f_r \in K[V]^G$ such $I = (f_1, \ldots, f_r)$. We will prove that $K[V]^G = K[f_1, \ldots, f_r]$. Suppose $h \in K[V]^G$ homogeneous of degree d. We will prove that $h \in K[f_1, \ldots, f_r]$ using induction on d. If $d = 0$, then $h \in K \subseteq K[f_1, \ldots, f_r]$. If $d > 0$, we can write

$$h = \sum_{i=1}^{r} g_i f_i \tag{2.2.1}$$

with $g_i \in K[V]$. Without loss of generality, we may assume that g_i is homogeneous of degree $d - \deg(f_i) < d$. The Reynolds operator $\mathcal{R} : K[V] \twoheadrightarrow K[V]^G$ maps $K[V]_d$ onto $K[V]_d^G$ (see Corollary 2.2.7(a)). Applying the Reynolds operator to (2.2.1) and using Corollary 2.2.7(b), we obtain

$$h = \mathcal{R}(h) = \sum_{i=1}^{r} \mathcal{R}(g_i f_i) = \sum_{i=1}^{r} \mathcal{R}(g_i) f_i.$$

Because $\mathcal{R}(g_i) \in K[V]^G$ is homogeneous of degree $< d$, we have by induction that $\mathcal{R}(g_i) \in K[f_1, \ldots, f_r]$ for all i. We conclude that $h \in K[f_1, \ldots, f_r]$. $\square$

Corollary 2.2.11. *If G is a linearly reductive group acting regularly on an affine variety X, then $K[X]^G$ is finitely generated.*

---

[1] "This is theology and not mathematics."

Proof. Combine Theorem 2.2.10 with Corollary 2.2.9.   □

Corollary 2.2.12. *Let G be a linearly reductive algebraic group and let $[G, G]$ be its commutator subgroup. Let X be an affine G-variety. Then $K[X]^{[G,G]}$ is spanned by G-semi-invariants.*

Proof. If $f \in K[X]^{[G,G]}$, then there exists a finite dimensional G-stable vector space $W \subset K[X]$ with $f \in W$. By replacing W with $W^{[G,G]}$, we may assume that $[G, G]$ acts trivially on W. We can write $W = W_1 \oplus W_2 \oplus \cdots \oplus W_r$ with W_i an irreducible representation of G (and of $G/[G, G]$). Since $G/[G, G]$ is abelian, all its irreducible representations are 1-dimensional, say $W_i = K \cdot f_i$. Clearly, every f_i is a G-semi-invariant and f is a linear combination of the f_i.   □

2.2.2 Other Notions of Reductivity

For a linear algebraic group G, the **unipotent radical** $R_u(G)$ is defined as the largest connected normal unipotent subgroup of G. G is called **reductive** or **group theoretically reductive** if $R_u(G)$ is trivial (see Definition A.3.6). There are many examples of reductive groups, for example GL_n, SL_n, O_n, SO_n, Sp_n, finite groups, tori and semisimple groups. For details on reductive groups we refer to Section A.3 in this book or to one of the books on linear algebraic groups (see Borel [23], Humphreys [118], Springer [233]).

For the proof of the following theorem we refer to Nagata and Miyata [174].

Theorem 2.2.13. *If $\mathrm{char}(K) = 0$, then a linear algebraic group is reductive if and only if it is linearly reductive.*

Definition 2.2.14. *A linear algebraic group is called **geometrically reductive** if for every rational representation V and every $v \in V^G \setminus \{0\}$, there exists a homogeneous $f \in K[V]^G$ of degree > 0 such that $f(v) \neq 0$.*

Clearly, linear reductivity implies geometric reductivity. The converse is not true, though. For example a non-trivial finite p-group in characteristic p is geometrically reductive, but not linearly reductive. We state the following useful result without proof.

Theorem 2.2.15. *For any characteristic of K, a group is geometrically reductive if and only if it is reductive.*

Proof. In Nagata and Miyata [174] it was proven that geometrically reductive groups are reductive. Haboush proved the converse which had been conjectured by Mumford (see Haboush [101]).   □

Hilbert's finiteness theorem has also been proven for geometrically reductive groups (see Nagata [173]) using some new ideas.

Theorem 2.2.16. *If X is an affine G-variety and G is geometrically reductive, then $K[X]^G$ is finitely generated.*

The converse is also true. Popov [189] proved the following.

Theorem 2.2.17. *If $K[X]^G$ is finitely generated for every affine G-variety X, then G must be reductive.*

Example 2.2.18. Let $\mathbb{G}_a = K$ be the additive group. We define a regular action on K^2 by

$$\sigma \cdot (x,y) = (x + \sigma y, y), \qquad \sigma \in \mathbb{G}_a, \ (x,y) \in K^2.$$

The invariant ring $K[x,y]^{\mathbb{G}_a}$ is equal to $K[y]$. If $v \in K \times \{0\} = (K^2)^{\mathbb{G}_a}$, then every invariant vanishes on v. The group $\mathbb{G}_a$ is therefore not geometrically reductive. $\triangleleft$

In positive characteristic there exist only few linearly reductive groups as the following theorem shows.

Theorem 2.2.19 (Nagata [172]). *Suppose that $\mathrm{char}(K) = p > 0$. A linear algebraic group is linearly reductive if and only if the connected component G° of the identity element is a torus and $|G/G^\circ|$ is not divisible by p.*

Example 2.2.20. Assume that K is a field of characteristic $p > 0$ and let C_p be the cyclic group of order p, generated by σ. As in the previous example, we define an action on K^2 by

$$\sigma \cdot (x,y) = (x + y, y), \qquad (x,y) \in K^2.$$

The subspace $K \times \{0\} \subseteq K^2$ does not have a C_p-stable complement. So C_p is not linearly reductive whenever $p = \mathrm{char}(K)$. However, C_p is geometrically reductive. For example, if v is the invariant point $(1,0)$, then $f(v) \neq 0$ where f is the homogeneous invariant polynomial $f = \prod_{i \in \mathbb{F}_p} (x - iy) = x^p - xy^{p-1}$. $\triangleleft$

2.3 Categorical Quotients

In this section we give a geometric interpretation of invariant rings.

2.3.1 Geometric Properties of Quotients

If G is reductive and X is a G-variety, then $K[X]^G$ is finitely generated and we can define $X /\!/ G$ as the affine variety corresponding to the ring $K[X]^G$. The map $\pi : X \twoheadrightarrow X /\!/ G$ is the morphism corresponding to the inclusion $K[X]^G \subseteq K[X]$ and it is called the **categorical quotient**. We will study its geometric properties. We first remark that the categorical quotient does not always separate all the orbits as the following example shows.

Example 2.3.1. Let $\mathbb{G}_m = K^*$ be the multiplicative group acting diagonally on $V = K^2$ by

$$\sigma \cdot (x,y) = (\sigma x, \sigma y), \qquad \sigma \in \mathbb{G}_m, \ (x,y) \in K^2.$$

It is easy to see that $K[V]^{\mathbb{G}_m} = K$, so $V /\!/ \mathbb{G}_m$ is just a point. On the other hand, there are infinitely many $\mathbb{G}_m$-orbits, so $\pi : V \to V /\!/ \mathbb{G}_m$ does not separate the orbits. ◁

A few general remarks about the correspondence between ideals and their zeroes are useful here. If $I \subset K[X]^G$ is an ideal and $Y = \mathcal{V}(I)$ is its zero set in $X /\!/ G$, then the zero set of the ideal $K[X]I \subset K[X]$ corresponds to the inverse image $\pi^{-1}(Y)$ (but $K[X]I$ does not have to be a radical ideal). On the other hand if $I \subset K[X]$ is an ideal, and $Y = \mathcal{V}(I) \subset X$ is its zero set, then $I \cap K[X]^G = I^G$ is equal to $I(\overline{\pi(Y)})$, the vanishing ideal of $\overline{\pi(Y)}$. For the proofs of the following lemmas we will assume that G is linearly reductive. The results, however, remain true if G is geometrically reductive (see Mumford et al. [169] or Newstead [182]).

Lemma 2.3.2. *If X is a G-variety, then the quotient map $\pi : X \to X /\!/ G$ is surjective.*

Proof. Let $\mathfrak{m}_x \subseteq K[X]^G$ be the maximal ideal corresponding to a point $x \in X /\!/ G$. If $\pi^{-1}(x) = \emptyset$, then $1 \in \mathfrak{m}_x K[X]$, so we can write $1 = \sum_{i=1}^r a_i f_i$ with $f_i \in \mathfrak{m}_x$ and $a_i \in K[X]$ for all i. Applying the Reynolds operator yields $1 = \mathcal{R}(1) = \sum_{i=1}^r \mathcal{R}(a_i) f_i$, so $1 \in \mathfrak{m}_x$, a contradiction. Therefore $\pi^{-1}(x) \neq \emptyset$. □

Lemma 2.3.3. *If $Y_1, Y_2 \subseteq X$ are G-stable, then $\overline{\pi(Y_1)} \cap \overline{\pi(Y_2)} = \overline{\pi(Y_1 \cap Y_2)}$.*

Proof. Let $I_1 = I(Y_1)$ and $I_2 = I(Y_2)$ be the vanishing ideals of Y_1 and Y_2, respectively. Then $(I_1 + I_2)^G = \mathcal{R}(I_1 + I_2) = \mathcal{R}(I_1) + \mathcal{R}(I_2) = I_1^G + I_2^G$. □

Corollary 2.3.4. *If $Y \subseteq X$ is G-stable and closed, then $\pi(Y) \subseteq X /\!/ G$ is closed.*

Proof. Take $x \in \overline{\pi(Y)} \setminus \pi(Y)$. Then $\pi^{-1}(x)$ and Y are G stable and closed, $\pi^{-1}(x) \cap Y = \emptyset$, so $x \in \pi(\pi^{-1}(x)) \cap \overline{\pi(Y)} = \emptyset$ by Lemma 2.3.2 and Lemma 2.3.3. □

Corollary 2.3.5. *The topology of $X /\!/ G$ is the quotient topology.*

Proof. $Y \subseteq X /\!/ G$ is closed if and only $\pi^{-1}(Y)$ is closed. □

Corollary 2.3.6. *For every $x \in X /\!/ G$, the fiber $\pi^{-1}(x)$ contains exactly one closed orbit. This orbit is contained in the Zariski-closure of all (other) orbits in $\pi^{-1}(x)$.*

Proof. Suppose that there is no closed orbit in $\pi^{-1}(x)$. Choose $y_1 \in \pi^{-1}(x)$. Since $G \cdot y_1$ is not Zariski closed, we can find $y_2 \in \overline{G \cdot y_1}$ such that the orbit closure $\overline{G \cdot y_2}$ is strictly contained in the orbit closure $\overline{G \cdot y_1}$. Similarly, we can construct for all i a point y_i such that

$$\overline{G \cdot y_1} \supset \overline{G \cdot Y_2} \supset \overline{G \cdot y_3} \supset \cdots$$

where all inclusions are strict. This contradicts the Noetherian property, and we conclude that there exists a $y \in \pi^{-1}(x)$ whose orbit is closed.

If $z \in \pi^{-1}(x)$, then by Lemma 2.3.3 and Corollary 2.3.4 we have $\{x\} = \pi(\overline{G \cdot z}) \cap \pi(\overline{G \cdot y}) = \pi(\overline{G \cdot z} \cap \overline{G \cdot y})$. So $\overline{G \cdot z} \cap \overline{G \cdot y} \neq \emptyset$, which means that the orbit $G \cdot y$ lies in the closure of $G \cdot z$. $\square$

Example 2.3.7. Let $\mathbb{G}_m = K^*$ be the multiplicative group acting on $V = K^2$ by

$$\sigma \cdot (x,y) = (\sigma x, \sigma^{-1} y), \qquad \sigma \in \mathbb{G}_m, \ (x,y) \in K^2.$$

It is easy to see that $K[V]^{\mathbb{G}_m} = K[xy]$, so $V /\!\!/ \mathbb{G}_m \cong K$. The quotient map $\pi : V \to V /\!\!/ \mathbb{G}_m \cong K$ is given by

$$(x,y) \mapsto xy.$$

For $a \neq 0$, the fiber $\pi^{-1}(a)$ is just a single closed orbit. The zero fiber $\pi^{-1}(0)$ is given by $xy = 0$ and consists of three orbits, namely

$$\{(0,0)\}, \quad \{(x,0) \mid x \neq 0\}, \quad \{(0,y) \mid y \neq 0\}.$$

The only closed orbit is $\{(0,0)\}$, and it lies in the closure of the other two orbits.   ◁

We call $\pi : X \to X /\!\!/ G$ a **geometric quotient** if there is a 1–1 correspondence between G-orbits in X and points in $X /\!\!/ G$. If G is a finite group, then all G-orbits are closed. It follows that every fiber of $\pi : X \twoheadrightarrow X /\!\!/ G$ is a single orbit by Corollary 2.3.6. So $X /\!\!/ G$ is a geometric quotient in this case.

As we have seen in Example 2.3.1 and Example 2.3.7, a categorical quotient is not always a geometric quotient. Suppose that $\pi : X \to X /\!\!/ G$ is a geometric quotient and X is a connected G-variety. We have

$$\dim \pi^{-1}(\pi(x)) = \dim G \cdot x$$

for all $x \in X$, since the fibers are the orbits. But

$$\dim G \cdot x = \dim G - \dim G_x$$

where G_x is the stabilizer of x. Furthermore

$$\dim G_x = \dim \gamma^{-1}(\gamma(e,x))$$

where

$$Z = \{(g, x) \in G \times X \mid g \cdot x = x\}$$

and $\gamma : Z \to X$ is the projection onto X. Now $\dim G_x$ and $\dim G \cdot x$ depend semicontinuously on x (see Hartshorne [102, Exercise II.3.22]). This means that $\{x \mid \dim G \cdot x \leq C\}$ is Zariski *closed* for all constants C. On the other hand, $\dim \pi^{-1}(\pi(x))$ also depends semicontinuously on x, but in the other direction. This means $\dim \pi^{-1}(\pi(x)) \leq C$ is a Zariski *open* subset of X for all C. We conclude that $\dim \pi^{-1}\pi(x) = \dim G \cdot x$ must be a constant function on $x \in X$ since the only subsets of X which are Zariski open *and* closed are $\emptyset$ and X itself. We have shown that if X admits a geometric quotient and X is connected, then all the orbits must have the same dimension. This is also sufficient. If all orbits have the same dimension, then no orbit lies in the closure of any other orbit, and by Corollary 2.3.6, π separates the orbits.

Although not all categorical quotients are geometric, they still have many nice properties. Quotients with certain nice geometric properties are sometimes called "**good quotients**". We will not give an axiomatic definition here. Nice quotients are often needed for finding so-called moduli spaces, spaces which parametrize certain geometric objects (all non-singular curves of a given genus, for example). Typically, one needs a "good" or geometric quotient of a projective variety with a G-action. For example, if V is a rational representation of G, one can ask for a "good" quotient for the projective space $\mathbb{P}(V)$. Since the coordinate ring of $V /\!\!/ G$ is homogeneous, we can take the candidate $\mathbb{P}(V /\!\!/ G)$ (the projective variety corresponding to the graded ring $K[V]^G$). Now $\pi : V \to V /\!\!/ G$ induces a map

$$\tilde{\pi} : \mathbb{P}(V) \rightsquigarrow \mathbb{P}(V /\!\!/ G)$$

which is not well-defined, because for some $v \in V \backslash \{0\}$ we could have $\pi(v) = 0$. But if we exclude these points, we get a well defined "good" quotient

$$\tilde{\pi} : \mathbb{P}(V \setminus \pi^{-1}(0)) \to \mathbb{P}(V /\!\!/ G).$$

The elements in $V \setminus \pi^{-1}(0)$ are called **semi-stable**. For more details on the subject of Geometric Invariant Theory we refer to Mumford et al. [169] and Newstead [182].

2.3.2 Separating Invariants

The categorical quotient can be formed if generating invariants for $K[X]^G$ are known. Such a generating set can be large, hard to compute, or even infinite (in the case of non-reductive groups). It is an interesting question whether a smaller set of invariants might also suffice to achieve the same separation properties. In this subsection X is an affine variety over a field $K = \bar{K}$, and G is any group of automorphisms of the coordinate ring $K[X]$. G need not be reductive, and, in fact, not even an algebraic group.

Definition 2.3.8. *A subset $S \subseteq K[X]^G$ is said to be* **separating** *if for any two points $x, y \in X$ we have: If there exists an invariant $f \in K[X]^G$ with $f(x) \neq f(y)$, then there exists a $g \in S$ with $g(x) \neq g(y)$.*

Clearly $S \subseteq K[X]^G$ is separating if and only if the subalgebra $K[S] \subseteq K[X]^G$ generated by S is separating. If G is a reductive group acting regularly on X, then the above definition amounts to saying that for two points $x, y \in X$ with distinct closed G-orbits there exists $g \in S$ with $g(x) \neq g(y)$.

Example 2.3.9. Consider the finite group

$$G = \langle \begin{pmatrix} \omega & 0 \\ 0 & \omega \end{pmatrix} \rangle \leq \mathrm{GL}_2(K)$$

with ω a primitive third root of unity. The invariant ring $K[x, y]^G$ is minimally generated by

$$f_1 = x^3, \quad f_2 = x^2 y, \quad f_3 = xy^2, \quad f_4 = y^3.$$

We claim that $S := \{f_1, f_2, f_4\}$ is a separating subset. Indeed, we have $f_3 = f_2^2 / f_1$, so for $(\xi, \eta) \in K^2$ we obtain

$$f_3(\xi, \eta) = \begin{cases} f_2(\xi, \eta)^2 / f_1(\xi, \eta) & \text{if } f_1(\xi, \eta) \neq 0 \\ 0 & \text{if } f_1(\xi, \eta) = 0 \end{cases}.$$

This means that the value of f_3 at any point (ξ, η) is determined by the values of f_1 and f_2. Therefore S is separating.

Having seen this, one might expect that for all $a, b \in K$ with $(a, b) \neq (0, 0)$ the set $S' := \{f_1, af_2 + bf_3, f_4\}$ should also be separating. But, surprisingly, this is only the case if $ab = 0$. Indeed, if $ab \neq 0$, then for the points

$$v := \big(b(\omega - 1), a(1 - \omega^2)\big) \quad \text{and} \quad w := \big(b(\omega^2 - \omega), a(1 - \omega^2)\big)$$

it is easily verified that $f(v) = f(w)$ for every $f \in S'$, but $G \cdot v \neq G \cdot w$. ◁

The following proposition gives a necessary condition for a subalgebra to be separating.

Proposition 2.3.10. *Suppose that X is irreducible and $K[X]^G$ is finitely generated. Let $A \subseteq K[X]^G$ be a finitely generated and separating subalgebra. Then $\mathrm{Quot}(K[X]^G)$ is a finite, purely inseparable field extension of $\mathrm{Quot}(A)$. In particular, if $\mathrm{char}(K) = 0$ then*

$$\mathrm{Quot}(A) = \mathrm{Quot}(K[X]^G).$$

Proof. Let Y be the variety associated to A and consider the morphism $\pi \colon X /\!\!/ G \to Y$ coming from the inclusion $A \subseteq K[X]^G$. By the hypothesis, the restriction of π to the image of X in $X /\!\!/ G$ (which is contains a non-empty open subset of $X /\!\!/ G$) is injective. The restriction is also dominant, since $X \to Y$ comes from the embedding $A \subseteq K[X]$. This implies that the induced map $\pi^*\colon K(Y) \to K(X /\!/ G)$ makes $K(X /\!/ G)$ into a finite, purely inseparable extension of $K(Y)$ (see Humphreys [118, Theorem 4.6]). This yields the result. $\qquad\square$

We can say more in the case of a reductive group acting linearly (see Theorem 2.3.12). We need the following proposition.

Proposition 2.3.11. *Let R be a normal integral domain and $G \le \mathrm{Aut}(R)$ a group of automorphisms of R. Then the invariant ring R^G is also a normal integral domain.*

Proof. It is clear that R^G is an integral domain. G also acts on $F := \mathrm{Quot}(R)$. Let $f \in \mathrm{Quot}(R^G)$ be integral over R^G. Then $f \in F^G$, and f is integral over R. By the hypothesis f lies in R, hence $f \in R \cap F^G = R^G$. This shows that R^G is indeed normal. $\qquad\square$

Let $A \subseteq K[V]$ be a subalgebra of a polynomial ring of positive characteristic p. Then we call the algebra

$$\widehat{A} := \{ f \in K[V] \mid f^{p^r} \in A \text{ for some } r \in \mathbb{N} \} \subseteq K[V]$$

the **purely inseparable closure** of A in $K[V]$. In case $\mathrm{char}(K) = 0$ we set $\widehat{A} := A$.

Theorem 2.3.12. *Let G be a reductive group and V a rational representation. Moreover, let $A \subseteq K[V]^G$ be a finitely generated, graded, separating subalgebra, and write $\widehat{\widetilde{A}}$ for the purely inseparable closure of the normalization of A. Then*

$$\widehat{\widetilde{A}} = K[V]^G.$$

Proof. We first prove $\widehat{\widetilde{A}} \subseteq K[V]^G$. Proposition 2.3.11 implies that $\widetilde{A} \subseteq K[V]^G$, so the claimed inclusion holds in characteristic 0. Suppose $\mathrm{char}(K) = p > 0$ and take $f \in K[V]$ such that $f^q \in \widetilde{A}$ for a p-power q. Then for $\sigma \in G$ we have

$$(\sigma \cdot f - f)^q = \sigma \cdot f^q - f^q = 0,$$

so $f \in K[V]^G$.

Next we prove that $K[V]^G$ is integral over A. Consider the ideal $A_+ K[V]$ in $K[V]$ generated by all homogeneous elements of positive degree in A. Take $v \in \mathcal{V}(A_+ K[V])$. Then for every $f \in A$ we have

$$f(v) - f(0) = (f - f(0))\,(v) = 0.$$

Since A is separating, this implies $f(v) = 0$ for all $f \in K[V]^G_+$, so $v \in \mathcal{V}(K[V]^G_+ \cdot K[V])$. The Nullstellensatz now yields

$$K[V]^G_+ \subseteq \sqrt{A_+ K[V]} \cap K[V]^G. \tag{2.3.1}$$

Let $I = A_+ K[V]^G$ be the ideal in $K[V]^G$ generated by A_+. Then $I K[V] = A_+ K[V]$, and by Newstead [182, Lemma 3.4.2] we have $\sqrt{I K[V]} \cap K[V]^G \subseteq \sqrt{I}$. (This uses the reductivity of G and holds for any ideal in $K[V]^G$.) Using (2.3.1), we obtain

$$K[V]_+^G \subseteq \sqrt{A_+ K[V]^G}.$$

Therefore $K[V]^G / \left(A_+ K[V]^G \right)$ has Krull dimension 0, and thus

$$\dim_K \left(K[V]^G / \left(A_+ K[V]^G \right) \right) < \infty.$$

By the graded version of Nakayama's lemma (see Lemma 3.5.1), this implies that $K[V]^G$ is finitely generated as a module over A, so $K[V]^G$ is integral over A, as claimed.

To show that $K[V]^G \subseteq \widehat{\widetilde{A}}$, take $f \in K[V]^G$. Proposition 2.3.10 shows that $f^q \in \mathrm{Quot}(A)$ with q a power of $\mathrm{char}(K)$ ($q = 1$ if $\mathrm{char}(K) = 0$). Moreover, $f^q \in K[V]^G$ is integral over A, hence $f^q \in \widetilde{A}$. This implies $f \in \widehat{\widetilde{A}}$. $\square$

The following trivial example shows that Theorem 2.3.12 would become false if we dropped the purely inseparable closure.

Example 2.3.13. Suppose K has characteristic $p > 0$, and let $G \leq \mathrm{GL}_1(K)$ be the trivial group. Then $K[V]^G = K[x]$, but $A := K[x^p]$ is an integrally closed, separating subalgebra, since the map $\xi \mapsto \xi^p$ is injective in characteristic p. Hence $K[V]^G \neq \widetilde{A}$. $\triangleleft$

The next example shows that the converse of Theorem 2.3.12 does not hold.

Example 2.3.14. Consider the group $G \cong C_2 \times C_2$ generated by the diagonal matrices

$$\begin{pmatrix} -1 & & & \\ & -1 & & \\ & & 1 & \\ & & & 1 \end{pmatrix} \quad \text{and} \quad \begin{pmatrix} 1 & & & \\ & 1 & & \\ & & -1 & \\ & & & -1 \end{pmatrix}$$

where $\mathrm{char}(K) = 0$. G acts on the polynomial ring $R := K[x_1, y_1, x_2, y_2]$, and R^G is clearly generated by the invariants

$$f_i := x_i^2, \quad g_i := y_i^2, \quad \text{and} \quad h_i := x_i y_i \quad (i = 1, 2).$$

Consider the subalgebra $A := K[f_1, g_1, f_2, g_2, h]$ with $h := h_1 + h_2$. We claim $\widetilde{A} = R^G$. Indeed, we have

$$h_1 = \frac{f_1 g_1 - f_2 g_2 + h^2}{2h} \quad \text{and} \quad h_2 = \frac{f_2 g_2 - f_1 g_1 + h^2}{2h},$$

so the h_i lie in $\mathrm{Quot}(A)$. But they are also integral over A, since $h_i^2 - f_i g_i = 0$. Thus $h_1, h_2 \in \widetilde{A}$, which proves our claim. Now consider the points

$$v = (1, -1, 1, 1) \quad \text{and} \quad w = (1, 1, -1, 1).$$

Clearly v and w are in distinct G-orbits, but nevertheless

$$f_i(v) = f_i(w), \quad g_i(v) = g_i(w), \quad \text{and} \quad h(v) = h(w).$$

Therefore A does not separate G-orbits. $\triangleleft$

It is not surprising that there exist examples like the one above, since a graded separating subalgebra $A \subseteq K[V]^G$ corresponds to a variety Y whose normalization $\widetilde{Y}$ coincides with $V /\!\!/ G$. But for A to be separating we need the additional hypothesis that all fibers of $\widetilde{Y} \to Y$ have size one. This hypothesis is rather strong, and it would be desirable to have a simple criterion for deciding whether for a given variety Y we have that all fibers of the normalization $\widetilde{Y} \to Y$ have size one.

Although the proof of the following theorem is very simple, the theorem may be surprising and shows that separating algebras are a useful concept.

Theorem 2.3.15. *Let X be an affine variety and $G \leq \mathrm{Aut}(K[X])$ a group of automorphisms of the coordinate ring $K[X]$. Then there exists a finite separating set $S \subseteq K[X]^G$.*

Proof. Set $R := K[X]$ and consider the ideal I in $R \otimes_K R$ generated by all $f \otimes 1 - 1 \otimes f$ with $f \in R^G$. Since $R \otimes_K R$ is Noetherian, there exist finitely many invariants $f_1, \ldots, f_m \in R^G$ such that

$$I = (f_1 \otimes 1 - 1 \otimes f_1, \ldots, f_m \otimes 1 - 1 \otimes f_m).$$

We claim that $\{f_1, \ldots, f_m\}$ is separating. Indeed, take $x, y \in X$ and assume that there exists an $f \in K[X]^G$ with $f(x) \neq f(y)$. We have $g_1, \ldots, g_m \in R \otimes_K R$ such that

$$f \otimes 1 - 1 \otimes f = \sum_{i=1}^{m} g_i(f_i \otimes 1 - 1 \otimes f_i).$$

Consider the homomorphism $\varphi \colon R \otimes_K R \to K$ of K-algebras sending $g \otimes h$ to $g(x)h(y)$. We have

$$0 \neq f(x) - f(y) = \varphi(f \otimes 1 - 1 \otimes f) = \sum_{i=1}^{m} \varphi(g_i)(f_i(x) - f_i(y)).$$

Thus $f_i(x) \neq f_i(y)$ for some i, which proves our claim. $\square$

The point about Theorem 2.3.15 is of course that $K[X]^G$ need not be finitely generated, for example if G is a non-reductive group. So the theorem says that if we are only interested in invariant theory for the sake of separating G-orbits (which is likely to have been the original motivation of invariant theory), then we need not worry about finite generation. Unfortunately, however, the proof of Theorem 2.3.15 is not constructive.

A defect of reductive groups in positive characteristic is that an epimorphism of representations does not in general remain surjective when restricted to the invariants. In particular, generating invariants are in general not mapped to generating invariants. However, this is true for separating invariants, as the next result shows.

Theorem 2.3.16. *Let G be a linear algebraic group. Then the following two statements are equivalent.*

(a) G is reductive.

(b) If G acts regularly on an affine variety X and if $Y \subseteq X$ is a G-stable, closed subvariety, then the restriction map $K[X] \to K[Y]$ takes every separating subset of $K[X]^G$ to a separating subset of $K[Y]^G$.

Proof. First assume that G is reductive and let $S \subseteq K[X]^G$ be a separating subset. If two points x, y of Y can be separated by an invariant in $K[Y]^G$, then the orbit closures do not meet: $\overline{G(x)} \cap \overline{G(y)} = \emptyset$ (see Newstead [182, Corollary 3.5.2]). But this also holds in X, therefore x and y can be separated by an invariant from $K[X]^G$ (again by Corollary 3.5.2 in [182]). Thus there is an $f \in S$ with $f(x) \neq f(y)$. The same inequality holds for the restriction of f to Y.

To prove the converse, let V be a rational representation of G. Then $V^G \subseteq V$ is G-stable and closed. $K[V^G]$ coincides with its invariant ring, thus two distinct points $v, w \in V^G$ can be separated by an invariant from $K[V^G]^G$. By the assumption (b), this implies that v and w can also be separated by an invariant from $K[V]^G$. Assume $v \neq 0$ and take $w = 0$. Then we have $f \in K[V]^G$ with $f(v) \neq f(0)$. The invariant $f_+ := f - f(0)$ has no constant term, and $f_+(v) \neq 0$. Hence there exists a homogeneous invariant g of positive degree with $g(v) \neq 0$. Thus G is geometrically reductive. $\qquad\square$

Further results on separating subalgebras can be found in Section 3.9.4.

2.4 Homogeneous Systems of Parameters

In this section we will assume that G is a reductive group and V is a rational representation. First, we need to introduce the nullcone.

2.4.1 Hilbert's Nullcone

In this section we give a criterion for a set of homogeneous polynomials $f_1, \ldots, f_r \in K[V]^G$ which ensures that $K[V]^G$ is a finite module over $K[f_1, \ldots, f_r]$. We will use Hilbert's notion of the nullcone.

Definition 2.4.1. *The* **nullcone** $\mathcal{N}_V \subseteq V$ *is the zero set of all homogeneous invariant polynomials of positive degree:*

$$\mathcal{N}_V := \{v \in V \mid f(v) = 0 \text{ for all } f \in K[V]_+^G\}.$$

There is also a more geometric description of the nullcone. If $\pi : V \to V /\!\!/ G$ is the categorical quotient, then $\mathcal{N}_V$ is exactly the zero fiber $\pi^{-1}(0)$.

Lemma 2.4.2. *The nullcone $\mathcal{N}_V$ is the set of all $v \in V$ such that the orbit closure $\overline{G \cdot v}$ contains 0.*

Proof. The only closed orbit in $\pi^{-1}(0) = \mathcal{N}_V$ must be 0 (see Corollary 2.3.6). All other orbits in $\mathcal{N}_V$ must have 0 in its closure. On the other hand, it is clear that every orbit which has 0 in its closure must be contained in $\pi^{-1}(0) = \mathcal{N}_V$. $\qquad\qquad\square$

For a finite group the nullcone only consists of 0.

The following theorem shows that if $v \in \mathcal{N}_V$, then not only does the Zariski closure of $G \cdot v$ contain 0, we also can find a multiplicative subgroup $\mathbb{G}_m \subseteq G$ such that $\mathbb{G}_m \cdot v$ contains 0 in its closure. This statement is known as the Hilbert-Mumford criterion. It was first proved by Hilbert [108] for SL_n, and later in a more general setting by Mumford (see Mumford et al. [169]).

Theorem 2.4.3. *Choose a maximal torus $T \subseteq G$. Let $\mathcal{N}_{V,T}, \mathcal{N}_{V,G} \subseteq V$ be the nullcones with respect to the actions of T and G respectively. Then $\mathcal{N}_{V,G} = G \cdot \mathcal{N}_{V,T}$.*

With the Hilbert-Mumford criterion it is often easy to decide which orbits lie in the nullcone, as the following example shows.

Example 2.4.4. Consider again the binary forms of degree d (see Example 2.1.2). We can choose a maximal torus

$$T = \left\{ \begin{pmatrix} \lambda & 0 \\ 0 & \lambda^{-1} \end{pmatrix} =: \sigma_\lambda \mid \lambda \in K^* \right\}$$

of all diagonal matrices in SL_2. Let $f = a_0 x^d + a_1 x^{d-1} y + \cdots + a_d y^d \in V_d$. For $\sigma_\lambda \in T$ we have

$$\sigma_\lambda \cdot f = a_0 x^d \lambda^d + a_1 x^{d-1} y \lambda^{d-2} + \cdots + a_d y^d \lambda^{-d}.$$

It is clear that $f \in \mathcal{N}_{V_d,T}$ if and only if f is divisible by x^e or y^e where $e = \lfloor \frac{1}{2} d \rfloor + 1$. It follows from Theorem 2.4.3 that $f \in \mathcal{N}_{V_d} = \mathcal{N}_{V_d,\mathrm{SL}_2}$ if and only if f is divisible by $(\sigma \cdot x)^e$ for some $\sigma \in \mathrm{SL}_2$. So $f \in \mathcal{N}_{V_d}$ if and only if f has a zero of multiplicity $> \frac{1}{2} d$. $\qquad\qquad\triangleleft$

Lemma 2.4.5. *If $f_1, \ldots, f_r \in K[V]^G$ are homogeneous and the zero set $\mathcal{V}(f_1, \ldots, f_r)$ is $\mathcal{N}_V$, then $K[V]^G$ is a finitely generated F-module where $F = K[f_1, \ldots, f_r]$.*

Proof. Suppose that $K[V]^G = K[h_1, \ldots, h_s]$ for some homogeneous invariants $h_1, \ldots, h_s$. Let $\mathfrak{m} = (h_1, \ldots, h_s)$ be the maximal homogeneous ideal of $K[V]^G$. The zero set of $\mathfrak{m}$ in $V /\!\!/ G$ is just $\pi(0)$ (which we also call 0). Let J be the ideal generated by $f_1, \ldots, f_r$ in $K[X]^G$ and I be the ideal generated by $f_1, \ldots, f_r$ in $K[X]$. Suppose $Y := \mathcal{V}(J) \subseteq V /\!\!/ G$ is the zero set of J. Then the zero set $\mathcal{V}(I) \subseteq V$ is equal to $\pi^{-1}(Y)$. On the other hand this zero set must be equal to the nullcone $\pi^{-1}(0)$ and because π is surjective (see Lemma 2.3.2), we get $Y = \{0\}$. By Hilbert's Nullstellensatz there exists an l such that $h_i^l \in J$ for all i. Let S be the set of all monomials in $h_1^{i_1} h_2^{i_2} \cdots h_s^{i_s}$ with $0 \leq i_1, \ldots, i_s < l$. It is clear that S generates $K[V]^G$ as an F-module. $\qquad\square$

2.4.2 Existence of Homogeneous Systems of Parameters

Definition 2.4.6. *Suppose that $R = \bigoplus_{d=0}^{\infty} R_d$ is a graded algebra over a field K such that $R_0 = K$. A set $f_1, \ldots, f_n \in R$ of homogeneous elements is called a* **homogeneous system of parameters** *if*

(a) $f_1, \ldots, f_r$ are algebraically independent and
(b) R is a finitely generated module over $K[f_1, \ldots, f_r]$.

If $f_1, \ldots, f_r \in K[V]^G$ is a homogeneous system of parameters, then we call the f_i **primary invariants**. The invariant ring $K[V]^G$ is a finite $K[f_1, \ldots, f_r]$-module, say

$$K[V]^G = F g_1 + F g_2 + \cdots + F g_s,$$

where F is the polynomial ring $K[f_1, \ldots, f_r]$ and $g_1, \ldots, g_s \in K[V]^G$ homogeneous. The invariants $g_1, \ldots, g_s$ are called **secondary invariants**. Some of the algorithms for computing invariant rings will compute primary invariants first, and then calculate a set of secondary invariants (see Sections 3.3 and 3.5).

Homogeneous systems of parameters always exist for invariant rings. To see this we first need the Noether Normalization Lemma (see Eisenbud [59, Theorem 13.3]).

Lemma 2.4.7. *Suppose that $R = \bigoplus_{d=0}^{\infty} R_d$ is a graded ring with $R_0 = K$. Suppose that $f_1, \ldots, f_r \in R_d$ and R is a finitely generated F-module where $F = K[f_1, \ldots, f_r]$. Then there exist $g_1, \ldots, g_s \in R_d$ which are linear combinations of $f_1, \ldots, f_r$ such that $g_1, \ldots, g_s$ is a homogeneous system of parameters.*

Finitely generated graded rings always have homogeneous systems of parameters as the following corollary shows. In particular invariant rings of reductive groups have homogeneous systems of parameters.

Corollary 2.4.8. *If $R = \bigoplus_{d=0}^{\infty} R_d$ is a finitely generated graded algebra with $R_0 = K$, then R has a homogeneous system of parameters.*

Proof. Take homogeneous generators $f_1, \ldots, f_r \in R$. Set $d_i = \deg(f_i)$ for all i. Let d be the least common multiple of $d_1, \ldots, d_r$ and define $f_i' = f_i^{d/d_i}$. Then $f_1', f_2', \ldots, f_r'$ are homogeneous of degree d. Now apply Lemma 2.4.7. $\square$

Example 2.4.9. Let $H \subseteq S_n$ act on K^n by permutations. The elementary symmetric polynomials $f_1, \ldots, f_n \in K[x_1, \ldots, x_n]^{S_n}$ (as defined in Example 2.1.1) are also H-invariant. The polynomial

$$\varphi(t) := (t - x_1)(t - x_2) \cdots (t - x_n) = t^n - f_1 t^{n-1} + f_2 t^{n-2} - \cdots + (-1)^n f_n$$

has exactly $x_1, \ldots, x_n$ as zero set. So if $f_1 = f_2 = \cdots = f_n = 0$, then $x_1 = x_2 = \cdots = x_n = 0$. But $\mathcal{V}(f_1, \ldots, f_n)$ is exactly Hilbert's nullcone $\mathcal{N}_V = \{0\}$. It follows that $K[x_1, \ldots, x_n]^H$ is a finite F-module, with $F = K[f_1, \ldots, f_n]$. Because $\dim F = \dim K[x_1, \ldots, x_n] = n$, we must have that $f_1, \ldots, f_n$ are algebraically independent. This shows that $f_1, \ldots, f_n$ is a system of primary invariants (for *any* permutation group $H \subseteq S_n$). $\triangleleft$

2.5 The Cohen-Macaulay Property of Invariant Rings

In this section K is an arbitrary field (not necessarily algebraically closed). First we will discuss the Cohen-Macaulay property in general. Then we will prove the important result of Hochster and Roberts about the Cohen-Macaulay property of invariant rings.

2.5.1 The Cohen-Macaulay Property

If $f_1, \ldots, f_n \in K[V]^G$ are primary invariants, $K[V]^G$ is finitely generated as a module over the subalgebra $A = K[f_1, \ldots, f_n]$. It would be very nice if $K[V]^G$ were in fact a *free* module over A. As we will see, this is the case precisely if $K[V]^G$ is Cohen-Macaulay, and the latter condition is always satisfied if G is linearly reductive.

Definition 2.5.1. *Let R be a Noetherian ring and M a finitely generated R-module.*

(a) *A sequence $f_1, \ldots, f_k \in R$ is called M-**regular** if $M/(f_1, \ldots, f_k)M \neq 0$ and multiplication by f_i induces an injective map on $M/(f_1, \ldots, f_{i-1})M$, for $i = 1, \ldots, k$.*

(b) *Let $I \subseteq R$ be an ideal with $IM \neq M$. Then the **depth** of I on M is the maximal length k of an M-regular sequence $f_1, \ldots, f_k$ with $f_i \in I$, denoted by $\mathrm{depth}(I, M) = k$.*

(c) *If R is a local or graded ring with maximal (homogeneous) ideal $\mathfrak{m}$, we write $\mathrm{depth}(M)$ for $\mathrm{depth}(\mathfrak{m}, M)$.*

(d) If R is a local Noetherian ring with maximal ideal $\mathfrak{m}$, then M is called **Cohen-Macaulay** *if* $\operatorname{depth}(M) = \dim(M)$, *where* $\dim(M)$ *is the Krull dimension of* $R/\operatorname{Ann}(M)$. *If R is not necessarily local, then M is called* **Cohen-Macaulay** *if for all maximal ideals* $\mathfrak{m} \in \operatorname{Supp}(M)$ *(i.e., $\mathfrak{m}$ containing* $\operatorname{Ann}(M)$*), $M_{\mathfrak{m}}$ is Cohen-Macaulay as an $R_{\mathfrak{m}}$-module.*

(e) R is called Cohen-Macaulay if it is Cohen-Macaulay as a module over itself.

We remark that always $\operatorname{depth}(I, M) \le \operatorname{ht}(I)$ (the height of I), and in the case of a local ring $\operatorname{depth}(M) \le \dim(M)$ (see Eisenbud [59, Proposition 18.2]). For more background about the Cohen-Macaulay property we refer the reader to the book of Bruns and Herzog [31].

Lemma 2.5.2. *A polynomial ring is Cohen-Macaulay.*

Proof. Let $R = K[x_1, \ldots, x_n]$ be a polynomial ring and $P \subseteq R$ a maximal ideal. Then R/P is a finite field extension of K, hence for each $i = 1, \ldots, n$ there exists a non-zero polynomial $f_i(x_i)$ which lies in P. It is elementary to check that $f_1(x_1), \ldots, f_n(x_n)$ is an R-regular sequence. Hence it is also R_P-regular, so

$$\dim(R_P) = n \le \operatorname{depth}(P_P, R_P) \le \operatorname{ht}(P) = n.$$

$\square$

The following proposition gives some important characterizations of the Cohen-Macaulay property for graded algebras.

Proposition 2.5.3. *Let R be a Noetherian graded algebra over a field K with $K = R_0$ the homogeneous part of degree 0. Then the following conditions are equivalent:*

(a) R is Cohen-Macaulay;

(b) every homogeneous system of parameters is R-regular;

(c) if $f_1, \ldots, f_n$ is a homogeneous system of parameters, then R is a free module over $K[f_1, \ldots, f_n]$;

(d) there exists a homogeneous system of parameters $f_1, \ldots, f_n$ such that R is a free module over $K[f_1, \ldots, f_n]$.

A proof of Proposition 2.5.3 can be found in Benson [18, Theorem 4.3.5] or Kemper [134].

Example 2.5.4 (Bruns and Herzog [31, Exercise 2.1.18]). Let $R = K[x^4, x^3y, xy^3, y^4] \subset K[x, y]$ and $S = K[x^4, y^4]$. It is clear that R is finite over S because $(x^3y)^4$ and $(xy^3)^4$ are in S. So x^4, y^4 is a homogeneous system of parameters. If R were Cohen-Macaulay, then x^4, y^4 would be an R-regular sequence. There is a relation $x^4(xy^3)^2 = y^4(x^3y)^2$, but $(x^3y)^2$ does not lie in the R-ideal (x^4). This shows that R is not Cohen-Macaulay. $\triangleleft$

It is the condition (c) in the above proposition that makes the Cohen-Macaulay property very relevant for the computation of invariant rings.

2.5.2 The Hochster-Roberts Theorem

In this section we will prove the following theorem.

Theorem 2.5.5 (Hochster and Roberts [113]). *If G is a linearly reductive group, then $K[V]^G$ is Cohen-Macaulay.*

It was shown in Hochster and Huneke [112] that the result still holds when V is not a representation but a smooth G-variety. Another important generalization can be found in Boutot [26] where it was proven that $K[V]^G$ has rational singularities (this implies that $K[V]^G$ is Cohen-Macaulay). It is hard to overemphasize the importance of the theorem of Hochster and Roberts. Apart from being an important tool for the computation of invariant rings of linearly reductive groups, it provides a good deal of information about the structure of invariant rings: They are finitely generated free modules over a subalgebra which is isomorphic to a polynomial ring. We also have a partial converse of Theorem 2.5.5.

Theorem 2.5.6 (Kemper [140]). *Suppose that G is a reductive group and that for every rational representation the invariant ring $K[V]^G$ is Cohen-Macaulay. Then G is linearly reductive.*

Of course this theorem is vacuous in characteristic 0 (see Theorem 2.2.13), but it shows, for example, that the classical groups in positive characteristic have rational representations with non-Cohen-Macaulay invariant ring. The hypothesis that G be reductive is necessary, as the following example shows.

Example 2.5.7. Consider the additive group $G := \mathbb{G}_a$ over the complex field $K := \mathbb{C}$. The action of G on a rational representation V can be extended to an action of $\mathrm{SL}_2(\mathbb{C})$ (see, for example, Kraft [152, III.3.9]), and we have an isomorphism

$$\mathbb{C}[V]^{\mathbb{G}_a(\mathbb{C})} \cong \mathbb{C}[V \oplus \mathbb{C}^2]^{\mathrm{SL}_2(\mathbb{C})}, \qquad (2.5.1)$$

where $\mathrm{SL}_2(\mathbb{C})$ acts naturally on $\mathbb{C}^2$ (see Example 4.2.12). This isomorphism was proved in 1861 by Roberts (see Springer [231, p. 69, Remark 4] or Schur [211, Satz 1.14]). Now Theorem 2.5.5 and the above isomorphism imply that $\mathbb{C}[V]^G$ is Cohen-Macaulay, although G is not (linearly) reductive.   ◁

We now embark on the proof of Theorem 2.5.5. The special case where G is a finite group whose order is not divisible by the characteristic of K will be proven in Section 3.4, since this proof is instructive and much more elementary than the general one. Our version of the general proof is based on an (almost) elementary proof due to Friedrich Knop, who used ideas from Hochster and Huneke [112]. Knop's proof can be found in Bruns and Herzog [31]. The proof we will give here is similar, and is also inspired by lectures of Mel Hochster at the University of Michigan. We will use methods in positive characteristic. Suppose that R is a finitely generated domain over a finite field K of characteristic $p > 0$. An important notion we will use is tight closure. If I is an ideal of R and q is a power of p, then we will write $I^{[q]}$ for the ideal generated by all q-powers of elements in I. Note that $I^{[q]} \subseteq I^q$, but they do not have to be equal.

Definition 2.5.8. *The* **tight closure** $I^\star$ *of an ideal I is the set of all elements $f \in R$ for which there exists a non-zero-divisor $a \in R$ such that for all sufficiently large p-powers q we have*

$$af^q \in I^{[q]}.$$

It is known that if R is regular (i.e., the coordinate ring of a smooth variety), then we have $I = I^\star$ for all ideals I (see Hochster and Huneke [112, Theorem 4.4]). Here we will prove a weaker statement.

Theorem 2.5.9. *If $R = K[x_1, \ldots, x_n]$ is a polynomial ring over a perfect field K of characteristic p and $I \subseteq R$ is an ideal then $I^\star = I$.*

Proof. Suppose that $f \in I^\star$, i.e., for some $a \in R$ we have

$$af^q \in I^{[q]}$$

for all p-powers $q \geq C$ where C is some constant. By Lemma 2.5.10 below we have

$$a \in \bigcap_{q \geq C} (I^{[q]} : f^q) = \bigcap_{q \geq C} (I : f)^{[q]} \subseteq \bigcap_{q \geq C} (I : f)^q.$$

If $(I : f) \neq R$, then we have $\bigcap_{q \geq C}(I : f)^q = 0$ which contradicts $a \neq 0$. We conclude that $(I : f) = R$ and $f \in I$. $\qquad\square$

Lemma 2.5.10. *If $R = K[x_1, \ldots, x_n]$ is a polynomial ring over a perfect field K of characteristic p, then we have*

$$(I^{[q]} : f^q) = (I : f)^{[q]}$$

for all ideals $I \subseteq R$ and all $f \in R$.

Proof. The inclusion $(I^{[q]} : f^q) \supseteq (I : f)^{[q]}$ is trivial. We will prove the reverse inclusion. Suppose that $a \in (I^{[q]} : f^q)$, so

$$af^q = \sum_{j=1}^{r} a_j f_j^q, \qquad (2.5.2)$$

where $I = (f_1, \ldots, f_r)$. $K[x_1, \ldots, x_n]$ is a free module over the subring $K[x_1^q, \ldots, x_n^q]$, in fact

$$K[x_1, \ldots, x_n] = \bigoplus_i K[x_1^q, \ldots, x_n^q] m_i, \qquad (2.5.3)$$

where the m_i runs over all monomials $x_1^{d_1} \cdots x_n^{d_n}$ with $0 \leq d_1, \ldots, d_n \leq q - 1$. In particular, we can write

$$a = \sum_i b_i m_i \tag{2.5.4}$$

with $b_i \in K[x_1^q, \ldots, x_n^q]$ for all i, and similarly

$$a_j = \sum_i b_{i,j} m_i$$

with $b_{i,j} \in K[x_1^q, \ldots, x_n^q]$ for all i and j. Using (2.5.3) and (2.5.2) it follows that

$$b_i f^q = \sum_j b_{i,j} f_j^q$$

holds for every i. We can write $b_i = c_i^q$ and $b_{i,j} = c_{i,j}^q$ with $c_i, c_{i,j} \in K[x_1, \ldots, x_n]$ for all i, j (note that we can take q-th roots of elements in K because K is perfect). We have

$$c_i^q f^q = \sum_j c_{i,j}^q f_j^q$$

and taking the q-th root gives

$$c_i f = \sum_j c_{i,j} f_j \in I.$$

So $c_i \in (I : f)$ for all i, and from (2.5.4) it follows that $a \in (I : f)^{[q]}$. $\square$

Lemma 2.5.11. *Suppose that V is a representation of a linearly reductive group G. If $I \subseteq K[V]^G$ is an ideal, then $IK[V] \cap K[V]^G = I$.*

Proof. The inclusion "$\supseteq$" is clear. Conversely, suppose that $f \in K[V]^G$ lies in the ideal $IK[V]$, i.e.,

$$f = \sum_{i=1}^r g_i f_i \tag{2.5.5}$$

with $g_1, \ldots, g_r \in K[V]$ and $f_1, \ldots, f_r \in I$. Applying the Reynolds operator to (2.5.5) yields

$$f = \mathcal{R}(f) = \sum_{i=1}^r \mathcal{R}(g_i f_i) = \mathcal{R}(g_i) f_i.$$

Since $\mathcal{R}(g_1), \ldots, \mathcal{R}(g_r) \in K[V]^G$, this proves that $f \in I$. $\square$

Because of the previous lemma, Theorem 2.5.5 follows from the theorem below.

Theorem 2.5.12. *Let $R := K[x_1, \ldots, x_n]$ be a polynomial ring over a field K. If $S \subseteq R$ is a finitely generated graded subalgebra with the property that $IR \cap S = I$ for every ideal $I \subseteq S$, then S is Cohen-Macaulay.*

Proof. Let $f_1, \ldots, f_s \in S$ be a homogeneous system of parameters and put $A := K[f_1, \ldots, f_s]$. S is a finitely generated A-module, say

$$S = \sum_{i=1}^{m} A g_i,$$

where we can assume $g_1 = 1$. We are going to prove that $f_1, \ldots, f_s$ is a regular sequence in S, which by Proposition 2.5.3 implies that S is Cohen-Macaulay. The idea is to reduce to a statement in positive characteristic which can be proven using the notion of tight closure. Suppose that

$$a_{r+1} f_{r+1} = a_1 f_1 + \cdots + a_r f_r \qquad (2.5.6)$$

with $a_1, \ldots, a_{r+1} \in S$. We would like to show that $a_{r+1} \in (f_1, \ldots, f_r)S$. By of the hypothesis we only have to prove that $a_{r+1} \in (f_1, \ldots, f_r)R$. So we need to find $b_1, \ldots, b_r \in R$ such that $a_{r+1} = b_1 f_1 + \cdots + b_r f_r$. If such b_i exist, we may assume their (total) degrees to be smaller than the degree of a_{r+1}. Thus the existence of $b_1, \ldots, b_r$ is equivalent to the solvability of a system of (inhomogeneous) linear equations for the coefficients of the b_i. This system is of the form

$$Mv = w, \qquad (2.5.7)$$

where the vector w and the matrix M have entries in K. Assume, by way of contradiction, that

$$a_{r+1} \notin (f_1, \ldots, f_r)R. \qquad (2.5.8)$$

This means that (2.5.7) has no solution. Thus there exists a non-zero sub-determinant $d \in K$ of the extended matrix $(M|v)$ of rank greater than the rank of M.

Let Z be the smallest subring of K containing all "necessary" elements:

(a) The ring Z contains all coefficients of $f_1, \ldots, f_s$ and $g_1, \ldots, g_m$ (as polynomials over K).

(b) For all i, j we can write

$$g_i g_j = \sum_{l=1}^{m} p_{i,j,l}(f_1, \ldots, f_s) g_l$$

with $p_{i,j,l}$ polynomials with coefficients in K. Let these coefficients also be contained in Z.

(c) For all i we can write

$$a_i = \sum_{j=1}^{m} q_{i,j}(f_1, \ldots, f_s) g_j$$

where $q_{i,j}$ are polynomials with coefficients in K. These coefficients should also lie in Z.

(d) Finally, $d^{-1} \in Z$.

Thus $Z \subseteq K$ is a finitely generated ring. Now we put $R_Z := Z[x_1, \ldots, x_n]$, $A_Z := Z[f_1, \ldots, f_s]$, and $S_Z := A_Z[g_1, \ldots, g_m]$. The condition (a) guarantees that $S_Z \subseteq R_Z$, and by (b) we have that $S_Z = \sum_{i=1}^{m} A_Z g_i$. Because of (c), the equation (2.5.6) involves only elements of S_Z. Moreover, the coefficients of (2.5.7) lie in Z. Let $\mathfrak{m} \subset Z$ be a maximal ideal. Since $d^{-1} \in Z$, (2.5.7) has no solutions modulo $\mathfrak{m}$. This means that

$$a_{r+1} \notin (f_1, \ldots, f_r)R_Z + \mathfrak{m}R_Z \tag{2.5.9}$$

for every maximal ideal $\mathfrak{m} \subset Z$.

For the rest of our argument we need to find a maximal ideal $\mathfrak{m}$ such that $A_Z/\mathfrak{m}A_Z \to S_Z/\mathfrak{m}S_Z$ is injective. For this end, choose a maximal A_Z-linearly independent subset of $\{g_1, \ldots, g_m\}$. Without loss, this subset can be assumed to be $\{g_1, \ldots, g_{m'}\}$ with $m' \leq m$. Set $F_Z := \sum_{i=1}^{m'} A_Z g_i \subseteq S_Z$. The sum is direct, and there exists a non-zero $c \in A_Z$ such that $c \cdot S_Z \subseteq F_Z$. Since R_Z is a finitely generated ring without zero divisors, the intersection of all maximal ideals of R_Z is zero (see Eisenbud [59, Theorem 4.19], but a much more elementary proof can be given here). Thus there exists a maximal ideal $\mathfrak{n} \subset R_Z$ with $c \notin \mathfrak{n}$. Set $\mathfrak{m} := Z \cap \mathfrak{n}$. $R_Z/\mathfrak{n}$ is a field and at the same time a finitely generated ring. From this it easily follows that $R_Z/\mathfrak{n}$ is a finite field. Hence the same is true for

$$k := Z/\mathfrak{m}.$$

In particular, $\mathfrak{m} \subset Z$ is a maximal ideal. Now take an element $f \in A_Z \cap \mathfrak{m}S_Z$. Then

$$cf \in A_Z \cap \mathfrak{m}F_Z = A_Z \cap \bigoplus_{i=1}^{m'} \mathfrak{m}A_Z g_i = \mathfrak{m}A_Z,$$

since $g_1 = 1$ and $g_1, g_2, \ldots, g_{m'}$ are independent. Moreover, $\mathfrak{m}A_Z \subset A_Z$ is a prime ideal, since the quotient ring is a polynomial ring over k. Since $c \notin \mathfrak{m}A_Z$, the above implies $f \in \mathfrak{m}A_Z$. Thus we have shown that $A_Z \cap \mathfrak{m}S_Z = \mathfrak{m}A_Z$, so $A_Z/\mathfrak{m}A_Z \to S_Z/\mathfrak{m}S_Z$ is indeed injective. Set

$$A_k := A_Z/\mathfrak{m}A_Z, \quad S_k := S_Z/\mathfrak{m}S_Z, \quad \text{and} \quad R_k := R_Z/\mathfrak{m}R_Z.$$

Denote the canonical map $S_Z \to S_k$ by $\bar{\ }$. Then $A_k = k[\overline{f}_1, \ldots, \overline{f}_s]$ and $S_k = \sum_{i=1}^{m} A_k \overline{g}_i$. The relation (2.5.6) reduces to

$$\overline{a}_{r+1}\overline{f}_{r+1} = \overline{a}_1\overline{f}_1 + \cdots + \overline{a}_r\overline{f}_r.$$

Let p be the characteristic of k and let q be a power of p. Raising the above equation to the q-th power and multiplying by $\overline{c}$ yields

$$\overline{c} \cdot \overline{a}_{r+1}^{q}\,\overline{f}_{r+1}^{q} = \overline{c} \cdot \overline{a}_1^{q}\overline{f}_1^{q} + \cdots + \overline{c} \cdot \overline{a}_r^{q}\overline{f}_r^{q} \subseteq \left(\overline{f}_1^{q}, \ldots, \overline{f}_r^{q}\right) F_k, \tag{2.5.10}$$

where we put $F_k := \sum_{i=1}^{m'} A_k \overline{g}_i \subseteq S_k$. The latter sum is direct, since $\sum_{i=1}^{m'} \alpha_i g_i \in \mathfrak{m} S_Z$ with $\alpha_i \in A_Z$ implies $c \cdot \sum_{i=1}^{m'} \alpha_i g_i \in \mathfrak{m} F_Z$. Thus $c\alpha_i \in \mathfrak{m} A_Z$ by the freeness of F_Z, and therefore $\alpha_i \in \mathfrak{m} A_Z$. Since $A_k = k[\overline{f}_1, \ldots, \overline{f}_s]$ is a polynomial ring (and therefore Cohen-Macaulay by Lemma 2.5.2), and F_k is a free A_k-module, it follows that $\overline{f}_1^q, \ldots, \overline{f}_s^q$ is F_k-regular. Hence from (2.5.10) we get

$$\overline{c} \cdot \overline{a}_{r+1}^q \in \left(\overline{f}_1^q, \ldots, \overline{f}_r^q\right) F_k \subseteq \left(\overline{f}_1^q, \ldots, \overline{f}_r^q\right) S_k.$$

Application of the canonical map $\widehat{} : R_Z \to R_k$ now yields

$$\widehat{c} \cdot \widehat{a}_{r+1}^q \in \left(\widehat{f}_1^q, \ldots, \widehat{f}_r^q\right) R_k.$$

Since this is true for every p-power q and since $\widehat{c} \neq 0$, this means that $\widehat{a}_{r+1} \in (\widehat{f}_1, \ldots, \widehat{f}_r)^\star$ (the tight closure of the ideal in R_k generated by the $\widehat{f}_i$). By Theorem 2.5.9 this implies $\widehat{a}_{r+1} \in (\widehat{f}_1, \ldots, \widehat{f}_r)$, which contradicts (2.5.9). This shows that the assumption (2.5.8) was false, and we are done. $\square$

2.6 Hilbert Series of Invariant Rings

An important tool for computing invariants is the Hilbert series. The Hilbert series of a ring contains a lot of information about the ring itself. For example, the dimension and other geometric invariants can be read off the Hilbert series (see Section 1.4).

In many cases we already know the Hilbert series $H(K[V]^G, t)$ before knowing generators of $K[V]^G$. For finite groups, $H(K[V]^G, t)$ can be computed using Molien's formula (see Theorem 3.2.2). For arbitrary linearly reductive groups, $H(K[V]^G, t)$ can also be computed (see Section 4.6). The Hilbert series often helps to find invariants more efficiently.

If $H(K[V]^G, t)$ is known, then we have a criterion for a set of homogeneous invariants $f_1, \ldots, f_r$ to generate $H(K[V]^G, t)$, namely

$$K[V]^G = K[f_1, \ldots, f_r] \quad \Longleftrightarrow \quad H(K[V]^G, t) = H(K[f_1, \ldots, f_r], t). \quad (2.6.1)$$

Algorithm 2.6.1. Criterion (2.6.1) gives us another strategy for computing invariants if $H(K[V]^G, t)$ is known. We start with $S := \emptyset$, an empty set of generators. Suppose we have already found a finite set of homogeneous invariants $S = \{f_1, \ldots, f_r\}$. We write $K[S]$ instead of $K[f_1, \ldots, f_r]$ for convenience. Compute $H(K[S], t)$ and compare it to $H(K[V]^G, t)$. If they are equal, we are done. If not, then look at the difference

$$H(K[V]^G, t) - H(K[S], t) = kt^d + \text{higher order terms}.$$

This means that $K[f_1, \ldots, f_r]$ contains all invariants of degree $< d$ and in degree d we are missing k invariants, i.e., $K[S]_d$ has codimension k in $K[V]_d$.

Using linear algebra (we do not want to be more specific at this point), find invariants $g_1, \ldots, g_k \in K[V]_d^G$, such that $K[V]_d^G$ is spanned by $K[f_1, \ldots, f_r]_d$ and $g_1, \ldots, g_k$. We add $g_1, \ldots, g_k$ to our generators, $S := S \cup \{g_1, \ldots, g_k\}$. We continue this process until $H(K[V]^G, t) = H(K[S], t)$ or equivalently $K[V]^G = K[S]$.

Let $v_t\big(H(K[V]^G, t) - H(K[S], t)\big)$ be the smallest integer d such that t^d appears in the power series $H(K[V]^G, t) - H(K[S], t)$. With each step, $v_t\big(H(K[V]^G, t) - H(K[S], t)\big)$ increases. Suppose that $K[V]^G$ is generated by invariants of degree $\leq D$. After D steps,

$$H(K[V]^G, t) - H(K[S], t) = kt^d + \text{higher order terms},$$

with $d > D$, or $H(K[V]^G, t) = H(K[S], t)$. This means that $K[S]$ contains all invariants of degree $\leq D$, so $K[V]^G = K[S]$. This shows that the algorithm terminates whenever $K[V]^G$ is finitely generated.

Notice that S will be a minimal set of generators for $K[V]^G$. From the above description of the algorithm, it is clear that none of the g_i can be omitted.

We will investigate the structure of $H(K[V]^G, t)$ when G is a linearly reductive group and V is a rational representation. First of all, we know that there is a homogeneous system of parameters $f_1, \ldots, f_r$ (or primary invariants) of $K[V]^G$ (see Corollary 2.4.8). So $K[V]^G$ is a free F-module, where $F \cong K[f_1, \ldots, f_r]$, because $K[V]^G$ is Cohen-Macaulay by Theorem 2.5.5 (see Proposition 2.5.3). So there is a decomposition

$$K[V]^G = F g_1 \oplus F g_2 \oplus \cdots \oplus F g_s \tag{2.6.2}$$

with $g_1, \ldots, g_s \in K[V]^G$ homogeneous. The decomposition 2.6.2 is often called a **Hironaka decomposition**. By Example 1.4.8, the Hilbert series of $K[f_1, \ldots, f_r]$ is equal to $\prod_{i=1}^{r}(1 - t^{d_i})^{-1}$ where $d_i := \deg(f_i)$ for all i. From (2.6.2) it follows that

$$H(K[V]^G, t) = \frac{\sum_{j=1}^{s} t^{e_j}}{\prod_{i=1}^{r}(1 - t^{d_i})}, \tag{2.6.3}$$

where $e_j = \deg(g_j)$ for all j.

The degree of a rational function $\deg(a(t)/b(t))$ is given by $\deg(a(t)) - \deg(b(t))$. Boutot proved that $K[V]^G$ has rational singularities if G is reductive and K has characteristic 0 (see Boutot [26]). A consequence is that $H(K[V]^G, t)$ has degree ≤ 0 if G is connected and semisimple. This inequality was proven earlier by Kempf (see Kempf [148]). It has been conjectured in Popov [192] that $H(K[V]^G, t)$ has degree $\geq -\dim(V)$. In fact, it was shown in Popov [192] that the degree of $H(K[V]^G, t)$ is equal to $-\dim(V)$ for "almost all" representations of a connected semisimple group G. Later, Knop and Littelmann classified all irreducible representations of connected semisimple groups G such that the degree of $H(K[V]^G, t)$ is greater than $-\dim(V)$ (see Knop and Littelmann [151]). For a more detailed exposition, see Popov [193]. In an attempt to prove Popov's conjecture, Knop proved the following inequality.

Theorem 2.6.2 (Knop [150]). *Let G be semisimple and connected and let K be a field of characteristic 0. The degree $H(K[V]^G, t)$ is $\leq -r$, where r is the Krull dimension of $K[V]^G$.*

For finite linearly reductive groups (in any characteristic) the same statement follows from Molien's formula (see Theorem 3.2.2).

The bound on the degree of $H(K[V]^G, t)$ can be used to find upper bounds for $\beta(K[V]^G)$ (see Popov [190], Popov [191]). Recall that $\beta(K[V]^G)$ is the smallest integer d such that the ring of invariants $K[V]^G$ is generated by invariants of degree $\leq d$ (see (2.1.1)). For the moment, we just note the following.

Corollary 2.6.3. *Suppose that G is semisimple and connected and $\mathrm{char}(K) = 0$, or that G is finite and $\mathrm{char}(K) \nmid |G|$. If $f_1, \ldots, f_r$ is a homogeneous system of parameters of $K[V]^G$, then*

$$\beta(K[V]^G) \leq \max\{d_1 + d_2 + \cdots + d_r - r, d_1, d_2, \ldots, d_r\} \tag{2.6.4}$$

where $d_i := \deg(f_i)$ for all i. In fact, the above degree bound holds for the secondary invariants.

Proof. This follows immediately from (2.6.3) and Theorem 2.6.2. $\qquad\square$

Remark 2.6.4. If at least two of the polynomials $f_1, \ldots, f_r$ are non-linear, (2.6.4) simplifies to

$$\beta(K[V]^G) \leq d_1 + d_2 + \cdots + d_r - r$$

$\lhd$

Example 2.6.5. Assume $\mathrm{char}(K) = 0$ and let $A_n \subset S_n$ be the alternating group of even permutations acting on $V = K^n$. We saw in Example 2.4.9 that the elementary symmetric polynomials $f_1, \ldots, f_n$ defined in Example 2.1.1 form a homogeneous system of parameters. By Corollary 2.6.3 we have $\beta(K[x_1, \ldots, x_n]^{A_n}) \leq 1 + 2 + \cdots + n - n = \binom{n}{2}$. Let us define $g = \prod_{i < j}(x_i - x_j)$. Then clearly g is A_n-invariant but not S_n-invariant. The degree of g is exactly $\binom{n}{2}$.

We will show that $K[x_1, \ldots, x_n]^{A_n} = F \oplus Fg$ where $F = K[x_1, \ldots, x_n]^{S_n}$ (which is generated by $f_1, \ldots, f_n$, see Example 2.1.1).

If h is an arbitrary homogeneous A_n-invariant, then $h = h_1 + h_2$ with

$$h_1 = \frac{h + \tau \cdot h}{2}, \qquad h_2 = \frac{h - \tau \cdot h}{2},$$

and $\tau = (1\ 2) \in S_n$ is the 2-cycle which interchanges 1 and 2. In this decomposition, h_1 is S_n-invariant and h_2 is an S_n-semi-invariant:

$$\sigma \cdot h_2 = \operatorname{sgn}(\sigma)h_2 \qquad \text{for all } \sigma \in S_n,$$

where $\operatorname{sgn}(\sigma)$ is the sign of the permutation $\sigma \in S_n$. It follows that whenever $x_i = x_j$ for $i \neq j$, we have

$$h_2(x_1,\ldots,x_n) = -((i\ j) \cdot h_2)(x_1,\ldots,x_n) = -h_2(x_1,\ldots,x_n),$$

so $h_2(x_1,\ldots,x_n) = 0$. Thus h_2 must be divisible by $x_i - x_j$ for all $i < j$, so in fact it is divisible by g. We conclude $h = h_1 + g \cdot (h_2/g)$ and $h_1, h_2/g \in F$.

Notice that g cannot be expressed in A_n-invariants of smaller degree, so the bound from Corollary 2.6.3 is sharp in this case. $\triangleleft$

3 Invariant Theory of Finite Groups

The invariant theory of finite groups has enjoyed considerable recent interest, as the appearance of the books by Benson [18] and Smith [225] and of many articles on the subject show. In this chapter we focus on computational aspects. As in Chapter 2, the central goal is the calculation of a finite set of generators for the invariant ring, but we will also address some interesting properties which invariant rings of finite groups may or may not have, and how they can be tackled algorithmically. Almost all algorithms treated in this chapter have implementations in various computer algebra systems. Here is an (almost certainly incomplete) list of computer algebra packages that are devoted to invariant theory mostly of finite groups, ordered chronologically.

- The Invariant Package [87] in MAS. A package for invariants of permutation groups written by Manfred Göbel in MAS (see Kredel [154]). The core procedure performs the reduction of invariants given in Algorithm 3.10.7. The Invariant Package is included in the standard distribution of MAS.
- INVAR [130]. A MAPLE package written by the second author for the computation of invariant rings of finite groups and their properties. An older version is part of the MAPLE share library which is shipped with the standard distribution of MAPLE (see Char et al. [42]). A new version also covers the modular case and can be obtained by anonymous ftp from the site `ftp.iwr.uni-heidelberg.de` under `/pub/kemper/INVAR2/`. For more information, please contact `Gregor.Kemper@iwr.uni-heidelberg.de`.
- FINVAR [106]. A package based on SINGULAR (see Greuel et al. [99]) written by Agnes E. Heydtmann. The scope of the package is similar to that of INVAR. For more information, please contact Wolfram Decker at `decker@math.uni-sb.de`.
- The MAGMA package for invariants [146]. Algorithms for invariant theory of finite groups have become part of the standard distribution of MAGMA (see Bosma et al. [24]). They were implemented by Allan Steel with the collaboration of the second author.
- SYMMETRY [76]. A MAPLE package written by Karin Gatermann and F. Guyard for invariants and equivariants of finite and continuous groups. The focus lies on the computation of invariants and equivariants of compact Lie groups. The package is available from the web-page `http://www.zib.de/gatermann/symmetry.html`.

– PerMuVAR [245]. A library written by Nicolas Thiéry for MuPAD (see Fuchssteiner et al. [72]) devoted to the computation of permutation invariants. The goals of PerMuVAR are similar to those of INVAR, but it is optimized for permutation groups in that it uses concise internal representations for invariants, for example. A beta-version of PerMuVAR is about to be released in the near future. For the current status, see at `http://www.mines.edu/~nthiery/PerMuVAR/`.

As a further tool for research, a database of invariant rings has been provided by the IWR in Heidelberg (see Kemper et al. [147]). This database focuses on modular invariant rings of finite groups and is meant as a research tool for making experiments, testing and setting up conjectures, and generally for finding interesting examples. The database, together with retrieval software and documentation, is available by anonymous ftp from the site `ftp.iwr.uni-heidelberg.de` in the directory `/pub/kemper/DataBase/`.

In this chapter K is a field and G is a finite group acting linearly on a finite dimensional vector space V over K. In other words, V is a module over the group ring KG. If the characteristic of K divides the group order $|G|$, we speak of the **modular case**. Otherwise, we are in the **non-modular case**, which includes $\mathrm{char}(K) = 0$. The number and degrees of generating invariants, as well as the properties of the invariant ring we will be interested in, do not change if we enlarge the ground field K. We may therefore assume that K is algebraically closed when this assumption is convenient, but we do not make it as a general assumption. Notice that if we regard G as a linear algebraic group, then every linear action on a vector space is rational. Since finite groups are (geometrically) reductive, it follows from Nagata's result (Theorem 2.2.16) that the invariant ring is finitely generated. There is, however, an older and much simpler proof for this result due to Emmy Noether [185].

Proposition 3.0.6. *Let R be a finitely generated algebra over a Noetherian commutative ring K, and let G be a finite group acting on R by automorphisms fixing K elementwise. Then R^G is finitely generated as a K-algebra.*

Proof. Let $x_1, \ldots, x_n$ be generators for R. The polynomial

$$\prod_{\sigma \in G} (T - \sigma \cdot x_i) = T^m + a_{i,1}T^{m-1} + \cdots + a_{i,m-1}T + a_{i,m} \in R^G[T]$$

provides an integral equation for x_i over R^G. Hence R is finitely generated as a module over the subalgebra $A := K[a_{1,1}, \ldots, a_{n,m}]$ generated by the coefficients of these equations. Since A is Noetherian, R is Noetherian as an A-module, and hence the same is true for the submodule R^G. Therefore R^G is a finitely generated module over A. $\qquad\square$

Notice that the above proof is not constructive. The main goal of this chapter can be rephrased by saying that we want to turn the proof into a constructive method for finding generators.

3.1 Homogeneous Components

The most basic task in computational invariant theory is to compute invariants of some given degree d. Recall that the invariant ring $K[V]^G$ is a graded algebra, where $K[V]_d^G$ is the space of homogeneous invariants of degree d (see on page 40). The monomials of degree d in the variables $x_1, \ldots, x_n$ form a basis of $K[V]_d$, hence $\dim(K[V]_d) = \binom{n+d-1}{n-1}$, where $n = \dim(V)$.

3.1.1 The Linear Algebra Method

Let $H \leq G$ be a subgroup of G whose invariants of degree d are known (typically, the trivial group), and take a set $S(G/H) \subseteq G$ such that H together with $S(G/H)$ generates G. Take the $|S(G/H)|$-fold direct sum of $K[V]$, whose components are indexed by the elements of $S(G/H)$, and consider the map

$$K[V]^H \to \bigoplus_{\sigma \in S(G/H)} K[V], \; f \mapsto (\sigma \cdot f - f)_{\sigma \in S(G/H)},$$

whose kernel is $K[V]^G$. This map is K-linear (in fact it is a homomorphism of modules over $K[V]^G$), and it preserves the grading. Restriction to the degree-d component yields a linear map $K[V]_d^H \to K[V]_d^{|S(G/H)|}$, whose kernel is $K[V]_d^G$. This map is explicitly given, so its kernel can be effectively calculated by solving a system of linear equations over K. For $H = 1$, the number of unknowns in this system is $\binom{n+d-1}{n-1}$, and the number of equations is $|S(G/H)| \cdot \binom{n+d-1}{n-1}$. This can become enormous for large values of n and d.

3.1.2 The Reynolds Operator

A further method for calculating the homogeneous component $K[V]_d^G$ is by means of the Reynolds operator, which is only available in the non-modular case. Recall that for finite groups, the Reynolds operator is simply averaging over the group (see Example 2.2.3).

Algorithm 3.1.1 (available in the non-modular case). Apply the Reynolds operator to all monomials in $K[V]$ of degree d. This yields a generating set of $K[V]_d^G$ as a vector space. By linear algebra, a basis can be extracted from this.

More generally, let $H \leq G$ be a subgroup such that the index $[G : H]$ is not divisible by the characteristic p of K. Then the **relative Reynolds operator** is defined as

$$\mathcal{R}_{G/H}\colon K[V]^H \to K[V]^G, \quad f \mapsto \frac{1}{[G:H]} \sum_{\sigma \in G/H} \sigma \cdot f,$$

where G/H denotes a set of left coset representatives of H in G. $\mathcal{R}_{G/H}$ is independent of the choice of the coset representatives, and it is easily checked that it is a projection of modules over $K[V]^G$. In particular, the images under $\mathcal{R}_{G/H}$ of a basis of $K[V]_d^H$ generate the desired vector space $K[V]_d^G$. Again, it is a problem of linear algebra to select a basis of $K[V]_d^G$ from a generating set. Carrying this idea further, we can also use a chain $H = G_1 < G_2 < \cdots < G_{r-1} < G_r = G$ of subgroups. Then $\mathcal{R}_{G/H} = \mathcal{R}_{G_r/G_{r-1}} \circ \cdots \circ \mathcal{R}_{G_2/G_1}$, and applying $\mathcal{R}_{G/H}$ to an $f \in K[V]^H$ by composing the $\mathcal{R}_{G_i/G_{i-1}}$ only requires $\sum_{i=2}^r [G_i : G_{i-1}]$ applications of group elements and even fewer summations. If we want to compute $K[V]_d^G$ with this method, we have the additional benefit that we can select a vector space basis of $K[V]_d^{G_i}$ after having applied $\mathcal{R}_{G_i/G_{i-1}}$ to a basis of $K[V]_d^{G_{i-1}}$, so we need to apply the subsequent relative Reynolds operators to fewer elements.

In the non-modular case there is a choice between using Algorithm 3.1.1 or the linear algebra method for calculating homogeneous invariants. In Kemper and Steel [146], the authors analyzed the computational cost of both approaches. The general tendency is that the Reynolds operator performs better for small $|G|$ and large d.

3.2 Molien's Formula

For the computation of the homogeneous invariants of degree d by using the Reynolds operator, it would be a great advantage to know the dimension of $K[V]_d^G$ a priori. These dimensions are encoded in the Hilbert series $H(K[V]^G, t)$ (see Definition 1.4.1). In later sections, especially 3.3 and 3.5, we will see many more uses of the Hilbert series. In this section we prove Molien's formula, which holds in the non-modular case and gives a way for computing the Hilbert series without touching a single invariant. Then we investigate generalizations to the modular case.

3.2.1 Characters and Molien's Formula

We start with the following definitions: If V is a finite dimensional module over the group ring KG, we set

$$\sigma_t(V) := \sum_{d=0}^{\infty} K[V]_d \cdot t^d,$$

which is a formal power series with KG-modules as coefficients. (More precisely, $\sigma_t(V)$ is a formal power series over the representation ring of KG).

Let p be the characteristic of K (which may be zero) and set $|G| = p^a m$ with $p \nmid m$. Choose an isomorphism between the m-th roots of unity in an algebraic closure $\bar{K}$ and in $\mathbb{C}$. For $\tau \in G_{p'}$, the subset of G consisting of the elements of order not divisible by p, we can use this isomorphism to lift the eigenvalues of the τ-action on V to characteristic 0. If $\lambda_1, \ldots, \lambda_n \in \mathbb{C}$ are the lifted eigenvalues, we define the **Brauer character** as

$$\Phi_\tau(V) := \lambda_1 + \cdots + \lambda_n$$

(see, for example, Curtis and Reiner [50, p. 420]), and moreover

$$\det_V^0 (1 - t\tau) := (1 - t\lambda_1) \cdots (1 - t\lambda_n) \in \mathbb{C}[t].$$

Of course we can take the usual trace and the usual determinant of $1 - t\tau$ if K has characteristic 0. We can now apply Φ_τ to $\sigma_t(V)$ coefficient-wise, which yields a formal power series over $\mathbb{C}$.

Lemma 3.2.1. *Let $\tau \in G_{p'}$ be an element of order not divisible by $p = \mathrm{char}(K)$. Then with the above notation we have*

$$\Phi_\tau \left(\sigma_t(V) \right) = \frac{1}{\det_V^0 \left(1 - t \cdot \tau^{-1} \right)}. \tag{3.2.1}$$

Proof. First note that neither side of (3.2.1) changes if we replace K by its algebraic closure $\bar{K}$. Hence we may assume K to be algebraically closed. As functions of V, both sides of the equality remain unchanged if we restrict V to the subgroup generated by τ. Hence we can assume that G is cyclic of order coprime to $\mathrm{char}(K)$. Also observe that for finite dimensional KG-modules U and V we have $K[U \oplus V] = K[U] \otimes_K K[V]$, hence $\sigma_t(U \oplus V) = \sigma_t(U) \cdot \sigma_t(V)$, where multiplication of the coefficients is given by the tensor product. Moreover, Φ_τ is additive and multiplicative. It follows that

$$\Phi_\tau(\sigma_t(U \oplus V)) = \Phi_\tau(\sigma_t(U)) \cdot \Phi_\tau(\sigma_t(V)).$$

The same rule holds for the right-hand side of (3.2.1). Hence we can also assume that V is an indecomposable KG-module, hence it is one-dimensional. But then (3.2.1) is clearly true. $\square$

We can now prove Molien's formula.

Theorem 3.2.2 (Molien's formula). *Let G be a finite group acting on a finite dimensional vector space V over a field K of characteristic not dividing $|G|$. Then*

$$H(K[V]^G, t) = \frac{1}{|G|} \sum_{\sigma \in G} \frac{1}{\det_V^0 (1 - t \cdot \sigma)}.$$

If K has characteristic 0, then $\det_V^0 (1 - t \cdot \sigma)$ can be taken as $\det_V (1 - t \cdot \sigma)$.

Proof. By character theory (see Curtis and Reiner [50, Theorem 18.23]), the operator

$$T(U) = \frac{1}{|G|} \sum_{\sigma \in G} \Phi_\sigma(U)$$

counts the multiplicity of the trivial representation in a KG-module U. Therefore

$$H(K[V]^G, t) = T(\sigma_t(V)),$$

from which Molien's formula follows by Lemma 3.2.1. $\qquad\square$

Because of the great importance of Molien's theorem, the Hilbert series of an invariant ring is sometimes also called the Molien series. Molien's formula is very easy to evaluate in practice. In fact, its summands only depend on the conjugacy class of σ, hence the summation over all elements of G is in fact a summation over the conjugacy classes. Moreover, for any $\sigma \in G$ the coefficients of $\det_V(1 - t \cdot \sigma)$ are (up to signs) the elementary symmetric polynomials in the (lifted) eigenvalues of σ, and these can be expressed in terms of the power sums of the eigenvalues by using Newton's formulae (see Curtis and Reiner [50, p. 314]). But the i-th power sum of lifted eigenvalues of σ is nothing but $\Phi_{\sigma^i}(V)$. Hence there exists a polynomial F_n over $\mathbb{Q}$ in $n + 1$ variables such that $\det_V^0(1 - t \cdot \sigma) = F_n(\Phi_\sigma(V), \ldots, \Phi_{\sigma^n}(V), t)$. In other words, we can evaluate Molien's formula if we only know the Brauer character associated to V and the power maps of G, and we do not even have to compute any determinants!

Remark 3.2.3. Suppose that W is another finite dimensional KG-module. Then the following modification of Molien's formula yields the Hilbert series of the module $(K[V] \otimes_K W)^G$ of equivariants (also called covariants):

$$H\left((K[V] \otimes_K W)^G, t\right) = \frac{1}{|G|} \sum_{\sigma \in G} \frac{\Phi_{\sigma^{-1}}(W)}{\det_V^0(1 - t \cdot \sigma)}.$$

This is proved as Theorem 3.2.2. $\qquad\triangleleft$

3.2.2 Extended Hilbert Series

It is interesting to ask if there is a version of Molien's formula which gives information about the homogeneous components $K[V]_d$ in the modular case. For a finite dimensional KG-module V, define $m(V, K)$ to be the multiplicity of the trivial KG-module in a composition series of V. In the non-modular case we have $m(V, K) = \dim(V^G)$. Consider the **extended Hilbert series**

$$\tilde{H}(K[V], K, t) := \sum_{d=0}^{\infty} m(K[V]_d, K)t^d.$$

Let $\psi\colon G_{p'} \to \mathbb{C}$ be the projective indecomposable character of the trivial module K (see Goldschmidt [90, p. 6-11]), which by definition is the Brauer character of the (unique) indecomposable projective module containing K. This can easily be calculated if the Brauer character table X of G is known, the main step being an inversion of X. The following theorem, which can be found in Mitchell [167, Proposition 1.2] (see also Smith [228, p. 218]), gives the appropriate generalization of Molien's formula in the modular case.

Theorem 3.2.4. *With the above notation, we have*

$$\tilde{H}(K[V], K, t) = \frac{1}{|G|} \cdot \sum_{\sigma \in G_{p'}} \frac{\psi(\sigma)}{\det_V^0(1 - t\sigma)}. \tag{3.2.2}$$

Proof. For an irreducible KG-module S we have by Goldschmidt [90, Theorem 6.10] that $1/|G| \sum_{\tau \in G_p} \psi(\tau^{-1})\Phi_\tau(S)$ equals 1 if $S \cong K$, and 0 otherwise. Since Brauer characters are additive along composition series, it follows that

$$\frac{1}{|G|} \cdot \sum_{\tau \in G_{p'}} \psi(\tau^{-1})\Phi_\tau(U) = m(K, U)$$

for any finite dimensional KG-module U. Therefore

$$\tilde{H}(K[V], K, t) = \frac{1}{|G|} \cdot \sum_{\tau \in G_{p'}} \psi(\tau^{-1})\Phi_\tau(\sigma_t(V)).$$

Now the result follows from Lemma 3.2.1. $\qquad\qquad\square$

Remark 3.2.5. If S is an irreducible KG-module, then (3.2.2) gives the generating function $\tilde{H}(K[V], S, t)$ counting the multiplicity of S as a composition factor in $K[V]_d$ if we substitute ψ by the projective indecomposable character of S. One can also easily obtain a formula for the extended Hilbert series $\tilde{H}(K[V] \otimes_K W, S, t)$ of the module of equivariants, where W is another finite dimensional KG-module. $\qquad\qquad\triangleleft$

We will see in Section 3.3.3 how Theorem 3.2.4 can be used to derive a priori constraints on the degrees of primary invariants.

3.2.3 Hilbert Series of some Modular Invariant Rings

The cyclic group C_p of order p acting on a vector space over $\mathbb{F}_p$ may be regarded as the simplest example of modular invariant theory. However, it is extremely difficult to give a generating set of invariants in this case. Shank [217] achieved this goal for the indecomposable representations of dimension at most 5, for general p. In spite of these difficulties, Almkvist and Fossum [9] were able to develop formulas giving the Hilbert series of the invariant ring of C_p for any indecomposable representation. In fact, Shank used these formulas to prove his result. Almkvist and Fossum's results were generalized by

Hughes and Kemper [115], who lifted the restriction to indecomposable modules, and also obtained formulas for the generating functions encoding the multiplicities of any indecomposable module as a direct summand of $K[V]_d$. This work was further generalized in [116] to any group G whose order is divisible by $p = \mathrm{char}(K)$, but not by p^2. The authors give a formula for the Hilbert series of the invariants for any representation of such a group, which is reminiscent of (though much more complicated than) Molien's formula. Denis Vogel wrote a program in MAGMA which evaluates this formula.

3.3 Primary Invariants

The first strategic goal in the calculation of an invariant ring of a finite group is the construction of a homogeneous system of parameters (see Section 2.4.2). The invariants occurring in a homogeneous system of parameters are called primary invariants. The existence of a homogeneous system of parameters is guaranteed by the Noether Normalization Theorem (Corollary 2.4.8). We have the following criterion for primary invariants.

Proposition 3.3.1. *Let $f_1, \ldots, f_n \in K[V]_+^G$ be homogeneous invariants of positive degree with $n = \dim_K(V)$. Then the following statements are equivalent:*

(a) $f_1, \ldots, f_n$ form a homogeneous system of parameters;
(b) $\mathcal{V}_{\bar{K}}(f_1, \ldots, f_n) = \{0\}$, where $\mathcal{V}_{\bar{K}}(f_1, \ldots, f_n)$ is defined as $\{v \in \bar{K} \otimes_K V \mid f_i(v) = 0$ for $i = 1, \ldots, n\}$ and $\bar{K}$ is an algebraic closure of K;
(c) the Krull dimension $\dim(K[V]/(f_1, \ldots, f_n))$ is zero;
(d) $\dim(K[V]/(f_1, \ldots, f_i)) = n - i$ for $i = 1, \ldots, n$.

Proof. Since $K[V]$ is integral over $K[V]^G$ (see the proof of Proposition 3.0.6), it follows by Eisenbud [59, Proposition 9.2] that

$$\dim(K[V]^G) = \dim(K[V]) = n.$$

Therefore any homogeneous system of parameters in $K[V]^G$ must have n elements, and (a) is equivalent to the condition that $K[V]^G$ is a finitely generated module over the subalgebra $A := K[f_1, \ldots, f_n]$, which in turn is equivalent to the condition that $K[V]$ is a finitely generated module over A. By the graded version of Nakayama's lemma (see Lemma 3.5.1 on page 89), this is equivalent to

$$\dim_K(K[V]/(f_1, \ldots, f_n)) < \infty.$$

A K-algebra is of finite K-dimension if and only if its Krull dimension is zero. So we have shown the equivalence of (a) and (c). By Krull's Principal Ideal Theorem (see Eisenbud [59, Theorem 10.2]), (c) and (d) are equivalent. Finally, (b) is equivalent to (c) since the ideal $(f_1, \ldots, f_n)$ is homogeneous. $\square$

Proposition 3.3.1 is the key to all algorithms for constructing primary invariants.

3.3.1 Dade's Algorithm

It is important to note that there are many choices of a homogeneous system of parameters. For example, one can substitute any element in a homogeneous system of parameters by a power of itself. As we will see in Section 3.5, it is crucial for the efficiency of subsequent calculations that a homogeneous system of parameters be chosen whose degrees are as small as possible. More precisely, one usually wants to minimize the product $\prod_{i=1}^{n} \deg(f_i)$ (see Theorem 3.7.1). An algorithm for the construction of a homogeneous system of parameters for $K[V]^G$ was given by Dade (see Stanley [236], Reiner and Smith [198]). It is based on the following observation.

Proposition 3.3.2. *Let* $n := \dim(V)$ *and suppose that* $l_1, \ldots, l_n \in V^* \setminus \{0\}$ *are linear forms such that*

$$l_i \notin \bigcup_{\sigma_1, \ldots, \sigma_{i-1} \in G} \langle \sigma_1 \cdot l_1, \ldots, \sigma_{i-1} \cdot l_{i-1} \rangle_{K\text{-vector space}} \quad for \quad i = 2, \ldots, n.$$

Let f_i *be the product over all* l *in the* G-*orbit of* l_i. *Then* $\{f_1, \ldots, f_n\}$ *is a homogeneous system of parameters of* $K[V]^G$.

Proof. We show that condition (b) from Proposition 3.3.1 is satisfied. Take $v \in \mathcal{V}_{\bar{K}}(f_1, \ldots, f_n)$. Then $(\sigma_i \cdot l_i)(v) = 0$ for some $\sigma_i \in G$. But the assumption says that $\sigma_1 \cdot l_1, \ldots, \sigma_n \cdot l_n$ form a basis of V^*, hence $v = 0$. $\qquad\square$

It is clear how Proposition 3.3.2 can be turned into an algorithm, provided that the ground field K is large enough to make the avoidance of a union of at most $|G|^{n-1}$ proper subspaces possible. This algorithm is simple and quick, but the main drawback is that it tends to produce invariants whose degrees are of the same order of magnitude as $|G|$. For some experimental data on this see Kemper [135]. In that paper, various other approaches for the calculation of a homogeneous system of parameters are explored, with the outcome that a computable criterion to decide whether a given degree vector $d_1, \ldots, d_n$ represents the degrees of a homogeneous system of parameters is required in order to obtain an algorithm which always produces an optimal homogeneous system of parameters.

3.3.2 An Algorithm for Optimal Homogeneous Systems of Parameters

The following provides a criterion for the existence of primary invariants of given degrees.

Theorem 3.3.3 (Kemper [135]). *Let $d_1, \ldots, d_k$ be positive integers, and assume that K is an infinite field. Then the following are equivalent:*

(a) There exist homogeneous $f_1, \ldots, f_k \in K[V]^G$ with $\deg(f_i) = d_i$ such that

$$\dim(K[V]/(f_1, \ldots, f_k)) = n - k;$$

(b) for each subset $\mathcal{I} \subseteq \{1, \ldots, k\}$ the inequality

$$\dim\left(K[V]/(K[V]_{d_i}^G \mid i \in \mathcal{I})\right) \leq n - |\mathcal{I}|$$

holds. Here $(K[V]_{d_i}^G \mid i \in \mathcal{I})$ denotes the ideal in $K[V]$ generated by the union of all homogeneous components $K[V]_{d_i}^G$ with $i \in \mathcal{I}$.

The implication "(a) $\Rightarrow$ (b)" also holds if K is a finite field.

Observe that the ideals $(K[V]_{d_i}^G \mid i \in \mathcal{I}) \subseteq K[V]$ for $\mathcal{I} \subseteq \{1, \ldots, n\}$ can be calculated since K-bases for the subspaces $K[V]_{d_i}^G$ can be obtained by the methods of Section 3.1. Moreover, the dimensions of these ideals can be computed by using Gröbner basis methods (see Section 1.2.5).

We obtain the following rough idea of an algorithm for the construction of an optimal homogeneous system of parameters.

Algorithm 3.3.4 (Optimal primary invariants, rough algorithm).

(1) Loop through all degree vectors $(d_1, \ldots, d_n) \in \mathbb{N}^n$, ordered by rising values of $\prod_{i=1}^{n} d_i$, until one is found which satisfies the conditions in (b) of Theorem 3.3.3.

(2) Loop through all $f_1 \in K[V]_{d_1}^G$ until f_1 is found such that $(d_2, \ldots, d_n)$ satisfies the conditions in (b) of Theorem 3.3.3, with $K[V]$ replaced by $K[V]/(f_1)$.

(3) By recursion, obtain $f_2, \ldots, f_n$ of degrees $d_2, \ldots, d_n$ such that $f_1, \ldots, f_n$ is the desired homogeneous system of parameters.

(4) If the loop through $K[V]_{d_i}^G$ fails at some level in the recursion (which by Theorem 3.3.3 can only happen if K is finite), go back into the loop (1) and choose a new degree vector $(d_1, \ldots, d_n)$.

To make the algorithm more precise, one has to specify a procedure to enumerate the (usually infinite) vector space $K[V]_{d_i}^G$ in such a way that for a non-zero polynomial f on $K[V]_{d_i}^G$, a vector where f does not vanish is found after finitely many steps. For details, we refer to [135] and remark here that there is no theoretical or practical difficulty involved in this task. While it is clear that the above algorithm terminates and produces a homogeneous system of parameters with a minimal degree product, it still appears quite appalling, since it involves up to 2^n Gröbner basis computations for testing the conditions from (b) of Theorem 3.3.3 for each degree vector, and a further minimum of 2^n Gröbner basis computations for the recursive construction of the f_i.

However, with a few modifications the algorithm becomes quite feasible. Most importantly, some strong and easily testable restrictions can be applied on degree vectors before they are passed to the recursive loops. We will discuss such restrictions below. Furthermore, in the recursive loops as few of the conditions from (b) of Theorem 3.3.3 as possible are applied. This way a refined algorithm is obtained, which is given in detail in [135]. The approach of trying to get along with testing only a minimal number of conditions from (b) of Theorem 3.3.3 is justified by the fact that the subset of $K[V]_{d_1}^G \times \cdots \times K[V]_{d_n}^G$ consisting of those $(f_1, \ldots, f_n)$ which form a homogeneous system of parameters is Zariski-open (see [135, Proposition 1]). This means that the refined algorithm probabilistically only requires one Gröbner basis computation. It is implemented in MAGMA (see Bosma et al. [24]), and experience shows that it works quite well.

3.3.3 Constraints on the Degrees of Primary Invariants

Let $f_1, \ldots, f_n$ be primary invariants of degrees $d_1, \ldots, d_n$. By Remark 1.4.3, the Hilbert series of $A := K[f_1, \ldots, f_n]$ is

$$H(A, t) = \frac{1}{(1 - t^{d_1}) \cdots (1 - t^{d_n})}. \tag{3.3.1}$$

By looking at a graded free resolution of $K[V]^G$ as a module over A (see Section 1.3.2), we conclude that the Hilbert series of $K[V]^G$ can be written as

$$H(K[V]^G, t) = \frac{f(t)}{(1 - t^{d_1}) \cdots (1 - t^{d_n})} \quad \text{with} \quad f(t) \in \mathbb{Z}[t]. \tag{3.3.2}$$

Multiplying this by $(1 - t)^n$ and substituting $t = 1$ yields the value $f(1)/(d_1 \cdots d_n)$, hence $H(K[V]^G, t)$ has a pole at $t = 1$ of order at most n. Therefore we have a Laurent expansion

$$H(K[V]^G, t) = \frac{a_0}{(1 - t)^n} + \frac{a_1}{(1 - t)^{n-1}} + \cdots$$

about $t = 1$, with $a_0 = f(1)/(d_1 \cdots d_n)$. Since $H(K[V]^G, t)$ is coefficient-wise bounded from below by $H(A, t) = \prod_{i=1}^n (1 - t^{d_i})^{-1}$, the coefficient a_0 must be non-zero. This coefficient is the degree of $K[V]^G$ (see Definition 1.4.7). Since $f(1)$ is an integer, the product $d_1 \cdots d_n$ is a multiple of $1/\deg(K[V]^G)$. On the other hand $\deg(K[V]^G) = 1/|G|$ by Smith [225, Theorem 5.5.3]. To summarize:

Proposition 3.3.5. *If $d_1, \ldots, d_n$ are the degrees of primary invariants of $K[V]^G$, then the product $d_1 \cdots d_n$ is divisible by $|G|$.*

An alternative proof using Galois theory is contained in the proof of Theorem 3.7.1. In addition, we know from Campbell et al. [36] and Kemper [133] that the least common multiple of the d_i is divisible by the exponent of G. These results pose restrictions on the degrees $d_1, \ldots, d_n$ which are applicable for any finite group. A stronger restriction is obtained by using Equation (3.3.2) directly in cases where the Hilbert series is known (e.g., if $\operatorname{char}(K)^2 \nmid |G|$). Indeed, picking the smallest d_i such that $H(K[V]^G, t) \cdot \prod_{i=1}^{n}(1 - t^{d_i})$ is a polynomial with integral coefficients often yields the actual degrees of an optimal homogeneous system of parameters. In the non-modular case, one even knows that the coefficients of $f(t)$ are non-negative (see Equation (3.5.1)).

Example 3.3.6. (a) We consider the permutation group G of order 4 generated by $(1\ 2)(3\ 4)$ and $(1\ 4)(2\ 3)$. Let V be the natural four-dimensional permutation module over $K = \mathbb{Q}$. Molien's formula yields

$$H(K[V]^G, t) = \frac{1}{4}\left(\frac{1}{(1-t)^4} + \frac{3}{(1-t^2)^2}\right) =$$

$$\frac{t^2 - t + 1}{(1-t)^2(1-t^2)^2} = \frac{1 + t^3}{(1-t)(1-t^2)^3},$$

so $(1, 2, 2, 2)$ is the smallest possible degree vector for primary invariants. Indeed, we find

$$f_1 = x_1 + x_2 + x_3 + x_4, \qquad f_2 = (x_1 - x_2 + x_3 - x_4)^2,$$
$$f_3 = (x_1 - x_2 - x_3 + x_4)^2, \quad f_4 = (x_1 + x_2 - x_3 - x_4)^2.$$

(b) Now take the abelian group G of order 8 generated by the matrices

$$\begin{pmatrix} 1 & & \\ & 1 & \\ & & i \end{pmatrix} \quad \text{and} \quad \begin{pmatrix} -1 & & \\ & -1 & \\ & & 1 \end{pmatrix} \in \operatorname{GL}_3(\mathbb{C}).$$

This is an example of Stanley (see Sloane [223]). Molien's formula yields

$$H(K[V]^G, t) = \frac{1}{(1-t^2)^3},$$

so the degree vector $(d_1, d_2, d_3) = (2, 2, 2)$ meets the restriction posed by Equation (3.3.2), and is minimal with that property. But $K[V]_2^G$ is generated by $x_1^2, x_1 x_2$ and x_2^2, so we obtain the Krull dimension $\dim(K[V]^G/(K[V]_2^G)) = 1$. But the condition in Theorem 3.3.3(b) for $\mathcal{I} = \{1, 2, 3\}$ is $\dim(K[V]^G/(K[V]_2^G)) \leq 3 - |\mathcal{I}| = 0$, hence there are no primary invariants of degrees $(2, 2, 2)$. The degree vector with the second-lowest product is $(d_1, d_2, d_3) = (2, 2, 4)$, and here our algorithm readily finds primary invariants

$$f_1 = x_1^2, \quad f_2 = x_2^2, \quad f_3 = x_3^4.$$

$\triangleleft$

Although one might argue that the extended Hilbert series $\tilde{H}(K[V], K, t)$ does not have much significance to invariant theory, it is still true that it poses the same type of constraints on the degrees of primary invariants as $H(K[V]^G, t)$, as the following theorem shows.

Theorem 3.3.7 (Kemper [137]). *Suppose that $f_1, \ldots, f_n \in K[V]^G$ are primary invariants of degrees $d_1, \ldots, d_n$. Then we can write $\tilde{H}(K[V], K, t)$ in the form*

$$\tilde{H}(K[V], K, t) = \frac{f(t)}{(1 - t^{d_1}) \cdots (1 - t^{d_n})} \tag{3.3.3}$$

with $f(t) \in \mathbb{Z}[t]$.

Proof. We write $A = K[f_1, \ldots, f_n]$. Then $K[V]$ is a finitely generated graded module over the group ring AG. We will prove the theorem for an arbitrary finitely generated graded AG-module M. Notice that the multiplication with any $f \in A$, when restricted to an irreducible submodule of M, becomes the zero map or a KG-monomorphism. Hence the socle $\mathrm{Soc}(M)$ is a graded AG-submodule of M. Now we construct a filtration of M by recursively defining $M^{(0)} = M$ and $M^{(i+1)} = M^{(i)} / \mathrm{Soc}(M^{(i)})$ for $i > 0$. There are canonical maps $M \to M^{(i)}$, whose kernels we denote by $\mathrm{Soc}^i(M)$. Then $\mathrm{Soc}^1(M) \subseteq \mathrm{Soc}^2(M) \subseteq \ldots \subseteq M$ is an ascending sequence of A-submodules of M, hence by Noetherianity $\mathrm{Soc}^{N+1}(M) = \mathrm{Soc}^N(M)$ for some N, so $M^{(i)} = 0$ for $i > N$. We form the graded module $\tilde{M} = \bigoplus_{i \geq 0} M^{(i)}$ associated to the filtration, which is a finitely generated graded AG-module. Now by the definition of the extended Hilbert series we have

$$\tilde{H}(M, K, t) = H(\tilde{M}^G, t),$$

where the right hand side is the ordinary Hilbert series. Since $\tilde{M}^G$ has a free resolution of finite length as an A-module, the result follows as Equation (3.3.2). $\qquad\square$

Remark 3.3.8. It is straightforward to generalize Theorem 3.3.7 to the series $\tilde{H}(K[V], S, t)$ for an irreducible KG-module S (see Remark 3.2.5).   ◁

Example 3.3.9. We consider two representations of the group $G = A_5$ in characteristic 2. In this example, the Brauer character tables are taken from the Atlas of Brauer Characters (Jansen et al. [119]), and the projective indecomposable character ψ of the trivial module is constructed from these. These enables us to use Theorem 3.2.4 to compute the extended Hilbert series.

(a) Since $A_5 \cong \mathrm{SL}_2(\mathbb{F}_4)$ there exists an irreducible two-dimensional module V defined over $K = \mathbb{F}_4$. We calculate the extended Hilbert series of $K[V]$ and obtain

$$\tilde{H}(K[V], K, t) = \frac{t^{14} + t^{12} + t^{10} + 2t^9 + 2t^7 + 2t^5 + t^4 + t^2 + 1}{(1 - t^6)(1 - t^{10})}$$

as the representation in the form (3.3.3) with the smallest possible degrees d_i. (More precisely, this is the representation in which the product $d_1 d_2$ is minimal, and among those representations with minimal $d_1 d_2$, the sum $d_1 + d_2$ is minimal.) It turns out that the smallest degrees which are possible for primary invariants are $d_1 = 5$ and $d_2 = 12$ here, which corresponds to the representation

$$\tilde{H}(K[V], K, t) =$$
$$\frac{t^{15} + t^{13} + t^{11} + t^{10} + t^9 + t^8 + t^7 + t^6 + t^5 + t^4 + t^2 + 1}{(1 - t^5)(1 - t^{12})}$$

with the second-lowest d_1 and d_2. So it can be said that $\tilde{H}(K[V], K, t)$ contains very strong restrictions on the degrees of a homogeneous system of parameters.

(b) Now we take V as the irreducible module of dimension 4, which occurs as a submodule of the natural permutation module. Again, the representation of $\tilde{H}(K[V], K, t)$ with the second-smallest degrees d_i actually gives the degrees of an optimal homogeneous system of parameters. It is

$$\tilde{H}(K[V], K, t) = \frac{t^{10} + t^8 + 3t^7 + 4t^6 + 6t^5 + 4t^4 + 3t^3 + t^2 + 1}{(1 - t^2)(1 - t^3)(1 - t^4)(1 - t^5)}.$$

We will compute the ordinary Hilbert series of this invariant ring in Example 3.10.16(b). ◁

3.4 Cohen-Macaulayness

The Cohen-Macaulay property was defined in Section 2.5, and it was shown in Proposition 2.5.3 that a graded algebra is Cohen-Macaulay if and only if it is free as a module over the subalgebra generated by a homogeneous system of parameters. As we will see in Section 3.5, knowing that an invariant ring $K[V]^G$ is Cohen-Macaulay reduces the computation of generators of $K[V]^G$ as a module over the subalgebra generated by primary invariants to pure linear algebra. In the case of finite groups we have:

Theorem 3.4.1 (Hochster and Eagon [111]). *If* $\mathrm{char}(K)$ *does not divide the group order* $|G|$, *then* $K[V]^G$ *is Cohen-Macaulay.*

Proof. Let $f_1, \ldots, f_n$ be primary invariants. By Proposition 3.3.1, $f_1, \ldots, f_n$ also is a homogeneous system of parameters for $K[V]$, hence $f_1, \ldots, f_n$ is a $K[V]$-regular sequence by Lemma 2.5.2 and Proposition 2.5.3. Suppose that for some $i \in \{1, \ldots, n\}$ we have

$$g_i f_i = g_1 f_1 + \cdots + g_{i-1} f_{i-1}$$

with $g_1, \ldots, g_i \in K[V]^G$. Then g_i lies in the ideal generated by $f_1, \ldots, f_{i-1}$ in $K[V]$, so

$$g_i = h_1 f_1 + \cdots + h_{i-1} f_{i-1}$$

with $h_j \in K[V]$. Now we apply the Reynolds operator and obtain

$$g_i = \mathcal{R}(g_i) = \mathcal{R}(h_1) f_1 + \cdots + \mathcal{R}(h_{i-1}) f_{i-1}.$$

Therefore g_i lies in the $K[V]^G$-ideal generated by $f_1, \ldots, f_{i-1}$. Thus $K[V]^G$ is Cohen-Macaulay by Proposition 2.5.3(b). $\qquad\square$

Remark 3.4.2. We have several generalizations of Theorem 3.4.1.

(a) Let W be another finitely generated KG-module. Then $(K[V] \otimes_K W)^G$ is Cohen-Macaulay as a module over $K[V]^G$ if $\mathrm{char}(K) \nmid |G|$. The proof is analogous to the proof of Theorem 3.4.1 and uses that $K[V] \otimes_K W$ is Cohen-Macaulay. It is interesting to remark that although Theorem 3.4.1 holds for linearly reductive groups G by Hochster and Roberts [113], the generalization to covariants becomes false in general (see Van den Bergh [20]).

(b) Suppose that $H \leq G$ is a subgroup such that $\mathrm{char}(K)$ does not divide the index $[G : H]$. Then if $K[V]^H$ is Cohen-Macaulay, so is $K[V]^G$. This is a result of Campbell et al. [34]. The proof is again analogous to the proof of Theorem 3.4.1 and uses the relative Reynolds operator. Together with a result of Ellingsrud and Skjelbred [63] it follows that $K[V]^G$ is Cohen-Macaulay if $\dim_K(V) \leq 3$ (see also Smith [227]). $\qquad\triangleleft$

The following is an example of a modular invariant ring which is not Cohen-Macaulay.

Example 3.4.3. Let $G = \langle \sigma \rangle \cong C_p$ be the cyclic group of order $p := \mathrm{char}(K) > 0$, and consider the action on $K[V] = K[x_1, x_2, x_3, y_1, y_2, y_3]$ by

$$\sigma \cdot x_i = x_i \quad \text{and} \quad \sigma \cdot y_i = y_i + x_i.$$

We have invariants x_i and $u_{i,j} := x_i y_j - x_j y_i$ $(1 \leq i < j \leq 3)$. By Proposition 3.3.1 the sequence x_1, x_2, x_3 can be extended to a homogeneous system of parameters for $K[V]^G$. But the relation

$$u_{2,3} x_1 - u_{1,3} x_2 + u_{1,2} x_3 = \begin{vmatrix} x_1 & x_2 & x_3 \\ x_1 & x_2 & x_3 \\ y_1 & y_2 & y_3 \end{vmatrix} = 0$$

shows that x_1, x_2, x_3 is not $K[V]^G$-regular, since $u_{1,2}$ does not lie in the $K[V]^G$-ideal generated by x_1 and x_2. Hence $K[V]^G$ is not Cohen-Macaulay by Proposition 2.5.3. Notice that in this example the KG-module V is the direct sum of three copies of a two-dimensional KG-module U. For a sum of one or two copies of U the invariant ring is still Cohen-Macaulay. $\qquad\triangleleft$

It is an open problem to give a simple characterization which tells us when the invariant ring of finite group together with a modular representation is Cohen-Macaulay. At least it is known that for a group G and a field K whose characteristic divides $|G|$, there exists a finite dimensional KG-module V such that $K[V]^G$ is not Cohen-Macaulay (Kemper [138]). More precisely, it is shown in [138] that if U is a faithful KG-module, then for the direct sum $V = U^k := \oplus_{i=1}^k U$ of sufficiently many copies of U the invariant ring $K[V]^G$ is not Cohen-Macaulay. The proof uses the cohomology $H^*(G, K[V])$ with values in the polynomial ring, which becomes a module over $K[V]^G = H^0(G, K[V])$ via the cup product. The relevance of $H^*(G, K[V])$ to the Cohen-Macaulay property is given by the following proposition.

Proposition 3.4.4 ([138]). *Suppose that for a non-negative integer r we have $H^i(G, K[V]) = 0$ for $0 < i < r$. (This condition is vacuous if $r \leq 1$.) Then a $K[V]$-regular sequence $f_1, \ldots, f_{r+2} \in K[V]^G$ is $K[V]^G$-regular if and only if multiplication by the f_i induces an injective map*

$$H^r(G, K[V]) \longrightarrow \bigoplus_{i=1}^{r+2} H^r(G, K[V]).$$

This prompts the study of the annihilator $I := \mathrm{Ann}_{K[V]^G}(H^+(G, K[V]))$, where $H^+(G, K[V]) := \oplus_{i>0} H^i(G, K[V])$, and its variety $\mathcal{V}(I) = \mathrm{Supp}_{K[V]^G}(H^+(G, K[V]))$. For simplicity, we assume that K is algebraically closed and write $p := \mathrm{char}(K)$.

Proposition 3.4.5 ([141]). *With the above notation, we have*

$$\mathcal{V}(I) = \bigcup_{\substack{\sigma \in G \\ \mathrm{ord}(\sigma)=p}} V^\sigma.$$

If $V = U^k$ with U a faithful KG-module, it follows that the height of the annihilator I is at least k. Together with Proposition 3.4.4 and the fact that $H^r(G, K) \neq 0$ for some positive r (see Benson [17, Theorem 4.1.3]), this yields the result about non-Cohen-Macaulayness of vector invariants mentioned above. If G is a p-group with $p = \mathrm{char}(K)$, then a closer look at the geometry of the annihilator of the elements in $H^1(G, K)$ yields the following theorem.

Theorem 3.4.6 ([138]). *Assume that G is a p-group with $p := \mathrm{char}(K)$ and that $K[V]^G$ is Cohen-Macaulay. Then G is generated by bireflections, i.e., by $\sigma \in G$ which fix a subspace $U \subseteq V$ of codimension 2 pointwise.*

It is interesting to compare the above theorem to a result of Kac and Watanabe [124], which says that any finite linear group whose invariant ring is a complete intersection (a much stronger property than being Cohen-Macaulay!) is generated by bireflections. A corollary from Theorem 3.4.6 is that if G is a p-group acting faithfully on V, then $K[V^3]^G$ is not Cohen-Macaulay. A more elementary proof for this result was given by Campbell et al. [37].

3.5 Secondary Invariants

In this section we assume that primary invariants $f_1, \ldots, f_n \in K[V]^G$ have been constructed, so $K[V]^G$ is generated by homogeneous invariants $g_1, \ldots, g_m$ as a module over $A := K[f_1, \ldots, f_n]$. Such generators g_i are called secondary invariants (see Section 2.4.2). Together with the primary invariants, the g_i generate $K[V]^G$ as an algebra over K. It should be emphasized that neither primary nor secondary invariants are uniquely determined, and that being a primary or a secondary invariant is not an intrinsic property of an invariant. This section is devoted to the task of finding secondary invariants. We will give entirely different algorithms for the non-modular and the modular case (the non-modular one being much easier). In Section 3.11.2 a third algorithm will be given, which works in all cases, but only performs well if the invariant ring has a nice structure. The following graded version of Nakayama's lemma is of crucial importance to the finding of secondary invariants.

Lemma 3.5.1 (graded Nakayama Lemma). *Let R be a (non-negatively) graded algebra over a field $K = R_0$ and M a non-negatively graded R-module. We write $R_+ := \bigoplus_{d>0} R_d$ for the unique homogeneous maximal ideal. Then for a subset $S \subseteq M$ of homogeneous elements the following two conditions are equivalent:*

(a) S generates M as an R-module;
(b) S generates M/R_+M as a vector space over K. Here R_+M is the submodule of M generated by the elements $a \cdot g$ with $a \in R_+$ and $g \in M$.

In particular, a generating set S for M is of minimal cardinality if no proper subset of S generates M.

Proof. Clearly if S generates M, it also generates M/R_+M as a K-vector space.

Conversely, suppose that S generates M/R_+M and let $g \in M$ be homogeneous of some degree d. Then by assumption

$$g = \sum_{i=1}^{m} \alpha_i g_i + \sum_{j=1}^{r} a_j h_j$$

with $g_1, \ldots, g_m \in S$, $\alpha_i \in K$, $a_j \in R_+$ and $h_j \in M$. By multiplying out homogeneous parts and omitting those summands which are not of degree d, we can assume that the a_j and h_j are homogeneous with $\deg(a_j h_j) = d$.

Hence $\deg(h_j) < d$ and h_j lies in the submodule spanned by S by induction on d. Hence g lies in the module spanned by S.

The last remark on minimality follows from the corresponding property of vector spaces. $\square$

It follows that homogeneous invariants $g_1, \ldots, g_m$ generate $K[V]^G$ as a module over A if and only if their images generate the quotient $K[V]^G/A_+K[V]^G$ as a vector space over K. Thus a non-redundant system of secondary invariants has minimal cardinality, and moreover the degrees of such a system are uniquely determined. We have entirely different algorithms for the modular and non-modular case.

3.5.1 The Non-modular Case

We assume that the characteristic of K is not a divisor of the group order $|G|$. As we shall see, this has several beneficial effects on the efficiency of our algorithms. First, the invariant ring is always Cohen-Macaulay by Theorem 3.4.1. From the above remark, it follows that any system $g_1, \ldots, g_m$ of secondary invariants from which none can be omitted is a system of free generators. Let $e_1, \ldots, e_m$ be the degrees of the g_i. Then it follows from the additivity of the Hilbert series with respect to direct sums and from Equation (3.3.1) that

$$H(K[V]^G, t) = \frac{t^{e_1} + \cdots + t^{e_m}}{(1 - t^{d_1}) \cdots (1 - t^{d_n})}, \tag{3.5.1}$$

where $d_i = \deg(f_i)$. In fact, this is a special case of Equation (2.6.3). Furthermore, we can easily calculate the Hilbert series by Molien's formula, at least if the group is of reasonably small order (where "reasonable" means such that primary invariants can be successfully computed). Thus comparing Equations (3.5.1) and Theorem 3.2.2 yields complete information about the degrees of the secondary invariants.

In order to find the g_i most efficiently, we use Lemma 3.5.1 again. Let $g_1, \ldots, g_m \in K[V]^G$ be homogeneous invariants, with $m = \prod_{i=1}^{n} d_i/|G|$. Then the g_i are secondary invariants if and only if they generate $K[V]^G/A_+K[V]^G$ as a vector space over K. Since the number of g_i is correct, this is equivalent to the condition that the g_i are linearly independent modulo $A_+K[V]^G$. The ideal $A_+K[V]^G$ in $K[V]^G$ is generated by $f_1, \ldots, f_n$, but one cannot calculate with an ideal in $K[V]^G$ before $K[V]^G$ itself is known. To circumvent this problem, consider the map

$$K[V]^G \to K[V]/(f_1, \ldots, f_n)K[V], \quad f \mapsto f + (f_1, \ldots, f_n)K[V].$$

Clearly $A_+K[V]^G$ lies in the kernel. Conversely, an element f in the kernel has the form $f = h_1 f_1 + \cdots + h_n f_n$, and applying the Reynolds operator yields $f = \mathcal{R}(f) = \mathcal{R}(h_1)f_1 + \cdots + \mathcal{R}(h_n)f_n \in A_+K[V]^G$. Therefore we have an embedding

$$K[V]^G/A_+K[V]^G \hookrightarrow K[V]/(f_1,\ldots,f_n)K[V],$$

and conclude that $g_1,\ldots,g_m$ are secondary invariants if and only if they are linearly independent modulo the ideal $I := (f_1,\ldots,f_n)$ in $K[V]$. Now let $\mathcal{G}$ be a Gröbner basis of I with respect to any monomial ordering, and denote the normal form with respect to $\mathcal{G}$ by $\mathrm{NF}_\mathcal{G}$. Such a Gröbner basis has already been calculated in the process of finding the primary invariants f_i (see Section 3.3), so there is no extra cost involved. Then for $\alpha_1,\ldots,\alpha_m \in K$ we have

$$\alpha_1 g_1 + \cdots + \alpha_m g_m \in I \quad \Longleftrightarrow \quad \alpha_1\,\mathrm{NF}_\mathcal{G}(g_1) + \cdots + \alpha_m\,\mathrm{NF}_\mathcal{G}(g_m) = 0,$$

so all we have to do is check the linear independence of the normal forms of the g_i.

We obtain the following algorithm:

Algorithm 3.5.2 (Secondary invariants in the non-modular case).

(1) Let $\mathcal{G}$ be a Gröbner basis of the ideal $(f_1,\ldots,f_n) \subseteq K[V]$ generated by the primary invariants. ($\mathcal{G}$ was already calculated when the f_i were constructed.)
(2) Calculate the degrees $e_1,\ldots,e_m$ by using Molien's formula (Theorem 3.2.2) and comparing to (3.5.1).
(3) For $i = 1,\ldots,m$ perform the following two steps:
(4) Calculate a basis of the homogeneous component $K[V]^G_{e_i}$ by using the methods from Section 3.1.
(5) Select an element g_i from this basis such that the normal form $\mathrm{NF}_\mathcal{G}(g_i)$ lies outside the K-vector space generated by the polynomials $\mathrm{NF}_\mathcal{G}(g_1),\ldots,\mathrm{NF}_\mathcal{G}(g_{i-1})$.
(6) The invariants $g_1,\ldots,g_m$ are secondary invariants.

Remark 3.5.3. Algorithm 3.5.2 can be substantially optimized in (at least) two ways.

(a) One can try to use products of secondary invariants of smaller degrees as new secondary invariants. This is very often successful and has two benefits: It can save the calculation of homogeneous components $K[V]^G_{e_i}$ for some large e_i, and it actually produces a minimal system of generators of $K[V]^G$ as an algebra over $A = K[f_1,\ldots,f_n]$.
(b) If all products of known secondary invariants have been exhausted and the computation of "new" invariants becomes necessary, and if furthermore the Reynolds operator $\mathcal{R}_G$ is used to produce them, then it is enough to apply $\mathcal{R}_G$ to generators of $K[V]$ as a module over $A = K[f_1,\ldots,f_n]$. But by Lemma 3.5.1, such generators are given by a basis of $K[V]/(f_1,\ldots,f_n)$, which in turn can be chosen to consist of those monomials which lie outside $L(f_1,\ldots,f_n)$. The leading ideal is known by the Gröbner basis $\mathcal{G}$. Thus it is enough to apply the Reynolds

operator to all monomials of degree e_i which are not divisible by the leading monomial of any polynomial in $\mathcal{G}$, and of course one can stop as soon as enough new invariants are produced. ◁

Example 3.5.4. (a) We can now finish the computation of the invariant ring from Example 3.3.6(a). From the Hilbert series we see that the secondary invariants are of degrees 0 and 3. Using the above algorithm yields secondary invariants

$$g_1 = 1, \quad g_2 = x_1^3 + x_2^3 + x_3^3 + x_4^3.$$

(b) In Aslaksen et al. [11], the authors considered the permutation representation of degree 6 of the symmetric group $G = S_4$ given by $(1\ 4\ 6\ 3)(2\ 5)$ and $(2\ 4)(3\ 5)$. (This is the action of S_4 on subsets of two elements in $\{1, \ldots, 4\}$.) The ground field is $K = \mathbb{Q}$. Molien's formula yields

$$H(K[V]^G, t) = \frac{1 + t^3 + t^4 + t^5 + t^6 + t^9}{(1-t)(1-t^2)^2(1-t^3)^2(1-t^4)}.$$

Indeed, we find primary invariants of degrees 1,2,2,3,3,4. As secondary invariants we obtain

$$1,\ g_3,\ g_4,\ g_5,\ g_3^2,\ g_4 g_5,$$

where each g_i has degree i. Note that we only had to compute invariants of degrees up to 5. The complete computation takes about one second in MAGMA on a Sun workstation, and confirms the results from [11].

(c) A three-dimensional representation of the group $G = A_5$ over $K = \mathbb{R}$ is given by

$$(1\ 2\ 4) \mapsto \begin{pmatrix} 1 & \alpha & 0 \\ 0 & 0 & 1 \\ 0 & -1 & -1 \end{pmatrix}, \quad (1\ 2\ 3\ 4\ 5) \mapsto \begin{pmatrix} -\alpha & -\alpha & 0 \\ 0 & 0 & -1 \\ \alpha & 1 & 1 \end{pmatrix}$$

with $\alpha = (1 + \sqrt{5})/2$. The Hilbert series is

$$H(K[V]^G, t) = \frac{1 + t^{15}}{(1-t^2)(1-t^6)(1-t^{10})},$$

and MAGMA finds primary invariants of degrees 2,6,10 and secondary invariants of degrees 0 and 15 in about half a second. ◁

3.5.2 The Modular Case

Almost everything we used in the non-modular case is missing in the modular case: The Cohen-Macaulay property fails in general, we do not have Molien's formula and no Reynolds operator. Neither is there any reasonable upper bound known on the degrees of secondary invariants (see Section 3.9.3). Therefore we need a completely different approach for the calculation of secondary invariants. As we will see, this approach is nevertheless very straightforward. The idea is the following.

First, choose a subgroup $H \leq G$ with $\operatorname{char}(K) \nmid |H|$ (for example, the trivial group). Then use the algorithm from Section 3.5.1 to calculate (a minimal set of) secondary invariants $h_1, \ldots, h_r \in K[V]^H$ *with respect to the primary invariants $f_1, \ldots, f_n$ that were chosen for $K[V]^G$*. Since $K[V]^H$ is Cohen-Macaulay, the h_i define an isomorphism $A^r \to K[V]^H$ between a free module over $A = K[f_1, \ldots, f_n]$ and the invariant ring of H. If we assign the degrees of the h_i to the free generators of A^r, this isomorphism becomes degree-preserving. Now take a set $S(G/H) \subseteq G$ which together with H generates G, and consider the map

$$K[V]^H \to \bigoplus_{\sigma \in S(G/H)} K[V], \ f \mapsto (\sigma \cdot f - f)_{\sigma \in S(G/H)},$$

whose kernel is $K[V]^G$ (see Section 3.1.1). Observe that this map is a homomorphism of A-modules. $K[V]$ is Cohen-Macaulay by Lemma 2.5.2 and hence a free A-module. Therefore $\oplus_{\sigma \in S(G/H)} K[V] \cong A^k$. (We can see from Theorem 3.7.1 that $k = |S(G/H)| \cdot \prod_{i=1}^{n} \deg(f_i)$, often an enormous number.) We obtain the following commutative diagram with exact rows, where the map $A^r \to A^k$ is defined by the commutativity and M is its kernel.

$$
\begin{array}{ccccccc}
0 & \longrightarrow & K[V]^G & \longrightarrow & K[V]^H & \longrightarrow & \bigoplus_{\sigma \in S(G/H)} K[V] \\
 & & \uparrow\wr & & \uparrow\wr & & \uparrow\wr \\
0 & \longrightarrow & M & \longrightarrow & A^r & \longrightarrow & A^k
\end{array}
\qquad (3.5.2)
$$

Observe that each map in the diagram is a degree-preserving homomorphism of graded A-modules. The map $A^r \to A^k$ is a homomorphism of free modules over the polynomial algebra A. Hence its kernel M can be calculated by Gröbner basis methods (see Section 1.3.1). Then secondary invariants can be calculated by applying the map $A^r \to K[V]^H$ to the generators of M. We obtain the following algorithm.

Algorithm 3.5.5 (Secondary invariants in the modular case).

(1) Choose a subgroup $H \leq G$ with $\operatorname{char}(K) \nmid |H|$ and calculate secondary invariants $h_1, \ldots, h_r \in K[V]^H$ for H with respect to the primary invariants $f_1, \ldots, f_n$ that were chosen for G. The h_i define a map $A^r \to K[V]^H$, where as usual $A = K[f_1, \ldots, f_n]$.

(2) Calculate generators for $K[V]$ as an A-module. These give an isomorphism $A^k \to \oplus_{\sigma \in S(G/H)} K[V]$, where $S(G/H) \subseteq G$ together with H generates G.

(3) Calculate the preimage under the map $A^k \to \oplus_{\sigma \in S(G/H)} K[V]$ of each $(\sigma \cdot h_i - h_i)_{\sigma \in S(G/H)}$ $(i = 1, \ldots, r)$. This is done by writing down a general element of A^k of degree equal to $\deg(h_i)$ with unknown coefficients,

mapping it into $\oplus_{\sigma \in S(G/H)} K[V]$, equating to $(\sigma \cdot h_i - h_i)_{\sigma \in S(G/H)}$, and solving for the unknown coefficients. This is a system of inhomogeneous linear equations over K.

(4) The preimages calculated in step (3) define the map $A^r \to A^k$ from the diagram (3.5.2). Calculate generators of its kernel M using the methods from Section 1.3.1.

(5) Use linear algebra to omit generators of M which are redundant. By Lemma 3.5.1, this will result in a system of generators with minimal cardinality.

(6) Apply the map $A^r \to K[V]^H$ to the generators of M. The result is a set of minimal secondary invariants for $K[V]^G$.

Remark 3.5.6. (a) It is worth noting that in the authors' experience, in many examples the linear algebra involved in step (3) is much more expensive than the calculation of the syzygy module in step (4).

(b) There is a variant of the algorithm which does not require $K[V]^H$ to be Cohen-Macaulay. The only difficulty is to obtain a minimal set of secondary invariants. For this the A-linear relations between the secondary invariants $h_1, \ldots, h_r$ for $K[V]^H$ have to be calculated. For details we refer the reader to Kemper and Steel [146].

(c) It is possible to derive an upper bound for the degrees of the secondary invariants from Algorithm 3.5.5. We will pursue this possibility in Section 3.9.3. ◁

Example 3.5.7. (a) Let G be the permutation group of order 2 on 6 symbols, generated by $(1\ 2)(3\ 4)(5\ 6)$, and take K to be $\mathbb{F}_2$. Denote the variables of $K[V]$ by $x_1, y_1, x_2, y_2, x_3, y_3$, so the action of G is by exchanging x_i and y_i. We find primary invariants

$$x_i + y_i \quad \text{and} \quad x_i y_i \quad (i = 1, 2, 3).$$

Using the above algorithm, MAGMA finds the following minimal set of secondary invariants:

$$1, \quad x_i y_j + x_j y_i \quad (1 \leq i < j \leq 3) \quad \text{and} \quad x_1 x_2 x_3 + y_1 y_2 y_3.$$

The complete computation takes less than $1/10$ seconds. Note that in this example the number of secondary invariants (5) exceeds the product of the degrees of the primary invariants divided by the group order (4). We will see in Theorem 3.7.1 that this implies that $K[V]^G$ is not Cohen-Macaulay. In fact, this is a special case of Example 3.4.3.

(b) Let G be the 3-modular reduction of the Weyl group of type H_4. This is a subgroup of order $14\,400$ of $\mathrm{GL}_4(\mathbb{F}_9)$. We will calculate the invariant ring of this group in Section 3.11. Here we look at a p-Sylow subgroup P of G, for $p = 3$. P has order 9 and can be generated by the matrices

$$\begin{pmatrix} 1 & 0 & 0 & 0 \\ w+1 & 1 & 0 & 0 \\ -1 & 0 & 1 & 0 \\ w & 0 & -1 & 1 \end{pmatrix}, \quad \begin{pmatrix} 1 & 0 & 0 & 0 \\ 1 & 1 & 0 & 0 \\ 1 & 0 & 1 & 0 \\ w & -w & w & 1 \end{pmatrix} \in \mathrm{GL}_4(\mathbb{F}_9),$$

where $w^2 - w - 1 = 0$. The computation of primary and secondary invariants by MAGMA takes about 4 seconds. The result are primary invariants of degrees 1,2,3,9 and secondary invariants of degrees 0,3,4,7,8,11. In this example, the number of secondary invariants equals $\prod_{i=1}^{n} \deg(f_i)/|P|$. As we will see in Theorem 3.7.1, this means that $K[V]^G$ is Cohen-Macaulay.
$\triangleleft$

3.6 Minimal Algebra Generators and Syzygies

Let R be a graded algebra over $K = R_0$ and $g_1, \ldots, g_r \in R_+$ homogeneous of positive degree. Then it is seen by a similar (but easier) argument as the one used in the proof of Theorem 2.2.10 that $g_1, \ldots, g_r$ generate R as an algebra over K if and only if they generate the ideal $R_+ \subseteq R$. Moreover, by Lemma 3.5.1, this is equivalent to the condition that the images of the g_i generate the quotient R_+/R_+^2 as a vector space over K. Hence a homogeneous system of algebra generators has minimal cardinality if no generator is superfluous, and then the number and degrees of the generators are uniquely determined. In particular the maximal degree $\beta(R)$ (see Equation (2.1.1)) of a generator is well defined. One also sees that $\beta(R)$ remains unchanged under extensions of the ground field, i.e., if we pass from R to $R \otimes_K L$ for $L \geq K$ a field extension. Equivalently, $\beta(R)$ is the minimal number d such that R is generated as a K-algebra by homogeneous elements of degrees $\leq d$.

It is clear that the invariant ring $K[V]^G$ is generated as a K-algebra by the primary and secondary invariants. Although the secondary invariants are minimal module-generators, they are not minimal algebra-generators. For example, 1 is always a secondary invariant, but it is redundant as an algebra-generator. To test whether a given generator f within a system S of homogeneous algebra-generators is redundant is a linear algebra problem. The procedure is to set up a general element of the same degree as f in the algebra generated by $S \setminus \{f\}$ with unknown coefficients, equating to f and extracting the corresponding system of linear equations by comparison of coefficients. The system is solvable if and only if f can be omitted from S. Starting with S as the union of the primary and secondary invariants, one thus gets a minimal system of algebra-generators, and by the above, $\beta(K[V]^G)$ is its maximal degree.

Example 3.6.1. In Example 3.5.4, we have $\beta(K[V]^G) = 3, 5, 15$ in part (a), (b), (c), respectively. In Example 3.5.7(a), it is easily checked that the secondary invariant $x_1 x_2 x_3 + y_1 y_2 y_3$ cannot be expressed in terms of invariants of lower degree, hence $\beta(K[V]^G) = 3$. This shows that Noether's degree bound (see Section 3.8) does not hold in the modular case. In Example 3.5.7(b), we obtain minimal algebra generators of degrees 1,2,3,3,4,9, so the secondary invariants of degrees 7,8,11 are redundant. We obtain $\beta(K[V]^G) = 9 = |G|$, which confirms a conjecture made by the second author that Noether's degree bound holds if the invariant ring is Cohen-Macaulay. $\lhd$

Suppose now that we have generators $h_1, \ldots, h_r$ of a K-algebra R. Then we have a presentation of R if we know the kernel I of the map

$$\Phi \colon K[t_1, \ldots, t_r] \to R, \ t_i \mapsto h_i,$$

where the t_i are indeterminates. It is one of the basic tasks in invariant theory to compute generators of I as an ideal in the polynomial ring $K[t_1, \ldots, t_r]$. The elements of I are usually called **syzygies**. Often the term "syzygies" is used for elements in the kernel of a map of modules, not algebras (see Section 1.3). But in fact we have a special case here, since R becomes a module over $K[t_1, \ldots, t_r]$ via Φ, and then Φ is a module-homomorphism. We have shown how kernels of maps between free modules over a polynomial ring can be computed. But here the situation is different since R is usually not free, so we need different methods. The first of these takes advantage of the fact that in our situation $R = K[V]^G$ is embedded into the polynomial ring $K[V]$. Therefore we can use the standard Gröbner basis method for calculating the relations between the polynomials h_i (see Section 1.2.2).

Another, usually more efficient method uses the homogeneity of the generating invariants. Indeed, I becomes a homogeneous ideal if we set $\deg(t_i) = \deg(h_i)$. We also use that in our situation the set $\{h_1, \ldots, h_r\}$ is the union of a homogeneous system of parameters $\{f_1, \ldots, f_n\}$ and a set of secondary invariants $\{g_1, \ldots, g_m\}$. Since the f_i are algebraically independent, we are looking for the kernel I of the map

$$A[t_1, \ldots, t_m] \to K[V]^G, \ t_i \mapsto g_i,$$

where $A = K[f_1, \ldots, f_n]$, and the t_i are again indeterminates. Suppose that $S \subseteq I$ is a set of relations containing

(a) generators for the A-module $I \cap (\oplus_{i=1}^m A \cdot t_i)$ of A-linear relations between the g_i, and
(b) for each $1 \leq i \leq j \leq m$ a relation of the form $t_i t_j - f_{i,j}$ with $f_{i,j} \in \oplus_{k=1}^m A \cdot t_k$.

Then it is easy to show that S generates I (see Kemper and Steel [146, Proposition 12]). In other words, all that we have to know are the linear relations between the g_i with coefficients in A and the representation of each product $g_i g_j$ as an element of $\oplus_{k=1}^m A \cdot g_k$. In the non-modular case there are no A-linear relations. On the other hand, if the secondary invariants were computed by Algorithm 3.5.5, then the choice of generators of the module M means that we constructed an epimorphism $F \to M$ with F a free A-module. Composing with the embedding $M \subseteq A^r$ yields a homomorphism

$$F \longrightarrow A^r$$

of free A-modules whose kernel is the module of A-linear relations between the g_i. This kernel can be calculated by the methods of Section 1.3.1.

The representation of a product $g_i g_j$ or, more generally, any homogeneous element $f \in R$ of degree d, say, as an element of $\oplus_{i=1}^m A \cdot g_i$ can be calculated by equating f to a general element of $\oplus_{i=1}^m A \cdot g_i$ of degree d with unknown coefficients and solving the resulting inhomogeneous system of linear equations over K. This approach usually performs better than the Gröbner basis method. Nevertheless, the computation of relations can sometimes be quite expensive.

It is often important to obtain a *minimal* system of generators for the ideal I. Since $K[V]^G$ is a graded algebra, Lemma 3.5.1 applies again and tells us that it is enough to omit superfluous generators. If the linear algebra method is used, one can go a bit further by detecting superfluous relations even before calculating them: It is quite easy to decide whether the ideal generated by the relations that have already been computed at some point contains a relation giving the desired representation for a product $g_i g_j$. In fact, this again comes down to the solution of a system of linear equations.

A Noetherian graded algebra R is said to be a **complete intersection** if the minimal number of generators minus the minimal number of generating relations between them is equal to the Krull dimension $\dim(R)$. In other words, the dimension of the variety $\mathcal{V}_{\bar{K}}(r_1, \ldots, r_i)$ decreases by 1 with each new generating relation r_i as it enters into the ideal. In Example 3.5.4(a) and (c) the invariant rings are complete intersections.

A complete classification of all finite linear groups over $\mathbb{C}$ whose invariant rings are complete intersections was given independently by Nakajima [178, 179] and Gordeev [96].

3.7 Properties of Invariant Rings

One of the main reasons to calculate generators of an invariant ring $K[V]^G$ is that one wants to understand its structural properties. This applies especially (but not only) in the modular case, where many questions are still unanswered. In this section we address various properties or quantities associated to $K[V]^G$, and give methods to calculate them. We have already dealt with the computation of $\beta(K[V]^G)$ and the complete intersection property in Section 3.6. The following Sections 3.7.1–3.7.3 are only relevant in the modular case.

3.7.1 The Cohen-Macaulay Property

After (a minimal system of) secondary invariants have been calculated, the Cohen-Macaulay property of $K[V]^G$ can be tested by simply counting their number, as the following result shows.

Theorem 3.7.1. *Assume that the action of G on V is faithful, let $f_1, \ldots, f_n$ $\in K[V]^G$ be primary invariants of degrees $d_1, \ldots, d_n$, and let $g_1, \ldots, g_m$ be a minimal system of secondary invariants. Then*

$$m \geq \frac{d_1 \cdots d_n}{|G|}$$

with equality if and only if $K[V]^G$ is Cohen-Macaulay.

Proof. Let $h_1, \ldots, h_r$ be minimal homogeneous generators for $K[V]$ as a module over $A := K[f_1, \ldots, f_n]$. Since $K[V]$ is Cohen-Macaulay by Lemma 2.5.2, Equation (3.5.1) yields

$$H(K[V], t) = \frac{t^{\deg(h_1)} + \cdots + t^{\deg(h_r)}}{(1 - t^{d_1}) \cdots (1 - t^{d_n})}.$$

On the other hand, we have

$$H(K[V], t) = \frac{1}{(1 - t)^n}$$

by Equation (3.3.1). Equating, multiplying by $(1 - t)^n$ and setting $t := 1$ yields

$$r = d_1 \cdots d_n.$$

The h_i generate the rational function field $K(V)$ as a vector space over $K(f_1, \ldots, f_n) = \mathrm{Quot}(A)$. By the Cohen-Macaulay property of $K[V]$, they are also linearly independent over $\mathrm{Quot}(A)$. Hence the degree of the extension is

$$[K(V) : \mathrm{Quot}(A)] = r = d_1 \cdots d_n.$$

By Galois theory it follows that

$$[K(V)^G : \mathrm{Quot}(A)] = \frac{d_1 \cdots d_n}{|G|}. \tag{3.7.1}$$

(This provides an alternative proof of Proposition 3.3.5.) Since $g_1, \ldots, g_m$ generate $K(V)^G$ as a vector space over $\mathrm{Quot}(A)$, the claimed inequality follows. Moreover, we have equality if and only if the g_i are linearly independent over $\mathrm{Quot}(A)$, which by Proposition 2.5.3 is equivalent to the Cohen-Macaulay property of $K[V]^G$. $\qquad\square$

3.7.2 Free Resolutions and Depth

In the modular case, secondary invariants are usually calculated by Algorithm 3.5.5, i.e., a generating set for the A-module M from the diagram (3.5.2) is computed, where again A is the subalgebra generated by primary invariants. $K[V]^G$ is isomorphic to M, which is embedded into the free A-module A^r. Choosing a generating set of M means specifying a homomorphism

$$F_0 \longrightarrow A^r$$

of free A-modules, whose image is $M \cong K[V]^G$. We can thus use Algorithm 1.3.4 to compute a graded free resolution

$$0 \longrightarrow F_r \longrightarrow \ldots \longrightarrow F_1 \longrightarrow F_0 \longrightarrow K[V]^G \longrightarrow 0 \tag{3.7.2}$$

of $K[V]^G$ (as an A-module). By the method given in Section 1.3.2 this can be transformed into a minimal free resolution, so let us assume that (3.7.2) is minimal. In particular, the number $r = \operatorname{hdim}_A(K[V]^G)$ is the homological dimension of $K[V]^G$ (as an A-module). Now we can compute the depth of $K[V]^G$ as an A-module by the Auslander-Buchsbaum formula (see Benson [18, Theorem 4.4.4]), which yields

$$\operatorname{depth}_A(K[V]^G) = \operatorname{depth}(A) - \operatorname{hdim}_A(K[V]^G) = n - r, \tag{3.7.3}$$

where $n = \dim_K(V)$. The following lemma shows that the depth of $K[V]^G$ as a module over A is the same as the depth as a module over itself.

Lemma 3.7.2. *Let M be a finitely generated graded module over a Noetherian graded algebra R over a field $K = R_0$, and let $A \subseteq R$ be a Noetherian graded subalgebra over which R is integral. Then*

$$\operatorname{depth}(R_+, M) = \operatorname{depth}(A_+, M).$$

Proof. The inequality $\operatorname{depth}(A_+, M) \leq \operatorname{depth}(R_+, M)$ is clear. To show the reverse inequality, let $f_1, \ldots, f_k \in A_+$ be a maximal M-regular sequence. By Bruns and Herzog [31, Theorem 1.2.5], it suffices to show that $f_1, \ldots, f_k$ is also a maximal M-regular sequence *in* R_+. Every element of A_+ is a zero divisor on $M/(f_1, \ldots, f_k)M$, hence by Eisenbud [59, Theorem 3.1], A_+ is contained in the union of the associated primes *in* R of $M/(f_1, \ldots, f_k)M$. By the prime avoidance lemma (see Eisenbud [59, Lemma 3.3]), A_+ lies in some $P \in \operatorname{Ass}_R(M/(f_1, \ldots, f_k)M)$. The same is true for the ideal A_+R generated by A_+ in R. But since R is integral over A, only one prime ideal of R contains A_+R, namely R_+. Hence $P = R_+$, and again by Eisenbud [59, Theorem 3.1], $f_1, \ldots, f_k$ is a maximal M-regular sequence in R_+. $\qquad\square$

Now Equation (3.7.3) and Lemma 3.7.2 allow the computation of the depth of $K[V]^G$ (as a module over itself). Lemma 3.7.2 and Equation (3.7.3) also imply that the homological dimension of $K[V]^G$ over A is independent of the choice of primary invariants. Usually the computation of the minimal free resolution required to evaluate Equation (3.7.3) is fairly easy once the secondary invariants have been found (see Section 1.3.2). However, there is no algorithm known for calculating the depth of the invariant ring of a finite group without first computing generators for the invariant ring.

Only in special cases do we have formulas for the depth of a modular invariant ring, or easy methods to calculate it. The first result in this direction is given by the celebrated formula of Ellingsrud and Skjelbred [63], which says that for G a cyclic p-group (with $p = \mathrm{char}(K)$) we have

$$\mathrm{depth}(K[V]^G) = \min\{\dim_K(V^G) + 2, \dim_K(V)\}.$$

This was generalized by Campbell et al. [38] to the class of so-called shallow groups. We will not include the definition of shallowness here but just remark that it is still a rather limited class of groups, which contains all abelian groups with cyclic Sylow p-subgroup ($p = \mathrm{char}(K)$). For a shallow group the above formula holds with V^G replaced by V^P with $P \leq G$ a Sylow p-subgroup. A further interesting result of Shank and Wehlau [219] says that for $G = \mathrm{SL}_2(p)$ acting on binary forms (see Example 2.1.2) of degree d with $1 < d < p$ and $\gcd(d, p-1) \leq 2$ the depth of the invariant ring is 3. Recently these results were generalized by Kemper [141]. One of the results is the following.

Theorem 3.7.3. *Suppose that $|G|$ is divisible by $p := \mathrm{char}(K)$ but not by p^2. Let r be the smallest positive number such that $H^r(G, K[V]) \neq 0$. Then*

$$\mathrm{depth}(K[V]^G) = \min\left\{\dim_K(V^P) + r + 1, \dim_K(V)\right\},$$

where $P \leq G$ is a Sylow p-subgroup.

Among other ingredients, Proposition 3.4.5 is used in the proof. Theorem 3.7.3 reduces the computation of the depth to the problem of determining the smallest $r > 0$ with $H^r(G, K[V]) \neq 0$. At first glance, this seems to be an even harder problem than computing the invariant ring $K[V]^G = H^0(G, K[V])$. But it is for the same class of groups ($p^2 \nmid |G|$) that Hughes and Kemper [116] recently developed a method for finding generating functions encoding the multiplicities of the indecomposable KG-modules in the symmetric powers $K[V]_d$. They obtain formulas which involve a summation over all (complex) $(2p)$-th roots of unity. Since it is known for which indecomposable KG-modules U and for which i one has $H^i(G, U) \neq 0$, it is quite easy to derive formulas for the Hilbert series

$$H\left(H^i(G, K[V]), t\right) := \sum_{d=0}^{\infty} \dim_K\left(H^i(G, K[V]_d)\right) t^d.$$

From these the smallest r with $H^r(G, K[V]) \neq 0$ can be found. This provides a procedure for determining the depth in the case $p^2 \nmid |G|$ which is roughly of the same computational difficulty as the evaluation of Molien's formula. This procedure was implemented in MAGMA by Denis Vogel. For the groups $G = \mathrm{SL}_2(p)$ and $G = \mathrm{GL}_2(p)$ acting on binary forms of degree n, it is also possible to calculate explicit formulas for the above Hilbert series which hold for a given n but for general p. A program for this was written by the second author in MAGMA. The resulting depths are printed in Tables 3.1 and 3.2, which are taken from [141].

	$p = 3$	$p = 5$	$p = 7$	$p = 11$	$p = 13$	$p = 17$	$p = 19$	$p = 23$	$p = 29$
$n = 1$	2	$\rightarrow$							
$n = 2$	3	$\rightarrow$							
$n = 3$	-	3	4	3	$\rightarrow$				
$n = 4$	-	5	3	$\rightarrow$					
$n = 5$	-	-	3	$\rightarrow$					
$n = 6$	-	-	7	3	$\rightarrow$				
$n = 7$	-	-	-	3	$\rightarrow$				
$n = 8$	-	-	-	3	5	3	$\rightarrow$		
$n = 9$	-	-	-	3	4	3	$\rightarrow$		
$n = 10$	-	-	-	11	3	$\rightarrow$			
$n = 11$	-	-	-	-	3	$\rightarrow$			
$n = 12$	-	-	-	-	13	5	3	$\rightarrow$	
$n = 13$	-	-	-	-	-	3	$\rightarrow$		
$n = 14$	-	-	-	-	-	3	$\rightarrow$		
$n = 15$	-	-	-	-	-	3	4	3	$\rightarrow$
$n = 16$	-	-	-	-	-	17	3	$\rightarrow$	
$n = 17$	-	-	-	-	-	-	3	$\rightarrow$	

Table 3.1. Depth of the invariant ring of $SL_2(p)$ acting on binary forms of degree n.

In Table 3.1, arrows ($\rightarrow$) indicate that from this position on all depths are equal to 3 (or 2 in the very first row). It was in fact shown by Kemper [141] that all rows (except the first one) in Table 3.1 continue with 3's, but this proof does not generalize to the case of $GL_2(p)$-invariants. Thus no arrows appear in Table 3.2.

3.7.3 The Hilbert Series

If a graded free resolution (3.7.2) of length r of the invariant ring $K[V]^G$ as a module over the subalgebra A generated by primary invariants has been calculated, then for the Hilbert series we obtain

$$H(R,t) = \sum_{i=0}^{r} (-1)^i H(F_i, t)$$

(see Remark 1.4.4). Observe that the free generators of F_i must be of the right degrees to make the maps in (3.7.2) degree-preserving. If these degrees are $e_{i,1}, \ldots, e_{i,s_i}$, then

$p =$	3	5	7	11	13	17	19	23	29	31	37	41	43	47	53
$n = 1$	2	2	2	2	2	2	2	2	2	2	2	2	2	2	2
$n = 2$	3	3	3	3	3	3	3	3	3	3	3	3	3	3	3
$n = 3$	-	4	4	4	4	4	4	4	4	4	4	4	4	4	4
$n = 4$	-	5	5	5	4	4	3	3	3	3	3	3	3	3	3
$n = 5$	-	-	6	4	4	3	3	3	3	3	3	3	3	3	3
$n = 6$	-	-	7	5	4	3	3	3	3	3	3	3	3	3	3
$n = 7$	-	-	-	8	4	3	3	3	3	3	3	3	3	3	3
$n = 8$	-	-	-	9	9	3	3	3	3	3	3	3	3	3	3
$n = 9$	-	-	-	10	10	4	3	3	3	3	3	3	3	3	3
$n = 10$	-	-	-	11	11	5	3	3	3	3	3	3	3	3	3
$n = 11$	-	-	-	-	12	8	3	3	3	3	3	3	3	3	3
$n = 12$	-	-	-	-	13	13	5	3	3	3	3	3	3	3	3
$n = 13$	-	-	-	-	-	14	12	3	3	3	3	3	3	3	3
$n = 14$	-	-	-	-	-	15	15	3	3	3	3	3	3	3	3
$n = 15$	-	-	-	-	-	16	16	4	3	3	3	3	3	3	3
$n = 16$	-	-	-	-	-	17	17	9	3	3	3	3	3	3	3
$n = 17$	-	-	-	-	-	-	18	18	3	3	3	3	3	3	3

Table 3.2. Depth of the invariant ring of $\mathrm{GL}_2(p)$ acting on binary forms of degree n.

$$H(F_i, t) = (t^{e_{i,1}} + \cdots + t^{e_{i,s_i}}) \cdot H(A, t) = \frac{t^{e_{i,1}} + \cdots + t^{e_{i,s_i}}}{(1 - t^{d_1}) \cdots (1 - t^{d_n})},$$

where $d_i = \deg(f_i)$ (see Equation (3.3.1)). Thus the Hilbert series of $K[V]^G$ is an easy by-product of the resolution.

Another method, which is slightly more efficient if no resolution has been computed, appears by considering diagram (3.5.2) on page 93 again. If $N \subseteq A^k$ is the image of the map $A^r \to A^k$, then

$$H(K[V]^G, t) = H(A^r, t) - H(N, t) = H(A^r, t) - H(A^k, t) + H(A^k/N, t).$$

But the Hilbert series of A^r and A^k are known, and $H(A^k/N, t)$ can be computed as in Section 1.4.1 (which also works for submodules of A^k) from a Gröbner basis of N. Now fortunately a Gröbner basis of N has already been calculated as the first step in the syzygy calculation leading to M, as described in Section 1.3.1. Observe that both algorithms presented here compute the Hilbert series of $K[V]^G$ a posteriori, i.e., after generators have been found.

One reason why it is interesting to know the Hilbert series of a graded algebra R is that one can use it to check the Gorenstein property, which may be defined as follows: A Noetherian graded domain R over a field $K = R_0$ is **Gorenstein** if and only if it is Cohen-Macaulay and the Hilbert series satisfies the identity

$$H(R, 1/t) = (-1)^{\dim(R)} t^l \cdot H(R, t)$$

for some $l \in \mathbb{Z}$ (see Stanley [234]). R is called **strongly Gorenstein** (also "graded Gorenstein") if in addition $l = \dim(R)$. For example, the invariant rings in Example 3.5.4(a)–(c) and Example 3.5.7(b) are strongly Gorenstein. In the non-modular case we have the following result, whose proof can be found in Stanley [236].

Theorem 3.7.4 (Watanabe [253, 254]). *In the non-modular case, $K[V]^G$ is strongly Gorenstein if and only if G acts on V by transformations which lie in* $\mathrm{SL}(V)$.

3.7.4 Polynomial Invariant Rings and Reflection Groups

The most simple structure that a graded algebra can have is that it is isomorphic to a polynomial ring, i.e., it is generated by algebraically independent elements. Since we know how to calculate a minimal system of generators for an invariant ring $K[V]^G$ of a finite group (see Section 3.6), there is a way to decide whether $K[V]^G$ is (isomorphic to) a polynomial ring. However, there is a much more efficient algorithm for this, which does not involve the computation of primary or secondary invariants and reduces the question to linear algebra. It is based on the following result.

Theorem 3.7.5 (Kemper [132]). *Assume that the action of G on V is faithful, and let $f_1, \ldots, f_n \in K[V]^G$ be homogeneous invariants with $n = \dim(V)$. Then the following statements are equivalent:*

(a) $K[V]^G = K[f_1, \ldots, f_n]$ (in particular, $K[V]^G$ is a polynomial algebra).
(b) The f_i are algebraically independent over K and

$$\prod_{i=1}^{n} \deg(f_i) = |G|. \tag{3.7.4}$$

Proof. Set $d_i := \deg(f_i)$. Assume that $K[V]^G = K[f_1, \ldots, f_n]$. Then $f_1, \ldots, f_n$ form a system of primary invariants. In particular, the f_i are algebraically independent. Moreover, there is only one secondary invariant (the constant 1). From this, Equation (3.7.4) follows by Theorem 3.7.1.

Conversely, assume that the conditions in (b) hold. If we can show that $f_1, \ldots, f_n$ form a system of primary invariants, then it follows by Proposition 3.11.2 (which appears on page 133) that $K[V]^G = K[f_1, \ldots, f_n]$. Therefore we have to show that the f_i form a system of primary invariants, which by Proposition 3.3.1 is equivalent to

$$\mathcal{V}_{\bar{K}}(f_1, \ldots, f_n) = \{0\}, \tag{3.7.5}$$

where $\bar{K}$ is an algebraic closure of K. We take additional indeterminates $t_1, \ldots, t_n$ and x_0 and an algebraic closure $\tilde{K}$ of $K(t_1, \ldots, t_n)$ which contains

$\tilde{K}$. By Bézout's theorem (see Fulton [73, Example 12.3.7], where no assumption is made on the dimension of the zero manifold), the projective algebraic set $\mathcal{V} \subseteq \mathbb{P}^n(\tilde{K})$ given by

$$f_1(x_1, \ldots, x_n) - t_1 x_0^{d_1} = \cdots = f_n(x_1, \ldots, x_n) - t_n x_0^{d_n} = 0$$

has at most $\prod_{i=1}^{n} d_i = |G|$ irreducible components. So (3.7.5) will follow if we can show that there are at least $|G|$ components of $\mathcal{V}$ with $x_0 \neq 0$. By the assumption, $K(x_1, \ldots, x_n)$ is a finite field extension of $K(f_1, \ldots, f_n)$, so each x_i has a minimal polynomial over $K(f_1, \ldots, f_n)$, say $g_i(x_i, f_1, \ldots, f_n) = 0$. If $(\xi_1, \ldots, \xi_n) \in \tilde{K}^n$ is a solution of

$$f_1(x_1, \ldots, x_n) - t_1 = \cdots = f_n(x_1, \ldots, x_n) - t_n = 0, \qquad (3.7.6)$$

then $g_i(\xi_i, t_1, \ldots, t_n) = 0$. Hence there are only finitely many solutions of (3.7.6), and each constitutes a component of its own in $\mathcal{V}$. We shall complete the proof by giving $|G|$ solutions of (3.7.6).

Via the isomorphism $K(f_1, \ldots, f_n) \to K(t_1, \ldots, t_n)$, $f_i \mapsto t_i$, form

$$L := K(t_1, \ldots, t_n) \otimes_{K(f_1, \ldots, f_n)} K(x_1, \ldots, x_n),$$

which is a finite field extension of $K(t_1, \ldots, t_n)$ and can therefore be assumed to lie inside $\tilde{K}$. Take a $\sigma \in G$ and set $\xi_i := 1 \otimes \sigma \cdot x_i \in L$. Then

$$f_i(\xi_1, \ldots, \xi_n) = 1 \otimes f_i(\sigma \cdot x_1, \ldots, \sigma \cdot x_n) = 1 \otimes \sigma \cdot f_i = 1 \otimes f_i = t_i \otimes 1.$$

Hence the elements of G give rise to $|G|$ distinct solutions of (3.7.6). This yields $|G|$ components of $\mathcal{V}$ with $x_0 \neq 0$, which completes the proof. $\qquad \square$

We obtain the following algorithm.

Algorithm 3.7.6 (Test if $K[V]^G$ is polynomial). Build a graded subalgebra $R \subseteq K[V]^G$ as follows: Loop through the degrees $d = 1, 2, \ldots$. For each degree, compute $K[V]_d^G$ by the methods of Section 3.1. Choose $f_{d,1}, \ldots, f_{d,m_d} \in K[V]_d^G$ which provide a basis of the quotient space $K[V]_d^G / R_d$. Then $f_{d,1}, \ldots, f_{d,m_d}$ are new generators of R. If the degree product of all generators does not divide $|G|$, or if the number of generators exceeds n, then $K[V]^G$ is not polynomial. On the other hand, if at some stage there are n generators such that the product of their degrees equals $|G|$, then compute their Jacobian determinant. It is non-zero if and only if $K[V]^G$ is polynomial.

The above algorithm is certain to terminate, since there must be at least n generators for $K[V]^G$ (since $\dim(K[V]^G) = n$). In the non-modular case, the question when an invariant ring of a finite group is polynomial is settled by the famous theorem of Shephard and Todd [221], Chevalley [43], and Serre [215], which says that $K[V]^G$ is a polynomial ring if and only if the group G is generated by elements which act on V as (pseudo-)reflections , i.e., by non-identity elements σ which fix a codimension-1 subspace of V. (Regarding G as a subgroup of $\mathrm{GL}(V)$, we simply say that G is a **reflection group**.) It was also proved by Serre that in the modular case G has to be a reflection group if $K[V]^G$ is a polynomial ring (see Benson [18, Theorem 7.1.2] for a proof). However, the converse is not true in general, as the following example shows.

Example 3.7.7 (Nakajima [175]). Consider the group

$$G := \left\{ \begin{pmatrix} 1 & 0 & a+b & b \\ 0 & 1 & b & b+c \\ 0 & 0 & 1 & 0 \\ 0 & 0 & 0 & 1 \end{pmatrix} \in \mathrm{GL}_4(\mathbb{F}_q) \mid a,b,c \in \mathbb{F}_q \right\},$$

which is a reflection group of order q^3. By way of contradiction, assume that $K[V]^G$ is a polynomial ring. There are two invariants of degree 1, namely $f_1 = x_1$ and $f_2 = x_2$. By Theorem 3.7.5 there are two further generators f_3 and f_4. If $\deg(f_3) \leq \deg(f_4)$, then $\deg(f_4) > q$, and both degrees are powers of $p := \mathrm{char}(\mathbb{F}_q)$. Hence any invariant of degree $q+1$ must have the form $g_1 \cdot f_3^k + g_2$ with $g_1, g_2 \in K[x_1, x_2]$. But it is easily verified that

$$x_1^q x_3 - x_1 x_3^q + x_2^q x_4 - x_2 x_4^q$$

is an invariant which is not of the above form. This contradiction shows that $K[V]^G$ is not a polynomial ring.

For $q = p$ a prime, this example appeared in Nakajima [175]. Nakajima achieved much further reaching results on the polynomiality of invariants of reflection groups in [176, 177]. In [177], he gave a characterization of all p-groups over $\mathbb{F}_p$ which have polynomial invariant rings. $\triangleleft$

The following result is a strong tool to prove that an invariant ring is not a polynomial ring.

Proposition 3.7.8 (Serre [215]). *Suppose that $K[V]^G$ is a polynomial ring. Then for all linear subspaces $W \subseteq V$ the invariant ring $K[V]^{G_W}$ of the point-wise stabilizer $G_W := \{\sigma \in G \mid W \subseteq V^\sigma\}$ is also a polynomial ring. In particular, G_W is a reflection group.*

Proposition 3.7.8 was used by Nakajima [175] to prove that the invariant rings of many classical groups (over finite fields) are not isomorphic to polynomial rings. Later Kemper and Malle [143] classified the irreducible linear groups (i.e., finite groups G with an irreducible KG-module V) for which $K[V]^G$ is a polynomial ring. This classification was obtained by going through a complete list of all finite irreducible modular reflection groups given by Kantor [127], Wagner [251, 252], Zalesskii and Serezkin [264, 265]. Table 3.3 gives an overview of the finite irreducible modular reflection groups, and which of then have polynomial invariants rings.

Group	$K[V]^G$ polynomial	$K[V]^G$ not polynomial
G imprimitive	all	-
$n = \dim(V) \leq 2$	all	-
$\mathrm{SL}_n(q) \leq G \leq \mathrm{GL}_n(q)$	all	-
$\Omega_n^{(\pm)}(q) < G \leq \mathrm{O}_n^{(\pm)}(q),$ $G \neq \mathrm{SO}_n^{(\pm)}(q)$	$\mathrm{O}_3(q)$, $\mathrm{R}^+\mathrm{O}_3(q)$, $\mathrm{O}_4^-(q)$	all others
$\mathrm{SU}_n(q^2) \leq G \leq \mathrm{U}_n(q^2)$	$\mathrm{U}_3(q^2)$	all others
$G = \mathrm{Sp}_n(q)$, $n \geq 4$ even	-	all
$G = S_{n+1}$, $p \nmid (n+1)$	all	-
$G = S_{n+2}$, $p \mid (n+2)$, $n \geq 5$	-	all
$G = W_p(G_i)$ $(i = 23, \ldots, 37),$ $G = \mathrm{EJ}_3(5)$, $G = J_4(4)$	all except $\rightarrow$	$W_7(G_{24})$, $W_3(G_{28})$, $W_3(G_{30})$, $W_5(G_{30})$, $W_3(G_{31})$, $W_5(G_{32})$, $W_3(G_{33})$, $W_2(G_{34})$, $W_3(G_{36})$, $W_3(G_{37})$, $W_5(G_{37})$

Table 3.3. Finite irreducible modular reflection groups.

Most of the notation used in Table 3.3 should be clear. The representations of the symmetric group S_n are the representation called V is Example 3.7.10 below for $p \nmid n$, and U from Example 3.7.10 for $p \mid n$. $W_p(G_i)$ means the p-modular reduction of the complex reflection group appearing as number i in the classification of Shephard and Todd [221]. The groups $\mathrm{EJ}_3(5)$, $J_4(4)$ only appear in positive characteristic (5 and 2, respectively). For the exceptional groups (i.e., those appearing in the last row) whose invariant rings turned out to be polynomial, this was in most cases proved by explicitly constructing generating invariants using a computer. In each instance where $K[V]^G$ is not polynomial, Proposition 3.7.8 was used, so the following result emerges:

Theorem 3.7.9 (Kemper and Malle [143]). *The invariant ring $K[V]^G$ of a finite group G acting irreducibly on V is a polynomial ring if and only if G is a reflection group and $K[V]^{G_W}$ is a polynomial ring for every non-zero linear subspace $W \leq V$.*

It may be worth noting that in the above theorem it does not suffice to demand that G_W be a reflection group for all subspaces W (see Campbell et al. [35] for a non-irreducible counter example and Kemper and Malle [143, Example 2.2] for an irreducible counter example).

Another noteworthy fact about the paper [143] is that five years after its appearance it was discovered by Kasper Andersen that the authors had used Proposition 3.7.8 incorrectly. In fact, the convention adopted in [143] is that the polynomial ring is $S(V)$ (the symmetric algebra of V, rather than its dual V^*). Then the authors applied Proposition 3.7.8 to point-wise stabilizers of subspaces $W \leq V$. But according to their convention, they should have used subspaces $W \leq V^*$. Drawing attention to this inconsistency is far from being pedantic. In fact, it was pointed out by Andersen that Kemper and Malle's way of applying Proposition 3.7.8 might very well have resulted in incorrectly sending the group $W_5(G_{29})$ to the right hand column of Table 3.3. It is precisely the symmetric group S_5 with its four-dimensional representation discussed in Example 3.7.10 below that occurs as a stabilizer here. What really happened is that the authors had already proved by computations that $W_5(G_{29})$ has a polynomial invariant ring, so they did not even start to apply Proposition 3.7.8 to it. After hearing about their error from Andersen, the authors had to reconsider all cases where they had applied Proposition 3.7.8. Luckily, it turned out that the final results summarized in Table 3.3 and Theorem 3.7.9 all remain correct. In most of the cases the point-wise stabilizers G_W found by the authors are not generated by reflections, which means that $S(V)^{G_W}$ as well as $S(V^*)^{G_W}$ are not polynomial rings; so no trouble can arise in these cases.

Qualitatively speaking, "most" invariant rings of modular reflection groups are not polynomial rings. This raises the question what the structure of these invariant rings is. Could it be, for instance, that the invariant ring of a modular reflection group is always a complete intersection? For example, the invariant rings of symplectic groups were determined by Carlisle and Kropholler [41], and found to be all complete intersections (see also Benson [18, Section 8.3]). Nevertheless, all hopes of finding nice properties which hold in general for such rings are crushed by examples such as the following.

Example 3.7.10. Let n be a multiple of $p := \text{char}(K)$ and consider the symmetric group $G = S_n$. Let $W = K^n$ be the natural KG-module with basis $e_1, \ldots, e_n$, and consider the modules

$$V := W/K \cdot (e_1 + \cdots + e_n) \quad \text{and}$$
$$U := \{\alpha_1 e_1 + \cdots + \alpha_n e_n + K(e_1 + \cdots + e_n) \in V \mid \alpha_1 + \cdots + \alpha_n = 0\}.$$

G acts on U and V as a reflection group, and U is irreducible (at least for $p > 2$). Now assume $p \geq 5$. Then by Kemper [138, Corollary 2.8], $K[V]^G$ is not Cohen-Macaulay, and if moreover $n > 5$, then also $K[U]^G$ is not Cohen-Macaulay. If we restrict further and consider $G = S_p$, we even obtain the depths of $K[V]^G$ and $K[U]^G$ by [141, Section 3.2]: The result is 3 in both cases.

Hence we have two series of modular reflection groups (one irreducible and the other a mod-p reduction from a characteristic 0 reflection group) such that the homological dimension of their invariant rings grows arbitrarily large! On the other hand, it is known from Kemper and Malle [144] that the invariant field $K(U)^G$ is purely transcendental over K, so it is the field of fractions of some polynomial ring. A further fact seems even more bizarre: The invariant ring $K[V^*]^G = S(V)^G$ of the dual of V is a polynomial ring, generated by the (residue classes of) the elementary symmetric polynomials of degrees ≥ 2 in the e_i. Thus we have modules whose invariant rings have arbitrarily bad properties, but the invariants of the dual modules have the nicest possible structure!   ◁

In Section 3.11.3 we will come back to the question of the structure of non-polynomial invariant rings of modular reflection groups.

3.8 Noether's Degree Bound

We discussed in Section 3.6 how the degree bound $\beta(K[V]^G)$ can be calculated *a posteriori*, i.e., after having found generators for the invariant ring. On the other hand, having an a priori upper bound for $\beta(K[V]^G)$ leads to a "naive" algorithm for calculating the invariant ring by computing all invariants up to this degree bound (using the methods from Section 3.1) and then deleting superfluous generators by linear algebra. All that is needed for this approach is linear algebra and polynomial arithmetic. Although this algorithm is far from being efficient in most cases, degree bounds have enjoyed considerable interest in all periods of invariant theory.

In this section we will prove that the invariant ring of a finite group G can be generated by homogeneous invariants of degree at most the group order $|G|$, provided that the characteristic of the ground field K does not divide $|G|$. This was proved by Noether [183] in the case that $p := \mathrm{char}(K)$ is zero or bigger than $|G|$. For the case that p is smaller than $|G|$ but $p \nmid |G|$, the question whether Noether's bound always holds (known as the "Noether gap") was open for quite a while. It was never seriously doubted that this is true, but the lack of a proof was for many years an irritating problem, since in all other aspects the invariant theory in coprime characteristic parallels that in characteristic zero. Partial results were obtained by Smith [226] and Richman [201], who independently proved the bound for solvable groups. See also Fleischmann and Lempken [69], Smith [229]. Only very recently Fleischmann [68] and Fogarty [71] independently found proofs for the general statement and thus resolved the Noether gap. Subsequently these proofs were substantially simplified by D. Benson [19]. We will present a version of this simplification. It is in many ways similar to Noether's original proof, in that it is elementary and surprisingly simple, with a little touch of ingenuity, and gives a constructive way to express any invariant in terms of invariants of degree at most $|G|$. We are grateful to David Benson for giving us the permission to include the proof here.

We start by the following lemma, which asserts as a special case that if $p \nmid |G|$, then every homogeneous polynomial of degree at least $|G|$ lies in the "Hilbert ideal" $\left(K[V]_+^G\right) K[V]$.

Lemma 3.8.1 (Benson [19], Fogarty [71, Equation (1)]). *Let A be a commutative ring with unit, G a finite group of automorphisms of A, and $I \subseteq A$ a G-stable ideal. If the order of G is invertible in A, then*

$$I^{|G|} \subseteq I^G A.$$

Proof. Let $\prod_{\sigma \in G} f_\sigma$ be a product of $|G|$ elements of I which are indexed by $\sigma \in G$. For every $\tau \in G$ we have

$$\prod_{\sigma \in G} \left((\tau\sigma) \cdot f_\sigma\right) - f_\sigma \right) = 0.$$

Multiplying this out and summing over all $\tau \in G$ yields

$$\sum_{M \subseteq G} (-1)^{|G \setminus M|} \left(\sum_{\tau \in G} \prod_{\sigma \in M} \tau(\sigma f_\sigma) \right) \cdot \left(\prod_{\sigma \in G \setminus M} f_\sigma \right) = 0. \qquad (3.8.1)$$

The summand for $M = \emptyset$ is $\pm|G| \cdot \prod_{\sigma \in G} f_\sigma$, and all other summands lie in $I^G A$. Thus $\prod_{\sigma \in G} f_\sigma \in I^G A$, and the lemma is proved. $\qquad\square$

If we put $A = K[V]$ and $I = K[V]_+$, then Lemma 3.8.1 tells us that every homogeneous polynomial of degree $\geq |G|$ lies in the Hilbert ideal $J := \left(K[V]_+^G\right) K[V]$, so J is generated in degrees $\leq |G|$. From this the Noether bound for $K[V]^G$ follows by Hilbert's classical argument (see Theorem 2.2.10). So the proof for the most basic version of the Noether bound is already complete! Nevertheless, we will work a little bit harder to derive a bound which also holds for equivariants. Recall that $(K[V] \otimes_K W)^G$ (with W another KG-module) is the module of equivariants, whose elements can be viewed as polynomial functions $V \to W$ which commute with the G-actions. In fact, we generalize even further and consider the following situation. Let R be a commutative ring with unit and let $A = \oplus_{d=0}^\infty A_d$ be a graded, commutative R-algebra with unit. As usual, we write $\beta(A)$ for the least integer k such that A is generated by homogeneous elements of degree at most k as an R-algebra, or $\beta(A) = \infty$ if no such k exists. Let G be a finite group acting on A by degree-preserving R-automorphisms, and let W be an RG-module. We make $A \otimes_R W$ into a graded A-module by setting $(A \otimes_R W)_d := A_d \otimes_R W$. Thus A^G is a graded R-algebra, and $(A \otimes_R W)^G$ is a graded A^G-module. We will prove:

Theorem 3.8.2. *In the above situation, suppose that $|G|$ is invertible in R. Then the following statements hold:*

(a) $\beta(A^G) \leq |G| \cdot \beta(A)$.

(b) $(A \otimes_R W)^G$ *is generated as an A^G-module by homogeneous elements of degree at most $(|G| - 1) \cdot \beta(A)$.*

Proof. Write $M := A \otimes_R W$, $\beta := \beta(A)$, and $g := |G|$. Let B be the subalgebra of A^G generated by all homogeneous elements of degree at most $g\beta$. We prove that M^G is generated as a B-module by homogeneous elements of degree at most $(g-1)\beta$. From this (a) (as the special case $W = R$) and (b) will follow.

Let $d > (g-1)\beta$ be an integer. Any element from M_d can be written as a sum of products of the form $t \otimes w$, where $w \in W$ and $t = f_1 \cdots f_k$ is a product of homogeneous elements $f_i \in A_{\leq \beta}$ (i.e., $\deg(f_i) \leq \beta$) with $\deg(t) = d$. For such a product t, the sequence $f_1, \ldots, f_k$ contains at least g elements lying in A_+, say $f_1, \ldots, f_g$. By Lemma 3.8.1, and since $\deg(f_1 \cdots f_g) \leq g\beta$, we obtain $f_1 \cdots f_g \in B_+ A$ and therefore $t \in B_+ A_{<d}$. We conclude that $M_d \subseteq B_+ M_{<d}$. By induction this yields $M = B M_{\leq (g-1)\beta}$. Now by applying the Reynolds operator $\mathcal{R} \colon M \to M^G$ we get

$$M^G = \mathcal{R}(M) = B\mathcal{R}(M_{\leq (g-1)\beta}) = B M^G_{\leq (g-1)\beta},$$

as claimed. $\qquad\qquad\qquad\qquad\qquad\qquad\qquad\qquad\qquad\qquad\qquad\qquad\quad\square$

In the situation where G acts on vector spaces V and W over a field K we obtain:

Corollary 3.8.3. *Let G be a group and V a finite dimensional representation of G over a field K. Moreover, let $N \trianglelefteq G$ be a normal subgroup of finite index $[G : N]$ such that $\mathrm{char}(K) \nmid [G : N]$. Then we have*

(a) $\beta\left(K[V]^G\right) \leq [G : N] \cdot \beta\left(K[V]^N\right)$.

(b) *If W is another finite dimensional representation over K with $W^N = W$, then the module $(K[V] \otimes_K W)^G$ of equivariants is generated over $K[V]^G$ by homogeneous elements of degree at most $([G : N] - 1) \cdot \beta\left(K[V]^N\right)$.*

Proof. G/N acts on $A := K[V]^N$ and on W, and we have

$$(K[V] \otimes_K W)^G = (A \otimes_K W)^{G/N}.$$

Now the result follows from Theorem 3.8.2. $\qquad\qquad\qquad\qquad\qquad\qquad\quad\square$

Corollary 3.8.4 (Noether, Fleischmann, Benson, Fogarty). *Suppose that* $\mathrm{char}(K) \nmid |G|$. *Then*

$$\beta\left(K[V]^G\right) \leq |G|.$$

Remark 3.8.5. (a) As observed by Peter Fleischmann, one can also consider the summand for $M = G$ in Equation (3.8.1). This summand is $\sum_{\tau \in G} \tau \left(\prod_{\sigma \in G} \sigma \cdot f_\sigma\right)$, and equating it to the negative of the remaining summands yields the inclusion

$$\mathrm{Tr}_G(I^{|G|}) \subseteq A^G I, \qquad\qquad\qquad (3.8.2)$$

where Tr_G denotes the transfer map (or trace) $A \to A^G$, $f \mapsto \sum_{\tau \in G} \tau \cdot f$. The inclusion (3.8.2) holds in any characteristic, and is of interest in itself. For example, in the non-modular case the Noether bound can be derived from (3.8.2) by applying the Reynolds operator to it.

(b) It is reasonable to ask if Corollary 3.8.3(a) also holds if N is no longer assumed to be normal. This is true if $|G|!$ is invertible in K (i.e., $\mathrm{char}(K) = 0$ or $> |G|$, see Schmid [209, Lemma 3.2] or Smith [225, Theorem 2.4.2]). Unfortunately, the proof given above and the one by Fleischmann [68] fail if N is not normal. So there is still a "baby Noether gap" left. $\triangleleft$

We add two conjectures which generalize Corollary 3.8.4.

Conjecture 3.8.6. *(a) If $K[V]^G$ is Cohen-Macaulay, then $\beta\left(K[V]^G\right) \leq |G|$.*

(b) Let $I = \left(K[V]_+^G\right) K[V]$ be the Hilbert ideal. Then I is generated by homogeneous elements of degree at most $|G|$.

We have plenty of computational evidence for both conjectures. It is particularly striking that we found Conjecture 3.8.6(b) to be true in cases where $\beta\left(K[V]^G\right)$ exceeded $|G|$ by far (see Section 3.9.1). For example, it can be seen by inspection of the generators given by Campbell and Hughes [33] that the conjecture is true for vector invariants of the indecomposable two-dimensional representation of the cyclic group of order p over $K = \mathbb{F}_p$. Moreover, it was shown by Fleischmann [68, Theorem 4.1] that Conjecture 3.8.6(b) is true if V is a permutation module or, more generally, a trivial source module.

Remark 3.8.7. One might be tempted to conjecture that Lemma 3.8.1 might also hold in the modular case. However, this is not true. As an example, consider the group

$$G := \left\langle \begin{pmatrix} 1\,0\,0\,0 \\ 1\,1\,0\,0 \\ 0\,0\,1\,0 \\ 0\,0\,1\,1 \end{pmatrix} \right\rangle \leq \mathrm{GL}_4(\mathbb{F}_2)$$

of order 2. The invariant ring is generated by $x_1, x_3, x_2(x_1 + x_2), x_4(x_3 + x_4)$, and $x_1 x_4 + x_2 x_3$. Thus the Hilbert ideal is $I := (x_1, x_3, x_2^2, x_4^2)$, hence $x_2 x_4 \notin I$. $\triangleleft$

Improvements of Noether's bound. Noether's degree bound is sharp in the sense that no better bound can be given in terms of only the group order—just consider the example of the cyclic group of order n, coprime to $\mathrm{char}(K)$, acting on a one-dimensional vector space by multiplication with a primitive n-th root of unity. However, it was shown by Schmid [209] that if G is not cyclic and $\mathrm{char}(K) = 0$, then $\beta\left(K[V]^G\right) < |G|$. Schmid's methods were sharpened by Domokos and Hegedűs, who were able to prove the following surprisingly explicit result:

Theorem 3.8.8 (Domokos and Hegedűs [57]). *Let G be a finite, non-cyclic group acting on a finite-dimensional vector space V over a field K of characteristic 0. Then*

$$\beta\left(K[V]^G\right) \leq \begin{cases} \frac{3}{4}|G| & \text{if } |G| \text{ is even,} \\ \frac{5}{8}|G| & \text{if } |G| \text{ is odd} \end{cases}.$$

3.9 Degree Bounds in the Modular Case

We have already seen in Example 3.5.7(a) that Noether's bound fails in the modular case. So the question is how badly it fails. One might imagine that there should still exist some bound which only depends on the group order. If no such bound existed, this would mean that Noether's bound can be "arbitrarily wrong" in the modular case. On a different line of thought, one may ask for degree bounds which depend on $|G|$ and on the dimension of the representation. In this section we will look at lower degree bounds given by Richman, and then discuss an upper bound given by Broer, which holds if $K[V]^G$ is Cohen-Macaulay. Finally, we will derive a general (but unrealistic) upper bound.

3.9.1 Richman's Lower Degree Bound

It was quite a surprise when in 1990 Richman [200] proved that for the two-dimensional indecomposable module V of the cyclic group $G := C_p$ over a field of characteristic p the vector invariants have the property $\beta\left(K[V^m]^G\right) \geq m(p-1)$, where V^m is the direct sum of m copies of V. (Later, equality was proved by Campbell and Hughes [33].) Richman proved similar results for another class of groups, which contains the general and special linear groups over a finite field. Then in [202], which was written in 1990 but not published until 1996, he proved that for any group G of order divisible by $p := \mathrm{char}(K)$ and for any faithful, finite dimensional KG-module V the beta number $\beta\left(K[V^m]^G\right)$ tends to infinity with m. We will present a short and elementary proof for this statement in the case that $K = \mathbb{F}_p$ is a prime field. This proof was shown to us by Ian Hughes.

We start by fixing an arbitrary element $\sigma \in G$ whose order is $p = \mathrm{char}(K)$. There exists a basis $x_1, \ldots, x_n \in V^*$ such that σ is in Jordan canonical form with respect to this basis. Re-ordering the x_i and writing $\{x_{i,j} \mid 1 \leq i \leq m, 1 \leq j \leq n\}$ for the corresponding basis of $(V^m)^*$, we obtain

$$\sigma \cdot x_{i,j} = x_{i,j} + \epsilon_j x_{i,j-1} \quad \text{with} \quad \epsilon_j \in \{0,1\}, \ \epsilon_n = 1, \ \epsilon_1 = 0. \tag{3.9.1}$$

Lemma 3.9.1. *Let $t = \prod_{i=1}^m x_{i,n}^{e_i}$ be a monomial in $K[x_{1,n}, \ldots, x_{m,n}] \subseteq K[V^m]$. If t occurs in an invariant, then all exponents e_i are divisible by p.*

Proof. For $f, g \in K[V^m]$ we have

$$(\sigma - 1)(fg) = \sigma \cdot f \cdot (\sigma - 1)g + g \cdot (\sigma - 1)f. \tag{3.9.2}$$

Let $J \subset K[V^m]$ be the ideal generated by all $x_{i,j}$ with $1 \le i \le m$ and $1 \le j \le n - 1$. Then (3.9.1), (3.9.2), and induction on the degree yield $(\sigma - 1)K[V^m] \subseteq J$. Now let $I \subset K[V^m]$ be the ideal generated by

$$\{x_{i,j} \mid 1 \le i \le m, 1 \le j \le n - 2\} \cup \{x_{i,n-1}^2 \mid 1 \le i \le m\}.$$

We have $\sigma \cdot I \subseteq I$ and $J^2 \subseteq I$. Thus $(\sigma - 1)f \cdot (\sigma - 1)g \in I$ for $f, g \in K[V^m]$, so (3.9.2) yields that $\sigma - 1$ acts as a derivation on $K[V^m]/I$.

Take any $f \in K[V^m]$. Then $f \equiv h + \sum_{i=1}^{m} x_{i,n-1} g_i \mod I$ with $h, g_i \in K[x_{1,n}, \ldots, x_{m,n}]$. Using that $\sigma - 1$ is a derivation modulo I, we easily obtain

$$(\sigma - 1)f \equiv (\sigma - 1)h \mod I.$$

For a monomial $t = \prod_{i=1}^{m} x_{i,n}^{e_i}$ in h we have

$$(\sigma - 1)t \equiv \sum_{i=1}^{m} e_i x_{i,n-1} \cdot t/x_{i,n} \mod I.$$

If f is an invariant and $h = \sum_{e_1, \ldots, e_n} \alpha_{e_1, \ldots, e_n} \prod_{i=1}^{m} x_i^{e_i}$ with $\alpha_{e_1, \ldots, e_n} \in K$, then

$$0 = (\sigma - 1)f \equiv (\sigma - 1)h \equiv \sum_{e_1, \ldots, e_n} \alpha_{e_1, \ldots, e_n} \sum_{i=1}^{m} e_i x_{i,n-1} x_{i,n}^{-1} \prod_{k=1}^{m} x_k^{e_k} \mod I.$$

From this we see that $\alpha_{e_1, \ldots, e_n} = 0$ if any e_i is not divisible by p. This concludes the proof. $\qquad\square$

We can now prove the lower degree bound.

Theorem 3.9.2 (Richman [202]). *Suppose that $K = \mathbb{F}_p$ and G is a group of order divisible by p. If G acts faithfully on the K-space V of dimension n, then*

$$\beta(K[V^m]^G) \ge \frac{m}{n-1}.$$

Proof. Consider the polynomial

$$f = \sum_{\alpha_1, \ldots, \alpha_n \in K} \prod_{i=1}^{m} (\alpha_1 x_{i,1} + \cdots + \alpha_n x_{i,n})^{p-1},$$

which clearly is an invariant. We may assume that $m \ge n$. The coefficient of the monomial

$$t = \prod_{i=1}^{n-1} x_{i,i}^{p-1} \cdot \prod_{i=n}^{m} x_{i,n}^{p-1}$$

in f is

$$\sum_{\alpha_1,\ldots,\alpha_n \in K} \prod_{i=1}^{n-1} \alpha_i^{p-1} \cdot \alpha_n^{(m-n+1)(p-1)} = \sum_{\alpha_1,\ldots,\alpha_n \in K^*} 1 = (p-1)^n \neq 0,$$

so t occurs in f. Suppose that we have a set of homogeneous generators of $K[V]^G$, then

$$t = t_1 \cdot t_2 \cdots t_r \tag{3.9.3}$$

with t_i monomials occurring in elements of this generating set. We are done if we can show that $\deg(t_k) \geq m/(n-1)$ for some k. Assume that a t_i lies in $K[x_{1,n},\ldots,x_{m,n}]$, then by Lemma 3.9.1 all exponents of t_i must be divisible by p, which is impossible since t_i is a divisor of t by (3.9.3). Hence

$$\deg_{n'}(t_i) > 0 \quad \text{for all} \quad i = 1,\ldots,r,$$

where we write $\deg_{n'}(\prod_{i,j} x_{i,j}^{e_{i,j}}) = \sum_{i=1}^m \sum_{j=1}^{n-1} e_{i,j}$. With $u_i := \deg(t_i)$ and $v_i := \deg_{n'}(t_i)$, we have that

$$\frac{u_k}{v_k} \geq \frac{\sum_{i=1}^r u_i}{\sum_{i=1}^r v_i} \quad \text{for some} \quad k = 1,\ldots,r, \tag{3.9.4}$$

since otherwise $u_k \cdot \sum_{i=1}^r v_i < v_k \cdot \sum_{i=1}^r u_i$ for all k, and summing over k would yield the contradiction

$$\sum_{k=1}^r u_k \cdot \sum_{i=1}^r v_i < \sum_{k=1}^r v_k \cdot \sum_{i=1}^r u_i.$$

But from (3.9.3) we get that

$$\sum_{i=1}^r u_i = \deg(t) = m(p-1) \quad \text{and} \quad \sum_{i=1}^r v_i = \deg_{n'}(t) = (n-1)(p-1),$$

hence (3.9.4) yields

$$\deg(t_k) = u_k \geq v_k \cdot \frac{m(p-1)}{(n-1)(p-1)} \geq \frac{m}{n-1}.$$

This completes the proof. $\square$

Let G be a finite group and K a field. We say that (G,K) has a **global degree bound** k if there exists an integer k such that $\beta(K[V]^G) \leq k$ holds for all finite dimensional KG-modules V.

Corollary 3.9.3. *Let G be a finite group and K a field. Then (G,K) has a global degree bound if and only if* $\operatorname{char}(K) \nmid |G|$.

Proof. One direction follows from Noether's bound (Corollary 3.8.4), and the other from Theorem 3.9.2, since G has a faithful representation defined over $\mathbb{F}_p$, for example the regular representation. $\qquad\square$

Remark 3.9.4. The largest part of Richman's paper [202] is devoted to obtaining a lower degree bound for vector invariants, where the hypothesis that K is the prime field is dropped. The result is

$$\beta(K[V^m]^G) \geq \frac{m(p-1)}{p^{|G|-1} - 1}.$$

What we have presented above is a simplified version of the first part of [202].

The above bound (and the bound from Theorem 3.9.2) was improved by Kemper [133] for the case that G acts by permuting basis of V: Then the bound

$$\beta(K[V^m]^G) \geq m(p-1)$$

holds. This bound is known to be sharp if $G = S_p$ is the symmetric group on $p = n$ letters and $m > 1$ (see Fleischmann [67]). $\qquad\triangleleft$

3.9.2 Broer's Degree Bound

In this section we will discuss a degree bound given by Broer [30] that holds under the assumption that $K[V]^G$ is Cohen-Macaulay. We assume that $f_1, \ldots, f_n \in K[V]^G$ are primary invariants and set $A := K[f_1, \ldots, f_n]$. We start with a proposition that has been known for a long time (see Stanley [236]).

Proposition 3.9.5. *If* $\mathrm{char}(K) \nmid |G|$ *in the above situation, then* $K[V]^G$ *is generated as an A-module by homogeneous invariants of degrees at most*

$$\deg(f_1) + \cdots + \deg(f_n) - n.$$

Proof. By Molien's formula $H(K[V]^G, t)$ has degree at most $-n$ as a rational function in t. Comparing to Equation (3.5.1) yields the result. $\qquad\square$

Our goal is to extend Proposition 3.9.5 to the modular case under the assumption that $K[V]^G$ is Cohen-Macaulay. The following lemma holds in general.

Lemma 3.9.6. *Let* $K(V) := \mathrm{Quot}(K[V])$ *be the rational function field on* V, *and* $K(V)^G := \{f \in K(V) \mid \sigma \cdot f = f \ \forall \sigma \in G\}$ *the invariant field. Then*

$$K(V)^G = \mathrm{Quot}(K[V]^G).$$

Proof. Clearly $\mathrm{Quot}(K[V]^G) \subseteq K(V)^G$. Conversely, take $f/g \in K(V)$ with $f, g \in K[V]$ such that $f/g \in K(V)^G$. Then

$$\frac{f}{g} = \frac{f \cdot \prod_{\sigma \in G \setminus \{\mathrm{id}\}} \sigma \cdot g}{\prod_{\sigma \in G} \sigma \cdot g} \in \mathrm{Quot}(K[V]^G),$$

since the denominator of the right hand fraction is G-invariant, and therefore also the numerator. $\qquad\square$

We will use the transfer map

$$\mathrm{Tr} = \mathrm{Tr}_G \colon K[V] \to K[V]^G, \quad f \mapsto \sum_{\sigma \in G} \sigma \cdot f,$$

which is a degree-preserving homomorphism of modules over $K[V]^G$ and in particular over A. We thus obtain a map

$$\mathrm{Tr}^* \colon \ \mathrm{Hom}_A(K[V]^G, A) \to \mathrm{Hom}_A(K[V], A), \quad \varphi \mapsto \varphi \circ \mathrm{Tr}.$$

For two graded modules M and N over A, a homomorphism $\varphi \in \mathrm{Hom}_A(M, N)$ is called homogeneous of degree d if $\varphi(M_e) \subseteq N_{e+d}$ for all $e \in \mathbb{Z}$. This gives $\mathrm{Hom}_A(M, N)$ the structure of a graded A-module. With this, the above map Tr^* becomes a degree-preserving homomorphism of graded A-modules.

Lemma 3.9.7. *If V is faithful, then the map Tr^* is injective.*

Proof. Since $K(V)$ is a separable field extension of $K(V)^G = \mathrm{Quot}(K[V]^G)$ (equality by Lemma 3.9.6), there exist $g, h \in K[V]$ such that $\mathrm{Tr}(g/h) = 1$. Since $K(V)$ is a finite field extension of $\mathrm{Quot}(A)$, we can assume that h lies in A, hence $\mathrm{Tr}(g) = h \in A \setminus \{0\}$. Now suppose that $\varphi \in \mathrm{Hom}_A(K[V]^G, A)$ lies in the kernel of Tr^*. Then for $f \in K[V]^G$ we have

$$h \cdot \varphi(f) = \varphi(hf) = \varphi(\mathrm{Tr}(g) \cdot f) = \varphi(\mathrm{Tr}(gf)) = \left(\mathrm{Tr}^*(\varphi)\right)(gf) = 0,$$

hence $\varphi(f) = 0$. So $\varphi = 0$, which was to be shown. $\qquad\square$

Theorem 3.9.8 (Broer [30])**.** *Let $f_1, \ldots, f_n \in K[V]^G$ be primary invariants and assume that $K[V]^G$ is Cohen-Macaulay. Then there are secondary invariants of degrees at most*

$$\deg(f_1) + \cdots + \deg(f_n) - n.$$

Proof. We can clearly assume that V is a faithful KG-module. Let $g_1, \ldots, g_m \in K[V]^G$ be free generators of $K[V]^G$ as an A-module, and let $e_1 \le e_2 \le \ldots \le e_m$ be their degrees. Then there exists a $\varphi \in \mathrm{Hom}_A(K[V]^G, A)$ which maps g_i to $\delta_{i,m}$. This φ is homogeneous of degree $-e_m$, and no homomorphism in $\mathrm{Hom}_A(K[V]^G, A)$ of smaller degree exists. By Lemma 3.9.7, $\mathrm{Tr}^*(\varphi) \in \mathrm{Hom}_A(K[V], A)$ also has degree $-e_m$. On the other hand, Proposition 3.9.5

implies that a homogeneous homomorphism in $\mathrm{Hom}_A(K[V], A)$ has degree at least $-(\deg(f_1) + \cdots + \deg(f_n) - n)$, hence

$$-e_m \geq -(\deg(f_1) + \cdots + \deg(f_n) - n),$$

from which the result follows. $\qquad\square$

Corollary 3.9.9. *If $K[V]^G$ is Cohen-Macaulay, then*

$$\beta(K[V]^G) \leq \max\{|G|, n(|G| - 1)\}. \tag{3.9.5}$$

Proof. We can assume that K is large enough so that Dade's algorithm (see Section 3.3.1) works. Hence there exist primary invariants of degrees at most $|G|$. By Theorem 3.9.8, the corresponding secondary invariants have degrees bounded by $n(|G| - 1)$, and the result follows. $\qquad\square$

Remark 3.9.10. It has been conjectured that the bound from Theorem 3.9.8 holds without the hypothesis that $K[V]^G$ be Cohen-Macaulay. All computational evidence known to the authors is in favor of this conjecture. For example, it holds if G is a permutation group and the elementary symmetric polynomials are taken as primary invariants (see Corollary 3.10.9). From the conjectured bound it would follow that (3.9.5) holds as a general a priory bound for finite groups. It was shown by Hughes and Kemper [115] that (3.9.5) holds if $|G|$ is not divisible by p^2, where $p = \mathrm{char}(K)$. $\qquad\triangleleft$

3.9.3 A General Degree Bound

We have seen in Section 3.5.2 that in the modular case secondary invariants can be found by calculating a syzygy module. Hermann [105] gave an upper bound for the degrees of generators of syzygy modules. Her bound leads to the following degree bound for invariants.

Theorem 3.9.11. *With $n = \dim(V)$ we have the degree bound*

$$\beta\left(K[V]^G\right) \leq n(|G| - 1) + |G|^{n \cdot 2^{n-1} + 1} \cdot n^{2^{n-1} + 1}.$$

Proof. For homogeneous invariants $g_1, \ldots, g_m \in K[V]_+^G$ it is equivalent that they generate $K[V]^G$ as a K-algebra and that their images generate $K[V]_+^G \big/ \left(K[V]_+^G\right)^2$ as a K-vector space (see Lemma 3.5.1 and the proof of Theorem 2.2.10). Thus $\beta\left(K[V]^G\right)$ is the maximal degree occurring in the graded vector space $K[V]_+^G \big/ \left(K[V]_+^G\right)^2$. It follows that $\beta\left(K[V]^G\right)$ remains unchanged if we extend K, so we may assume that K is an infinite field. Therefore Dade's algorithm (see Section 3.3.1) works, so there exist primary invariants $f_1, \ldots, f_n$ with $\deg(f_i) = |G|$.

Set $A := K[f_1, \ldots, f_n]$ and consider the diagram (3.5.2), where we choose H to be the trivial group. If the map $A^r \to A^k$ in (3.5.2) is denoted by Φ, then generating solutions $(z_1, \ldots, z_r) \in A^r$ of the equation

$$\sum_{i=1}^{r} z_i \Phi(e_i) = 0 \tag{3.9.6}$$

(with e_i the free generators of A^r) lead to the secondary invariants $\sum_{i=1}^{r} z_i h_i$, where $K[V] = \bigoplus_{i=1}^{r} A h_i$. In order to use Hermann's bound, we need to make the coefficients of (3.9.6) more explicit. For $\sigma \in G$ and $1 \le i \le r$ write

$$\sigma \cdot h_i - h_i = \sum_{j=1}^{r} g_{i,j,\sigma} h_j$$

with $g_{i,j,\sigma} h_j \in A$. Then

$$\Phi(e_i) = \sum_{j_1} \sum_{\sigma \in S(G)} g_{i,j,\sigma} e_{j,\sigma},$$

where $S(G) \subseteq G$ is a generating system for G and the $e_{j,\sigma}$ are free generators of A^k. Let q be an upper bound for the degrees of $g_{i,j,\sigma}$ as polynomials in $f_1, \ldots, f_n$. Then Hermann [105, Satz 2] tells us that the z_i in the generating solutions of (3.9.6) are bounded by $\sum_{l=0}^{n-1} (qr)^{2^l}$. (As stated, Hermann's bound uses the number k of equations instead of the number r of unknowns, but in the proof she deletes linearly dependent equations until $k < r$.) A bound for q can be obtained as follows. We know from Proposition 3.9.5 that $\deg(h_i) \le n(|G| - 1)$. Thus $\deg(g_{i,j,\sigma}) \le n(|G| - 1) \le n|G|$, and this entails the bound $q = n$ for the degree of the $g_{i,j,\sigma}$ as polynomials in $f_1, \ldots, f_n$. We obtain the bound

$$\sum_{l=0}^{n-1} (nr)^{2^l} \le n(nr)^{2^{n-1}}$$

for the generating solutions $(z_1, \ldots, z_r)$. These are the degrees as polynomials in $f_1, \ldots, f_n$, so to obtain the proper degrees as elements of A, we need to multiply this bound by $|G|$. The degrees of the secondary invariants are therefore bounded by

$$n(|G| - 1) + |G| \cdot n(nr)^{2^{n-1}}.$$

Finally, we use Theorem 3.7.1 (which is proved in Section 3.7.1) to obtain $r = |G|^n$. Substituting this into the above expression yields the desired bound. $\square$

Remark 3.9.12. The degree bound in Theorem 3.9.11 is truly astronomical, and light-years away from being sharp. It strikes the eye that the bound displays a doubly-exponential behavior in n, which is also typical for Gröbner bases (see Remark 1.1.9).

Clearly Theorem 3.9.11 should be taken as a theoretical result, in the sense of "there exists a degree bound which only depends on $|G|$ and $n = \dim(V)$."
◁

3.9.4 Separating Invariants

Let us pick up the topic of separating subalgebras, as introduced in Section 2.3.2, and ask for degree bounds for separating invariants. The next theorem gives a way for calculating separating invariants *explicitly* for any finite group. All that is required is multiplying out a (large) polynomial.

Theorem 3.9.13. *Let $x_1, \ldots, x_n$ be a basis of V^*, and form the polynomial*

$$F := \prod_{\sigma \in G} \left(T - \sum_{i=1}^{n} A_i \cdot \sigma \cdot x_i \right) \in K[V]^G[A_1, \ldots, A_n, T]$$

with $T, A_1, \ldots, A_n$ indeterminates. Then the coefficients of F (as a polynomial in $T, A_1, \ldots, A_n$) generate a separating subalgebra of $K[V]^G$.

Proof. Take $x, y \in V$ with distinct orbits. There exist coefficients $\alpha_1, \ldots, \alpha_n$ in a field extension L of K such that the linear form $l := \sum_{i=1}^{n} \alpha_i x_i$ satisfies

$$l(y - \sigma \cdot x) \neq 0 \quad \text{for all} \quad \sigma \in G.$$

Thus with

$$f := \prod_{\sigma \in G} (l(y) - \sigma \cdot l) \in L[V]^G$$

we have $f(y) = 0$ and $f(x) \neq 0$. But f can be obtained by specializing $A_i = \alpha_i$ and $T = l(y)$ in F. Hence f is an L-linear combination of the coefficients of F. Therefore there exists a coefficient c of F with $c(x) \neq c(y)$. $\qquad\square$

Theorem 3.9.13 entails the following degree bound.

Corollary 3.9.14. *Let $A \subseteq K[V]^G$ be the subalgebra generated by all homogeneous invariants of degree $\leq |G|$. Then A is separating.*

Proof. This follows immediately from Theorem 3.9.13. $\qquad\square$

We obtain the following corollary.

Corollary 3.9.15. *Let $A \subseteq K[V]^G$ be the subalgebra generated by all homogeneous invariants of degree $\leq |G|$. Then*

$$K[V]^G = \widehat{\overline{A}},$$

where $\widehat{\overline{A}}$ denotes the purely inseparable closure of the normalization of A (see the definition before Theorem 2.3.12).

Proof. This follows immediately from Corollary 3.9.14 and Theorem 2.3.12.
$\square$

Example 3.9.16. Let us reconsider Example 3.5.7(a), which is the first example we came across where Noether's degree bound was violated (see Example 3.6.1). So G is the cyclic permutation group of order 2 generated by $(1\ 2)(3\ 4)(5\ 6)$ acting on the polynomial ring $R = \mathbb{F}_2[x_1, y_1, x_2, y_2, x_3, y_3]$. A minimal set of generating invariants is given by

$$f_i := x_i + y_i, \quad g_i := x_i y_i \quad (i = 1, 2, 3),$$
$$u_{i,j} := x_i y_j + x_j y_i \quad (1 \le i < j \le 3),$$
$$\text{and} \quad h := x_1 x_2 x_3 + y_1 y_2 y_3.$$

By Corollary 3.9.15, h should be integral over the subalgebra A generated by f_i, g_i and $u_{i,j}$. Indeed, we find

$$h^2 + (f_1 f_2 f_3 + f_2 u_{1,3} + f_3 u_{1,2})\, h +$$
$$f_2^2 f_3^2 g_1 + f_2^2 g_1 g_3 + f_2 f_3 g_1 u_{2,3} + f_3^2 g_1 g_2 + f_1^2 g_2 g_3 = 0.$$

In this case h also lies in the quotient field of A:

$$h = (f_1 f_3 + u_{1,3})\, f_2 + (f_1 f_3 g_2 + u_{1,2} u_{2,3}) / f_2.$$

So in this example the computation of the purely inseparable closure turns out to be unnecessary.   ◁

The next example generalizes the previous one.

Example 3.9.17. Let p be a prime number and consider the cyclic group $G = \langle \sigma \rangle \cong C_p$ acting on $R := \mathbb{F}_p[x_1, y_1, x_2, y_2, \ldots, x_n, y_n]$ by $\sigma \cdot x_i = x_i$ and $\sigma \cdot y_i = x_i + y_i$. The invariant ring R^G was studied by Richman [200], Campbell and Hughes [33], and Shank and Wehlau [220]. Richman [200] conjectured that the invariants

$$x_i, \quad N_i := \prod_{\tau \in G} \tau \cdot y_i = y_i^p - x_i^{p-1} y_i \quad (1 \le i \le n),$$
$$u_{i,j} := x_i y_j - x_j y_i \quad (1 \le i < j \le n), \quad \text{and}$$
$$\mathrm{Tr}_G\left(y_1^{e_1} \cdots y_n^{e_n}\right) \quad (0 \le e_i < p)$$

generate R^G, where Tr_G denotes the transfer. This conjecture was proved by Campbell and Hughes [33]. Recently Shank and Wehlau [220] showed that a *minimal* generating set of invariants is given by the x_i, N_i, $u_{i,j}$, and those $\mathrm{Tr}_G\left(y_1^{e_1} \cdots y_n^{e_n}\right)$ with $e_1 + \cdots + e_m > 2(p-1)$. It follows that the subalgebra $A \subseteq R^G$ generated by the homogeneous invariants of degree $\le |G|$ is just

$$A = \mathbb{F}_p[x_i, N_i, u_{i,j}].$$

By Corollary 3.9.14, A is a separating subalgebra, and Corollary 3.9.15 lets us hope that the normalization $\widetilde{A}$ coincides with R^G.

As it turns out, a subalgebra that is even smaller than A has $K[V]^G$ as its normalization. Indeed, for the algebra A' generated by the x_i, N_i ($i = 1, \ldots, n$), and $u_{1,j}$ ($j = 2, \ldots, n$) we have $\widetilde{A'} = R^G$, since $u_{i,j}$ is integral over A' (like any other element in R, due to the equations $y_i^p - x_i^{p-1} y_i - N_i = 0$), and

$$u_{i,j} = \frac{x_i u_{1,j} - x_j u_{1,i}}{x_1} \in \mathrm{Quot}(A').$$

But A' is not separating. Indeed, it is easy to see that the generators of A form a minimal set of separating invariants. For example, the vectors

$$v := (0, \ldots, 0, 1, 1, 1, 0) \quad \text{and} \quad w := (0, \ldots, 0, 1, 1, 1, 1) \in \mathbb{F}_p^{2n}$$

have the property that $u_{n-1,n}(v) \neq u_{n-1,n}(w)$, but $f(v) = f(w)$ for all other generators f of A.

So, among other things, we have seen that $\mathrm{Quot}(A) = \mathrm{Quot}(A') = \mathrm{Quot}(R^G)$. But concerning quotient fields, or more generally localization, something even stronger can be said. In fact, a result of Richman [200] states that

$$R_{x_1}^G = \mathbb{F}_p[N_1, x_1, \ldots, x_n, u_{1,2}, \ldots, u_{1,n}]_{x_1}, \tag{3.9.7}$$

where the localization is dividing by powers of x_1. Thus $R_{x_1}^G$ is isomorphic to a localized polynomial ring, although R^G itself is very complicated (e.g., not Cohen-Macaulay for $n \geq 3$, see Ellingsrud and Skjelbred [63]). The following short proof of (3.9.7) was communicated to us by Ian Hughes. The field $L := \mathbb{F}_p\left(N_1, x_1, \ldots, x_n, u_{1,2}, \ldots, u_{1,n}\right)$ is contained in the invariant field $\mathbb{F}_p\left(x_1, y_1, x_2, y_2, \ldots, x_n, y_n\right)^G$, and y_1 is algebraic of degree p over L. The equation

$$y_i = \frac{x_i y_1 + u_{1,i}}{x_1} \tag{3.9.8}$$

shows that $y_i \in L(y_1)$, thus $L(y_1) = \mathbb{F}_p\left(x_1, y_1, x_2, y_2, \ldots, x_n, y_n\right) =: N$. Therefore N is of degree p over L, which, by Galois theory, shows that

$$L = N^G = \mathrm{Quot}(R^G) = \mathrm{Quot}(R_{x_1}^G). \tag{3.9.9}$$

Substituting (3.9.8) into the definition of N_i yields

$$N_i = \left(\frac{x_i}{x_1}\right)^p N_1 + \left(\frac{u_{1,i}}{x_1}\right)^p - \frac{x_i^{p-1} u_{1,i}}{x_1} \in B,$$

where we write $B := \mathbb{F}_p[N_1, x_1, \ldots, x_n, u_{1,2}, \ldots, u_{1,n}]_{x_1}$ for simplicity. Since R is integral over $\mathbb{F}_p[x_1, \ldots, x_n, N_1, \ldots, N_n]$, it follows that R_{x_1} and also $R_{x_1}^G$ are integral over B. By (3.9.9), $R_{x_1}^G$ and B also have the same fields of fractions. Since B is a localization of a polynomial ring, it is normal, and therefore $R_{x_1}^G = B$ as claimed. $\lhd$

3.10 Permutation Groups

In this section we deal with the special case that G acts on V by permutations of a basis B of V. If $x_1, \ldots, x_n \in V^*$ form a dual basis of B, then G also permutes the x_i, and $K[V] = K[x_1, \ldots, x_n]$. Clearly all facts and methods from the previous section hold for this case as well, but we shall see that much more can be said.

3.10.1 Direct Products of Symmetric Groups

Let us first look at the example of symmetric groups and, more generally, direct products of such.

Theorem 3.10.1. *Let $G = S_{n_1} \times \cdots \times S_{n_r}$ be a direct product of symmetric groups acting on $R := K[x_{1,1}, \ldots, x_{1,n_1}, \ldots, x_{r,1}, \ldots, x_{r,n_r}]$ by*

$$(\sigma_1, \ldots, \sigma_r) \cdot x_{i,j} = x_{i,\sigma_i(j)}. \tag{3.10.1}$$

Then R^G is generated by the elementary symmetric polynomials

$$s_{i,j} := \sum_{\substack{\mathcal{I} \subseteq \{1,\ldots,n_i\} \\ |\mathcal{I}|=j}} \prod_{k \in \mathcal{I}} x_{i,k} \quad (1 \leq i \leq r,\ 1 \leq j \leq n_i). \tag{3.10.2}$$

R^G is isomorphic to a polynomial ring. This holds independent of the characteristic of K.

Proof. We give two alternative proofs. First, observe that

$$\prod_{j=1}^{n_i}(T - x_{i,j}) = T^{n_i} - s_{i,1}T^{n_i-1} + \cdots + (-1)^{n_i-1}s_{i,n_i-1}T + (-1)^{n_i}s_{i,n_i}.$$

If follows that the $s_{i,j}$ form a homogeneous system of parameters. Moreover, the product of their degrees is equal to the group order. Now the result follows from Theorem 3.7.5.

The second proof gives an algorithm for expressing any invariant in terms of the $s_{i,j}$. Indeed, fix the lexicographic monomial ordering with

$$x_{1,1} > x_{1,2} > \cdots > x_{1,n_1} > x_{2,1} > \cdots > x_{r,n_r}.$$

If $\prod_{i,j} x_{i,j}^{e_{i,j}}$ is the leading monomial of a non-zero invariant f, then $e_{i,1} \geq e_{i,2} \geq \cdots \geq e_{i,n_i}$ for all i. The leading monomial of $s_{i,j}$ is $x_{i,1} \cdots x_{i,j}$, therefore f has the same leading monomial as

$$\prod_{i=1}^{r} \prod_{j=1}^{n_i} s_{i,j}^{e_{i,j}-e_{i,j-1}},$$

where we set $e_{i,0} := 0$. We obtain the following algorithm.

Algorithm 3.10.2. Let f be an $(S_{n_1} \times \cdots \times S_{n_r})$-symmetric polynomial in R^G.

(1) Set $F := 0$ and $g := f$.
(2) While $g \neq 0$, perform the steps (3)–(5).
(3) Let $\mathrm{LM}(g) =: \prod_{i,j} x_{i,j}^{e_{i,j}}$ and $\mathrm{LC}(g) =: c$.
(4) Set

$$g := g - c \cdot \prod_{i=1}^{r} \prod_{j=1}^{n_i} s_{i,j}^{e_{i,j} - e_{i,j-1}}.$$

(5) Set

$$F := F + c \cdot \prod_{i=1}^{r} \prod_{j=1}^{n_i} T_{i,j}^{e_{i,j} - e_{i,j-1}}$$

with indeterminates $T_{i,j}$.
(6) When $g = 0$ is reached, we have

$$f = F(s_{1,1}, \ldots, s_{r,n_r}).$$

The termination of the algorithm is guaranteed by the fact that each new g is an invariant of smaller leading term than the previous one, and of maximal total degree at most the maximal total degree d of f. But there are only finitely many monomials of degree $\leq d$. $\qquad\square$

Remark 3.10.3. A third proof of Theorem 3.10.1 would be to use the fact that for $G_1 \leq \mathrm{GL}(V_1)$ and $G_2 \leq \mathrm{GL}(V_2)$ we have

$$K[V_1 \oplus V_2]^{G_1 \times G_2} \cong K[V_1]^{G_1} \otimes_K K[V_2]^{G_2},$$

and that $K[x_1, \ldots, x_n]^{S_n}$ is generated by the elementary symmetric polynomials in $x_1, \ldots, x_n$. But then we would have to prove this latter fact, which amounts to more or less the same as proving Theorem 3.10.1. $\qquad\lhd$

The second proof tells us that the elementary symmetric polynomials form a SAGBI basis of the invariant ring, according to the following definition.

Definition 3.10.4. *Let $>$ be a monomial ordering on the polynomial ring $K[x_1, \ldots, x_n]$. For a subalgebra $A \subseteq K[x_1, \ldots, x_n]$, write $L(A)$ for the algebra generated by all leading monomials of non-zero elements from A. A subset $S \subseteq A$ is called a* **SAGBI basis** *of A if the algebra generated by all leading monomials of elements from S equals $L(A)$:*

$$K[\mathrm{LM}(f) \mid f \in S] = L(A).$$

If S is a SAGBI basis of A then it also generates A as an algebra, and we have an algorithm such as Algorithm 3.10.2 to rewrite an element of A as a polynomial in the elements of S. SAGBI bases were invented independently by Robbiano and Sweedler [203] and Kapur and Madlener [128]. Unfortunately, even finitely generated subalgebras of $K[x_1, \ldots, x_n]$ often do not have finite SAGBI bases. We shall see such examples in Section 3.10.3.

3.10.2 Göbel's Algorithm

It follows from Theorem 3.10.1 that the elementary symmetric polynomials can serve as a homogeneous system of parameters for any permutation group $G \leq S_n$. However, it is not clear how the fact that G is a permutation group can be used to facilitate the computation of secondary invariants. It is the algorithm of Göbel [86] which gives a surprising answer to this question. The upshot is that secondary invariants can be chosen among the orbit sums of so-called special monomials (see below). Göbel's algorithm itself gives a representation of an invariant in terms of elementary symmetric polynomials and such orbit sums. In particular, a degree bound emerges which is independent of the ground field. We present a generalization of Göbel's algorithm to the case were G is a subgroup of the direct product $S_{n_1} \times \cdots \times S_{n_r}$. The elementary symmetric polynomials $s_{i,j}$ defined by (3.10.2) will act as primary invariants. We will need the following notation.

G will be a subgroup of the direct product $S_{n_1} \times \cdots \times S_{n_r}$ acting on $R := K[x_{1,1}, \ldots, x_{1,n_1}, \ldots, x_{r,1}, \ldots, x_{r,n_r}]$ as in (3.10.1). We write

$$M := \{\prod_{i=1}^{r} \prod_{j=1}^{n_i} x_{i,j}^{e_{i,j}} \mid e_{i,j} \in \mathbb{N}_0\}, \quad \text{and} \quad T := \{a \cdot t \mid 0 \neq a \in K,\ t \in M\}$$

for the set of monomials and terms, respectively. For $f \in R$ with $f = \sum_{t \in M} a_t \cdot t$ let $T(f) := \{a_t \cdot t \mid a_t \neq 0\} \subset T$. For $t \in T$, set

$$\mathrm{orb}_G(t) := \sum_{t' \in \{\sigma \cdot t \mid \sigma \in G\}} t' \ = \ \sum_{\sigma \in G/G_t} \sigma \cdot t \in R^G,$$

where G_t is the stabilizer. Note that every element of the orbit enters only once into the sum, so $\mathrm{orb}_G(t)$ is in general not equal to the transfer of t. A monomial $t = \prod_{i=1}^{r} \prod_{j=1}^{n_i} x_{i,j}^{e_{i,j}} \in M$ is called **special** if for each $i \in \{1, \ldots, r\}$ we have

$$\{e_{i,1}, \ldots, e_{i,n_i}\} = \{0, \ldots, k_i\} \quad \text{with} \quad k_i \in \mathbb{N}_0,$$

where mutually equal $e_{i,j}$ are counted only once, as we are considering the *set* of the $e_{i,j}$. In other words, the exponents in each block cover a range without gaps starting from 0. We write M_{spec} for the set of all special monomials. Observe that for a special term t the total degree is bounded by

$$\deg(t) \leq 1 + 2 + \cdots + (n_1 - 1) + \cdots + 1 + 2 + \cdots + (n_r - 1)$$

$$= \binom{n_1}{2} + \cdots + \binom{n_r}{2}. \qquad (3.10.3)$$

In particular, M_{spec} is a finite set. To each term $t \in T$ we associate a special term $\tilde{t}$ by applying the following step iteratively: Let $k \in \mathbb{N}_0$ be minimal with $k \notin \{e_{i,1}, \ldots, e_{i,n_i}\}$ but $k < \max\{e_{i,1}, \ldots, e_{i,n_i}\}$ for some i. Then lower all $e_{i,j}$ with $e_{i,j} > k$ by 1. Repeat this step until a special term $\tilde{t}$ is obtained.

We write $\tilde{t} =: \mathrm{Red}(t)$. Loosely speaking, $\mathrm{Red}(t)$ is obtained by shoving the exponents in each block together.

Lemma 3.10.5. *For $t \in T$ and $\sigma \in S_{n_1} \times \cdots \times S_{n_r}$ we have:*

(a) $\sigma \cdot \mathrm{Red}(t) = \mathrm{Red}(\sigma \cdot t)$,
(b) $\sigma \cdot \mathrm{Red}(t) = \mathrm{Red}(t) \quad \Leftrightarrow \quad \sigma \cdot t = t$.

Proof. Part (a) follows directly from the definition of $\mathrm{Red}(t)$. Hence if $\sigma \cdot t = t$, then $\sigma \cdot \mathrm{Red}(t) = \mathrm{Red}(t)$. If, on the other hand, $\sigma \cdot \mathrm{Red}(t) = \mathrm{Red}(t)$, then $\mathrm{Red}(\sigma \cdot t) = \mathrm{Red}(t)$, so $\sigma \cdot t$ and t have the same coefficient a. As the application of Red does not change the ordering of the exponents, we conclude $\sigma \cdot t = t$. $\square$

We introduce a relation $\succ$ on T as follows: For $t = \prod_{i=1}^{r} \prod_{j=1}^{n_i} x_{i,j}^{e_{i,j}}$ and $t' = \prod_{i=1}^{r} \prod_{j=1}^{n_i} x_{i,j}^{e'_{i,j}}$ choose permutations $\sigma_1, \sigma_1' \in S_{n_1}, \ldots, \sigma_r, \sigma_r' \in S_{n_r}$ such that for all $i \in \{1, \ldots, r\}$ and for $1 \leq j_1 \leq j_2 \leq n_i$ we have

$$e_{i,\sigma_i(j_1)} \geq e_{i,\sigma_i(j_2)} \quad \text{and} \quad e'_{i,\sigma_i'(j_1)} \geq e'_{i,\sigma_i'(j_2)}.$$

Then we say that $t \succ t'$ if there exists an $i_0 \in \{1, \ldots, r\}$ and a $j_0 \in \{1, \ldots, n_{i_0}\}$ such that

$$e_{i,\sigma_i(j)} = e'_{i,\sigma_i'(j)} \quad \text{if } i < i_0 \text{ or if } i = i_0 \text{ and } j < j_0,$$

and furthermore

$$e_{i_0,\sigma_{i_0}(j_0)} > e'_{i_0,\sigma_{i_0}'(j_0)}.$$

Less formally, one can think of the relation $\succ$ as first comparing the biggest exponent in the first block, then the second-biggest, and so on. If the exponents in the first block agree (up to permutations), the same comparison is performed on the second block and so on. We write $t \preceq t'$ if $t \succ t'$ does not hold. Note that $\preceq$ is not an order even when restricted to M, since $t \preceq t'$ and $t' \preceq t$ fail to imply $t = t'$. But $\succ$ and $\preceq$ are transitive relations, and there exist no infinite, strictly decreasing chains of monomials. We are ready to prove the central lemma (compare Göbel [86, Lemma 3.10]).

Lemma 3.10.6. *Let $t \in T$ be a term and set $u := t/\mathrm{Red}(t) \in M$. Then with $\widehat{G} := S_{n_1} \times \cdots \times S_{n_r}$ we have*

(a) $t \succ s$ for all $s \in T(\mathrm{orb}_{\widehat{G}}(u) \cdot \mathrm{Red}(t) - t)$,
(b) $t \succ s$ for all $s \in T\left(\mathrm{orb}_{\widehat{G}}(u) \cdot \mathrm{orb}_G(\mathrm{Red}(t)) - \mathrm{orb}_G(t)\right)$.

Proof. (a) Obviously t lies in $T(\mathrm{orb}_{\widehat{G}}(u) \cdot \mathrm{Red}(t))$. We have to show that $t \succ s$ for all other $s \in T(\mathrm{orb}_{\widehat{G}}(u) \cdot \mathrm{Red}(t))$. So take such an s and assume that $t \preceq s$. Clearly $s = (\sigma_1, \ldots, \sigma_r) \cdot u \cdot \mathrm{Red}(t)$ with $(\sigma_1, \ldots, \sigma_r) \in \widehat{G}$. If we can show that $(\sigma_1, \ldots, \sigma_r) \cdot u = u$, then $s = t$ and (a) is proved.

Let $\mathrm{Red}(t) = a \prod_{i=1}^{r} \prod_{j=1}^{n_i} x_{i,j}^{e_{i,j}}$ and $u = \prod_{i=1}^{r} \prod_{j=1}^{n_i} x_{i,j}^{d_{i,j}}$. Thus t and s have exponents $e_{i,j} + d_{i,j}$ and $e_{i,j} + d_{i,\sigma_i(j)}$, respectively. By renumbering the $x_{i,j}$, we may assume that $e_{i,j} \geq e_{i,j'}$ for $j \leq j'$, and then also $d_{i,j} \geq d_{i,j'}$ by the construction of $\mathrm{Red}(t)$. Let k be maximal with $d_{1,1} = d_{1,2} = \ldots = d_{1,k}$. Then $t \preceq s$ implies that σ_1 must permute the set $\{1, \ldots, k\}$, since the biggest exponents in the first block of t and s must coincide. Proceeding analogously for the second-highest exponents $d_{1,k+1}, \ldots, d_{1,l}$, we conclude that σ_1 also permutes $\{k + 1, \ldots, l\}$, and finally arrive at $\sigma_1 \cdot \prod_{j=1}^{n_1} x_{1,j}^{d_{1,j}} = \prod_{j=1}^{n_1} x_{1,j}^{d_{1,j}}$. The same argument shows that σ_2 fixes the second block of u, and so on. Thus indeed $(\sigma_1, \ldots, \sigma_r) \cdot u = u$.

(b) For $\sigma \in G$ we have by (a)

$$\sigma \cdot t \succ s \quad \text{for all } s \in T \left(\mathrm{orb}_{\widehat{G}} \left(\frac{\sigma \cdot t}{\mathrm{Red}(\sigma \cdot t)} \right) \cdot \mathrm{Red}(\sigma \cdot t) - \sigma \cdot t \right),$$

hence also $t \succ s$ for these s. But by Lemma 3.10.5(a)

$$\mathrm{orb}_{\widehat{G}} \left(\frac{\sigma \cdot t}{\mathrm{Red}(\sigma \cdot t)} \right) \cdot \mathrm{Red}(\sigma \cdot t) = \mathrm{orb}_{\widehat{G}}(\sigma \cdot u) \cdot \sigma \cdot \mathrm{Red}(t) = \mathrm{orb}_{\widehat{G}}(u) \cdot \sigma \cdot \mathrm{Red}(t).$$

Therefore

$$t \succ s \quad \text{for all } s \in T \left(\mathrm{orb}_{\widehat{G}}(u) \cdot \sigma \cdot \mathrm{Red}(t) - \sigma \cdot t \right).$$

Now by Lemma 3.10.5(b), the $\sigma \in G$ fixing t are the same that fix $\mathrm{Red}(t)$, so summation over coset representatives σ of the stabilizer of t in G yields the result. $\qquad \square$

We can give Göbel's algorithm now. The purpose of the algorithm is to write an invariant as a linear combination of orbits sums of special monomials, where the coefficients of this linear combination are polynomials in the elementary symmetric polynomials $s_{i,j}$.

Algorithm 3.10.7 (Göbel's algorithm). Let $G \leq \widehat{G} := S_{n_1} \times \cdots \times S_{n_r}$ be a subgroup acting on $R := K[x_{1,1}, \ldots, x_{1,n_1}, \ldots, x_{r,1}, \ldots, x_{r,n_r}]$ as in (3.10.1). Given an invariant $f \in R^G$, perform the following steps.

(1) Set $g := f$, and for each $t \in M_{\mathrm{spec}}$ set $F_t := 0$.
(2) While $g \neq 0$, perform steps (i)–(iv).
    i)   Choose $t \in T(g)$ which is maximal with respect to the relation $\succ$.
    ii) Compute $\mathrm{Red}(t) =: a \cdot \tilde{t}$ with $\tilde{t} \in M_{spec}$ and set $u := t / \mathrm{Red}(t)$.
    iii) Use Algorithm 3.10.2 to represent $\mathrm{orb}_{\widehat{G}}(u)$ in terms of the elementary symmetric polynomials (3.10.2):

$$\mathrm{orb}_{\widehat{G}}(u) = F(s_{1,1}, \ldots, s_{1,n_1}, \ldots, s_{r,1}, \ldots, s_{r,n_r})$$

    with F a polynomial.
    iv) Set

$$g := g - \mathrm{orb}_{\widehat{G}}(u) \cdot \mathrm{orb}_G(\mathrm{Red}(t))$$

    and $F_{\tilde{t}} := F_{\tilde{t}} + aF$.

(3) When $g = 0$ is reached, we have

$$f = \sum_{t \in M_{\mathrm{spec}}} F_t(s_{1,1}, \ldots, s_{1,n_1}, \ldots, s_{r,1}, \ldots, s_{r,n_r}) \cdot \mathrm{orb}_G(t). \qquad (3.10.4)$$

It is clear that Algorithm 3.10.7 is correct if it terminates.

Theorem 3.10.8. *Algorithm 3.10.7 terminates after a finite number of steps.*

Proof. Each pass through the while-loop (i)–(iv) replaces g by

$$g - \mathrm{orb}_G(t) + \left(\mathrm{orb}_G(t) - \mathrm{orb}_{\widehat{G}}(u) \cdot \mathrm{orb}_G(\mathrm{Red}(t))\right).$$

Since G permutes the terms in $T(g)$, each term of $\mathrm{orb}_G(t)$ is also a term of g. Therefore and by Lemma 3.10.6, some maximal terms are removed from g, and only strictly smaller terms are added. Hence after a finite number of steps the maximum of $T(g)$ with respect to $\succ$ decreases strictly, and iterating further eventually yields $g = 0$, since every strictly decreasing sequence of terms is finite. $\qquad\square$

Algorithm 3.10.7 was implemented by Göbel [87] in the computer algebra system MAS (see Kredel [154]). This forms a package in MAS which is now included in the standard distribution.

It is clear from (3.10.4) that the orbit sums of the special monomials form a (usually non-minimal) system of secondary invariants. The number of orbit sums of special monomials was studied by Göbel [88]. Since the degree of a special monomial is bounded from above by (3.10.3), we obtain the following degree bound.

Corollary 3.10.9. *Let $G \leq S_{n_1} \times \cdots \times S_{n_r}$ be a subgroup acting on $R := K[x_{1,1}, \ldots, x_{1,n_1}, \ldots, x_{r,1}, \ldots, x_{r,n_r}]$ as in (3.10.1). If one chooses the elementary symmetric polynomials $s_{i,j}$ $(1 \leq i \leq r, 1 \leq j \leq n_r)$ given by (3.10.2) as primary invariants, then the secondary invariants have degrees at most*

$$\binom{n_1}{2} + \cdots + \binom{n_r}{2}.$$

In particular,

$$\beta(R^G) \leq \max\left\{\binom{n_1}{2} + \cdots + \binom{n_r}{2}, n_1, \ldots, n_r\right\}.$$

Remark 3.10.10. (a) The bound in Corollary 3.10.9 for the secondary invariants matches Broer's bound (Theorem 3.9.8) exactly: It is the sum of the degrees of the primary invariants minus the dimension of the representation. However, Corollary 3.10.9 holds over any ground field K (and, in fact, over any commutative ring with unit). Hence there are many examples where R^G is not Cohen-Macaulay (such as cyclic groups of order divisible by $p = \mathrm{char}(K)$, where R^G can only be Cohen-Macaulay if $p \leq 3$ by Campbell et al. [38]), but Corollary 3.10.9 still applies.

(b) With an appropriate generalization of the notion of special monomials, Göbel's algorithm and degree bound can be generalized to monomial groups, i.e., groups consisting of permutation matrices where the 1's may be replaced by roots of unity (see Kemper [137]). Again, the degree bound obtained for the secondary invariants is like Broer's bound.   ◁

3.10.3 SAGBI Bases

We have seen in Section 3.10.1 that the invariant ring of a direct product of symmetric groups has a finite SAGBI basis. The following example shows that invariant rings of finite groups do not always have finite SAGBI bases, even in characteristic 0.

Example 3.10.11 (Göbel [86]). Consider the action of the cyclic group $G = \langle \sigma \rangle \cong C_3$ of order 3 on $R := K[x_1, x_2, x_3]$ by $\sigma\colon x_1 \mapsto x_2 \mapsto x_3 \mapsto x_1$. Fix the lexicographic monomial ordering with $x_1 > x_2 > x_3$. In order to show that R^G has no finite SAGBI basis, we have to convince ourselves that the leading algebra $L(R^G)$ is not finitely generated as an algebra over K. So assume, by way of contradiction, that $L(R^G)$ is finitely generated. Then the quotient $L(R^G)_+/(L(R^G)_+)^2$ is finitely generated as a K-vector space, so there exists an integer N such that $\mathrm{LM}(f) \in (L(R^G)_+)^2$ for every homogeneous invariant f of degree at least N. But we have

$$\mathrm{LM}(x_1^{i-1} x_2^i + x_2^{i-1} x_3^i + x_1^i x_3^{i-1}) = x_1^i x_3^{i-1},$$

which does not lie in $(L(R^G)_+)^2$, since there is no monomial $x_1^j x_3^k$ with $j \le k$ in $L(R^G)_+$. Hence there is no bound N, and $L(R^G)$ is indeed not finitely generated.   ◁

Even more discouraging is a result of Göbel which gives a converse to Theorem 3.10.1

Theorem 3.10.12 (Göbel [89]). *Let $G \le S_n$ be a permutation group acting on $R := K[x_1, \dots, x_n]$ by $\sigma \cdot x_i = x_{\sigma(i)}$. Then there exists a finite SAGBI basis of the invariant ring R^G with respect to the lexicographic monomial ordering if and only if G is a direct product of symmetric groups acting as in (3.10.1).*

The "if" direction is Theorem 3.10.1, and the "only if" direction is the main result of [89]. We do not give the proof here.

It may seem that Theorem 3.10.12 shatters all hope that SAGBI bases can be of any use in invariant theory of finite groups. However, whether or not a finite SAGBI basis exists depends very much on the basis of V that is chosen. Indeed, Shank [217] gave a SAGBI basis for the invariant ring of the cyclic group of order p acting on an indecomposable KG-module of dimension up to 5 (for a general $p = \mathrm{char}(K)$). He used a basis such that the action assumes upper triangular form, and the graded reverse lexicographic monomial ordering. Obtaining a SAGBI basis is a crucial element in Shank's approach, since this enables him to compute the Hilbert series of a subalgebra of the invariants and comparing this to the "target" Hilbert series given by Almkvist and Fossum [9] to prove that the subalgebra is in fact equal to the invariant ring. More generally, we have the following result.

Proposition 3.10.13 (Shank [218], Shank and Wehlau [220]). *Let G be a finite group and V a KG-module whose dual V^* has a basis $x_1, \ldots, x_n$ such that*

$$\sigma \cdot x_i \in Kx_i \oplus \cdots \oplus Kx_n$$

for every $\sigma \in G$ and every $i = 1, \ldots, n$. Then $K[V]^G$ has a finite SAGBI basis with respect to any monomial ordering with $x_1 > x_2 > \cdots > x_n$.

Proof. By the special form of the action, the invariant $N_i := \prod_{\sigma \in G}(\sigma \cdot x_i)$ has the leading monomial

$$\mathrm{LM}(N_i) = x_i^{|G|}.$$

Hence $A := K[x_1^{|G|}, \ldots, x_n^{|G|}]$ is contained in $L(K[V]^G)$. But $K[V]$ is finitely generated as a module over A and hence a Noetherian module. Therefore the submodule $L(K[V]^G)$ is also Noetherian over A. Thus $L(K[V]^G)$ is finitely generated. $\qquad\square$

Remark 3.10.14. If G is a p-group with $p = \mathrm{char}(K)$, then a basis as required in Proposition 3.10.13 always exists. For example, the invariant ring considered in Example 3.10.11 does have a finite SAGBI basis in the case $\mathrm{char}(K) = 3$ (but only after changing the vector space basis). $\qquad\triangleleft$

SAGBI bases are also used by Stillman and Tsai [238] to compute invariants of tori and finite abelian groups.

3.10.4 Hilbert Series for Trivial Source Modules

Göbel's algorithm makes essential use of the fact that a permutation group G also permutes the monomials in $x_1, \ldots, x_n$. This also has consequences for the Hilbert series of the invariant ring $K[V]^G = K[x_1, \ldots, x_n]^G$. In fact, a basis of the homogeneous component $K[V]_d^G$ of degree d is given by the sums over the G-orbits of monomials of degree d. The number of such G-orbits is independent of the characteristic of K. Therefore one can use Molien's formula pretending to be in characteristic 0, and get the correct result even in the modular case (see Smith [225, Proposition 4.3.4]). If $\sigma \in G \leq S_n$ has a disjoint cycle representation with cycles of lengths $l_1, \ldots, l_k$, then the contribution $1/\det(1 - t \cdot \sigma)$ in Molien's formula (Theorem 3.2.2) is

$$\frac{1}{(1 - t^{l_1}) \cdots (1 - t^{l_k})}.$$

More generally, we can use Molien's formula to get the Hilbert series of the invariant ring $K[V]^G$ in the case that V is a trivial source module, i.e., a direct summand of a permutation module. More precisely, assume that K has positive characteristic p and let $(\widetilde{K}, R, K)$ be a p-modular system (see Curtis and Reiner [50, p. 402]), so $\mathrm{char}(\widetilde{K}) = 0$. By Benson [16, Corollary 3.11.4], V lifts to an RG-module $\hat{V}$ which is also a trivial source module. Now $\widetilde{V} := \widetilde{K} \otimes_R \hat{V}$ is a $\widetilde{K}G$-module, and we have the following result.

Proposition 3.10.15 (Kemper [137, Theorem 5]). *In the above situation we have*
$$H(K[V]^G, t) = H(\widetilde{K}[\widetilde{V}]^G, t) = \frac{1}{|G|} \sum_{\sigma \in G} \frac{1}{\det_{\widetilde{V}}(1 - t \cdot \sigma)}.$$

Example 3.10.16. We present a few examples of trivial source modules.

(a) If the order of G is not divisible by $p := \mathrm{char}(K)$ (for which we say that G is a p'-group), then every KG-module is a trivial source module by the theorem of Maschke. Furthermore, if W is a trivial source module for a subgroup $H \le G$, then the induced module $\mathrm{Ind}_H^G(W)$ is also a trivial source module (see Thévenaz [244, Corollary 27.4]). Hence every module which is induced from a p'-subgroup is a trivial source module. An important class of examples are monomial representations, i.e., modules induced from a one-dimensional module of a subgroup.

(b) Examples of trivial source modules which are not induced from p'-subgroups can be obtained by taking a permutation module $P = \oplus_{i=1}^n K e_i$ with $p \nmid n$ and $V := \{\alpha_1 e_1 + \cdots + \alpha_n e_n \in P \mid \alpha_1 + \cdots + \alpha_n = 0\}$. Then the sequence
$$0 \longrightarrow V \longrightarrow P \longrightarrow K \longrightarrow 0$$
of KG-modules splits, so V is a trivial source module. As a concrete example, consider the natural permutation representation of A_5 over a field of characteristic 2, and take V as above. We calculated the extended Hilbert series in Example 3.3.9(b). For the ordinary Hilbert series we now obtain
$$H(K[V]^G, t) = \frac{t^{10} + 1}{(1 - t^2)(1 - t^3)(1 - t^4)(1 - t^5)}.$$

A comparison with Example 3.3.9(b) illustrates the differences and similarities between the ordinary and the extended Hilbert series.   ◁

3.11 Ad Hoc Methods

It quite often happens that the algorithms given in the above sections (especially 3.3 and 3.5) are not feasible due to large time or memory requirements. In such cases one has to put one's hope in ad hoc methods, which may work and produce the invariant ring. The chances of success are particularly high if one already has a guess of a nice structure of the invariant ring. In any case, ad hoc methods depend on a mixture of experience, luck and optimism.

The methods that we employ in this section are far from covering all the ad hoc methods for computing invariants that are around. For example, Shank and Wehlau [220] provide an assortment of methods which are well suited for the computation of invariants of p-groups in characteristic p. We will give an account of some methods in this section, and use the following example throughout. Let $G := W_3(H_4)$ be the 3-modular reduction of the Weyl group of type H_4 of order $14\,400$. According to the classification of Shephard and Todd [221], $G = W_3(G_{30})$ is the 3-modular reduction of the irreducible complex reflection group number 30. Thus G is generated by reflections, but since its order is a multiple of the characteristic of K, the invariant ring need not be isomorphic to a polynomial ring (see Section 3.7.4). So the question is whether $K[V]^G$ is a polynomial ring or not, and if not, what structure it has then. G is a subgroup of $\mathrm{GL}_4(\mathbb{F}_9)$ and can be generated by the full permutation group S_4 together with the matrices

$$\begin{pmatrix} -1 & 0 & 0 & 0 \\ 0 & -1 & 0 & 0 \\ 0 & 0 & 1 & 0 \\ 0 & 0 & 0 & 1 \end{pmatrix} \quad \text{and} \quad \begin{pmatrix} 0 & -1 & -1 & \zeta^2 \\ -1 & 0 & -1 & \zeta^2 \\ -1 & -1 & 0 & \zeta^2 \\ \zeta^2 & \zeta^2 & \zeta^2 & -1 \end{pmatrix},$$

where $\zeta \in \mathbb{F}_9$ is an element of order 8. Trying the standard algorithms on this group quickly shows that even the construction of primary invariants runs into serious memory problems. In Example 3.5.7(b) we have calculated the invariant ring of a Sylow p-subgroup of G, for $p = 3$, and seen that this invariant ring is Cohen-Macaulay. It follows by Remark 3.4.2(b) that the invariant ring $K[V]^G$ of G is also Cohen-Macaulay.

All explicit calculations in this section were done in MAGMA.

3.11.1 Finding Primary Invariants

The following observation gives lower bounds for the degrees of primary invariants.

Proposition 3.11.1. *For a finite group G and an n-dimensional KG-module V, let $J_d \subseteq K[V]$ be the ideal generated by the homogeneous invariants $K[V]_k^G$ with $0 < k \le d$. If $d_1 \le \cdots \le d_n$ are the degrees of primary invariants, then*

$$d_i \ge \min\{d \mid \dim(R/J_d) \le n - i\}.$$

Proof. By Proposition 3.3.1(d) we have

$$n - i = \dim\left(K[V]/(f_1, \ldots, f_i)\right) \ge \dim\left(K[V]/J_{d_i}\right).$$

This yields the result. $\qquad\qquad\square$

We now apply Proposition 3.11.1 to the group $G = W_3(H_4)$. The computation of the homogeneous components $K[V]_d^G$ is feasible up to degrees around $d = 40$, and Proposition 3.11.1 yields

$$d_1 \geq 2, \; d_2 \geq 10, \quad \text{and} \quad d_3 \geq 36,$$

but no information on the last degree d_4 (except that $d_4 > 40$). By trying random invariants f_2, f_{10}, f_{36} of degrees 2, 10 and 36 (computed by the method of Section 3.1.1), we are lucky enough to arrive at an ideal (f_2, f_{10}, f_{36}) of dimension 1, so there is only one further primary invariant missing. We have

$$f_2 = x_1^2 + x_2^2 + x_3^2 + x_4^2, \quad f_{10} = x_1^{10} + x_2^{10} + x_3^{10} + x_4^{10},$$

but f_{36} is much more complicated. By Proposition 3.3.5 the degree d_4 of the missing primary invariant must be a multiple of $|G|/(d_1 d_2 d_3) = 20$. Since $d_4 > 40$, we might hope to find the last primary invariant of degree 60. However, it is impossible due to time and storage problems to compute all invariants of degree 60. Instead, we try to construct a polynomial f_{60} of degree 60 by using Steenrod operations.

Steenrod operations are a very helpful tool in modular invariant theory, and we briefly explain their definition following Smith [225]. Suppose $K = \mathbb{F}_q$ is a finite field. We take an additional indeterminate T and define a homomorphism $P \colon K[V] \to K[V][T]$ of K-algebras by sending an $x \in V^*$ to $x + x^q \cdot T \in K[V][T]$. It is easily checked that P commutes with the action of $\mathrm{GL}(V)$ on $K[V]$. For $f \in K[V]$, write

$$P(f) = \sum_{i \geq 0} \mathcal{P}^i(f) \cdot T^i.$$

Then $\mathcal{P}^i(f)$ is the i-th Steenrod operation on f. It follows from the $\mathrm{GL}(V)$-compatibility of P that Steenrod operations applied to invariants yield invariants again. It is also easy to check that for a homogeneous f

$$\deg(\mathcal{P}^i(f)) = \deg(f) + i(q - 1)$$

if $\mathcal{P}^i(f)$ is non-zero, and $\mathcal{P}^i(f) = 0$ if $i > \deg(f)$. Let us note here that the Steenrod operations provide a modular invariant ring with an additional structure of an unstable algebra over the Steenrod algebra (see Smith [225, p. 296]). This structure has been successfully used to prove results in modular invariant theory (see Adams and Wilkerson [5], Bourguiba and Zarati [25], Neusel [180], to mention just a few). For an approach to the Steenrod algebra which is different from the one presented above, the reader might turn to Wood [261].

In this book we only use Steenrod operations to produce new invariants from old ones. In the case $G = W_3(H_4)$, we first note that $f_{10} = -\mathcal{P}^1(f_2)$. Now $f_{60} := \mathcal{P}^3(f_{36})$ has the desired degree 60, and indeed we are lucky enough to find that the ideal generated by f_2, f_{10}, f_{36} and f_{60} has dimension 0. Thus a complete system of primary invariant is found. Since this system is optimal in the sense that no primary invariants of smaller degrees could be chosen, we conclude already here that $K[V]^G$ is not a polynomial ring, since

$$\frac{\deg(f_2) \cdot \deg(f_{10}) \cdot \deg(f_{36}) \cdot \deg(f_{60})}{|G|} = 3, \qquad (3.11.1)$$

and this quotient should be 1 if $K[V]^G$ were polynomial by Theorem 3.7.5.

3.11.2 Finding Secondary Invariants

We start by presenting a whole new algorithm for computing secondary invariants. The reason why we put this into the section on ad hoc methods is that we believe that the algorithm will only perform better than the ones given in Section 3.5 if the invariant ring has a nice structure (see Remark 3.11.4(e)). The algorithm is based on the following result.

Proposition 3.11.2. *Suppose that $R \subseteq K[V]^G$ is a subalgebra of an invariant ring of a finite group. Then $R = K[V]^G$ if and only if all of the following three conditions hold:*

(a) R contains a system of primary invariants $f_1, \ldots, f_n$.

(b) $[\mathrm{Quot}(R) : K(f_1, \ldots, f_n)] = \dfrac{\deg(f_1) \cdots \deg(f_n)}{|G|}$.

(c) R is normal.

In (b), $[\mathrm{Quot}(R) : K(f_1, \ldots, f_n)]$ denotes the degree of $\mathrm{Quot}(R)$ as a field extension of the subfield generated by the f_i.

Proof. First suppose that $R = K[V]^G$. Then (a) holds, and Lemma 3.9.6 shows that $\mathrm{Quot}(R) = K(V)^G$, hence (b) follows from Equation (3.7.1). Since the polynomial ring $K[V]$ is normal, (c) follows from Proposition 2.3.11.

Now suppose that (a)–(c) hold for R, and take any invariant $f \in K[V]^G$. It follows from (b) and Equation (3.7.1) that $\mathrm{Quot}(R) = K(V)^G$, hence $f \in \mathrm{Quot}(R)$. Moreover, f is integral over R by (a), hence (c) implies that $f \in R$. $\qquad\square$

It is quite easy to turn Proposition 3.11.2 into an algorithm. Indeed, if $R = K[f_1, \ldots, f_n, g_1, \ldots, g_m]$, we have Gröbner basis methods for computing the relations between the f_i and g_j (see Section 1.2.2). From these, the degree of $\mathrm{Quot}(R)$ over $K(f_1, \ldots, f_n)$ can easily be determined. (Sweedler [242] would be a reference for the relevant algorithms, but in our special situation it is almost immediately clear what to do.) Moreover, the relations can be used to check whether R is normal (see Section 1.6), or to extend R if it is not. We obtain the following algorithm. It computes generators for $K[V]^G$ as an algebra over K. If needed, secondary invariants can then be constructed from these.

Algorithm 3.11.3 (Calculate generators of $K[V]^G$). Let G be a finite group and V a finite dimensional KG-module, where K is a perfect field. Suppose we are given primary invariants $f_1, \ldots, f_n$. To obtain homogeneous invariants $g_1, \ldots, g_m$ which together with $f_1, \ldots, f_n$ generate $K[V]^G$ as a K-algebra, proceed as follows.

(1) If $\deg(f_1)\cdots\deg(f_n) = |G|$, then by Theorem 3.7.5 $K[V]^G$ is generated by the f_i and we are done. Otherwise, set $m := 1$.

(2) Let g_m be a homogeneous invariant of minimal degree not contained in $R_{m-1} := K[f_1,\ldots,f_n,g_1,\ldots,g_{m-1}]$. Invariants can be found by the methods of Section 3.1, and membership in R_{m-1} can be tested by linear algebra.

(3) Using a monomial ordering ">" with

$$Y_j \gg T_1,\ldots,T_n,Y_1,\ldots,Y_{j-1}$$

for all $j = 1,\ldots,m$, compute a Gröbner basis $\mathcal{G}$ of the kernel I of the map

$$\Phi\colon K[T_1,\ldots,T_n,Y_1,\ldots,Y_m] \to R_m,\ \ T_i \mapsto f_i,\ Y_j \mapsto g_j.$$

This can be done by the methods of Section 1.2.2. Assigning degrees $\deg(T_i) := \deg(f_i)$ and $\deg(Y_j) := \deg(g_j)$ makes I into a homogeneous ideal, and we can assume that the polynomials from $\mathcal{G}$ are also homogeneous.

(4) Let e_j be the minimal Y_j-degree of a polynomial in $\mathcal{G}$ involving the variable Y_j but none of $Y_{j+1},\ldots,Y_m$. If

$$e_1\cdots e_m < \frac{\deg(f_1)\cdots\deg(f_n)}{|G|},$$

then set $m := m+1$ and go back to step (2).

(5) Choose an $(m \times m)$-minor $f \in K[T_1,\ldots,T_n,Y_1,\ldots,Y_m]$ of the Jacobian matrix of the polynomials in $\mathcal{G}$ such that $f \notin I$, and set $J := \sqrt{I + (f)}$.

(6) Choose a homogeneous $g \in J \setminus I$ (e.g., $g = f$), and compute the quotient ideal $(g \cdot J + I) : J$. If this ideal is contained in $(g) + I$, then we are done.

(7) Otherwise, choose a homogeneous $h \in ((g \cdot J + I) : J) \setminus ((g) + I)$ and set $m := m+1$ and $g_m := \Phi(h)/\Phi(g)$.

(8) Compute a homogeneous generating set $\mathcal{G}$ of the kernel I of the map

$$\Phi\colon K[T_1,\ldots,T_n,Y_1,\ldots,Y_m] \to R_m,\ \ T_i \mapsto f_i,\ Y_j \mapsto g_j$$

(possibly using Lemma 1.3.2), and go back to step (5).

The following remark contains a justification why the algorithm is correct.

Remark 3.11.4. (a) In step (4), a polynomial of minimal positive Y_j-degree involving only $T_1,\ldots,T_m$ and $Y_1,\ldots,Y_j$ provides a minimal polynomial for g_j over the field $K(f_1,\ldots,f_n,g_1,\ldots,g_{j-1})$. Therefore

$$e_j = [K(f_1,\ldots,f_n,g_1,\ldots,g_j) : K(f_1,\ldots,f_n,g_1,\ldots,g_{j-1})],$$

and it follows that the product of the e_j is the degree of $K(f_1,\ldots,f_n,g_1,\ldots,g_n)$ over $K(f_1,\ldots,f_n)$. Thus when we get to step (5), the properties (a) and (b) of Proposition 3.11.2 are already satisfied.

(b) By Remark 1.6.5, the ideal J formed in step (5) satisfies the assumption of Theorem 1.6.4(b). Moreover, f is homogeneous since the elements of $\mathcal{G}$ are homogeneous.

(c) Steps (6) and (7) are applications of Theorem 1.6.4. Note that $\Phi(h)$ will automatically be divisible by $\Phi(g)$ (as polynomials in $K[V]$).

(d) Algorithm 3.11.3 is not guaranteed to produce a minimal set $\{g_1, \ldots, g_m\}$ of invariants which together with the f_i generates $K[V]^G$. But such a set can be obtained by eliminating redundant generators. Moreover, secondary invariants can be constructed by taking products of the g_j.

(e) Experience shows that in cases where the invariant ring is not a complete intersection, it is usually more expensive to compute relations between generating invariants than to compute the invariant ring itself using the "standard" algorithms from Section 3.5. Since Algorithm 3.11.3 requires the computation of relations in each loop, we conclude that its performance is worse than that of the standard algorithms unless $K[V]^G$ has a nice structure. $\qquad\qquad\lhd$

We return to the example $G = W_3(H_4)$. In Section 3.11.1 we have found primary invariants of degrees 2,10,36 and 60. We have also seen in Example 3.5.7(b) that the invariant ring of a Sylow 3-subgroup of G is Cohen-Macaulay. Hence by Remark 3.4.2(b) the invariant ring $K[V]^G$ of G is also Cohen-Macaulay. Thus by Theorem 3.7.1 and Equation (3.11.1) the number of secondary invariants is 3. The first secondary invariant is always 1. In order to find the next one, we search for a homogeneous invariant of minimal degree not contained in $A := K[f_2, f_{10}, f_{36}, f_{60}]$. The membership test is done by equating an invariant from a basis of $K[V]^G_d$ (computed by the methods of Section 3.1) to a general element of A of degree d and checking solvability of the corresponding linear system. We find the next secondary, g_{22}, in degree 22. Now we boldly make the guess that the secondary invariants are 1, g_{22} and g_{22}^2. In other words, we wish to show that $K[V]^G$ is generated as an algebra by the f_i and g_{22}. The property (b) of Proposition 3.11.2 is clearly satisfied, since $g_{22} \notin K(f_2, f_{10}, f_{36}, f_{60})$ and $[K(V)^G : K(f_2, f_{10}, f_{36}, f_{60})] = 3$.

In order to show property (c) of Proposition 3.11.2, we have to get a presentation for the algebra generated by the f_i and g_{22}. It is hopeless to get this by the standard methods of Section 1.2.2. But if it is true that 1, g_{22} and g_{22}^2 form a system of secondary invariants, then g_{22}^3 will be an A-linear combination of 1, g_{22} and g_{22}^2. Equating g_{22}^3 to such a linear combination of degree 66 with unknown coefficients, we see that the corresponding linear system is solvable indeed and obtain the relation

$$g_{22}^3 + (f_2^{11} - f_2 f_{10}^2)g_{22}^2 +$$
$$(f_2^{17} f_{10} - f_2^7 f_{10}^3 - f_2^{12} f_{10}^2 + f_2^4 f_{36} + f_2^2 f_{10}^4)g_{22} +$$
$$+ f_2^{18} f_{10}^3 - f_2^{28} f_{10} + f_{10}^3 f_{36} + f_2^{15} f_{36} -$$
$$f_2^{13} f_{10}^4 - f_2^5 f_{10}^2 f_{36} + f_2^8 f_{10}^5 - f_2^3 f_{60} = 0. \quad (3.11.2)$$

We claim that this relation generates the kernel I of

$$\Phi\colon K[T_2, T_{10}, T_{36}, T_{60}, Y_{22}] \to K[f_2, f_{10}, f_{36}, f_{60}, g_{22}], \quad T_i \mapsto f_i, \ Y_{22} \mapsto g_{22}.$$

Indeed, if there existed a relation not divisible by (3.11.2), then the division algorithm would yield a relation of Y_{22}-degree smaller than 3, in contradiction to $[K(f_2, f_{10}, f_{36}, f_{60}, g_{22}) : K(f_2, f_{10}, f_{36}, f_{60})] = 3$.

We can now go into step (5) of Algorithm 3.11.3. As a (1×1)-minor of the Jacobian matrix we choose the derivative of the relation with respect to T_{60}, which is $-T_2^3$. Hence $T_2 \in J = \sqrt{I + (T_2^3)}$. On the other hand,

$$I + (T_2) = (T_2, Y_{22}^3 + T_{10}^3 T_{36})$$

is a prime ideal, so it is equal to J. Choose $g = T_2$. We have to show that $(g \cdot J + I) : J \subseteq (g) + I$, so take any $h \in (g \cdot J + I) : J$. Then $T_2 h \in T_2 \cdot J + I$, so, writing r for the relation (3.11.2), we have

$$T_2 h = h_1 T_2^2 + h_2 r$$

with polynomials h_1, h_2. Thus T_2 divides $h_2 r$, but it does not divide r, so $h_2 = h_3 T_2$ with another polynomial h_3. Dividing by T_2, we obtain

$$h = h_1 T_2 + h_3 r \in (T_2) + I.$$

It follows that indeed $(g \cdot J + I) : J \subseteq (g) + I$, and we conclude from Proposition 3.11.2 that $K[V]^G = K[f_2, f_{10}, f_{36}, f_{60}, g_{22}]$. Therefore even though $K[V]^G$ is not a polynomial ring, it is a complete intersection (more specifically, a hypersurface, i.e., a K-algebra of Krull dimension n which is generated by $n + 1$ elements).

3.11.3 The Other Exceptional Reflection Groups

The previous calculation showed us that the invariant ring of $W_3(G_{30})$ is a hypersurface. But what about the other finite irreducible reflection group that appear in Table 3.3 on page 106 and have non-polynomial invariant rings? The second author used methods just like those described above to determine the structure of the invariant rings of the *exceptional* reflection groups, i.e., those groups appearing in the last row of Table 3.3. The results are summarized in Table 3.4.

In Table 3.4 we list the relevant finite irreducible complex reflection groups according to the classification of Shephard and Todd [221]. For each of these, we give the dimension of the action, the group order, and the structures of the invariant rings of the p-modular reduction $W_p(G_i)$ for those primes p where non-polynomial invariants rings occur. A "–" in the table signifies that the corresponding reduction becomes reducible or isomorphic to a classical

G	$\dim(V)$	$\lvert G \rvert$	$p = 2$	$p = 3$	$p = 5$	$p = 7$
G_{24}	3	336	PR	PR	PR	HS
G_{28}	4	1152	–	HS	PR	PR
G_{30}	4	14400	–	HS	HS	PR
G_{31}	4	46080	HS	HS	PR	PR
G_{32}	4	155520	–	–	nCM	PR
G_{33}	5	51840	–	HS	PR	PR
G_{34}	6	39191040	HS	–	PR	PR
G_{36}	7	2903040	–	nCM	PR	PR
G_{37}	8	696729600	–	nCM	nCM	PR

Table 3.4. Invariant rings of exceptional modular reflection groups.

group. "PR", "HS", and "nCM" mean that the invariant ring is a polynomial ring, a hypersurface, or not Cohen-Macaulay, respectively. It is probably a coincidence that in this table nothing appears between hypersurface and non-Cohen-Macaulay. For the results on non-Cohen-Macaulayness, Theorem 3.7.3 (with the algorithm for computing cohomology described after Theorem 3.7.3) and the following proposition was used.

Proposition 3.11.5 (Kemper [136, Korollar 5.14, or 142]). *Suppose that $K[V]^G$ is Cohen-Macaulay. Then for every linear subspace $W \subseteq V$ the invariant ring $K[V]^{G_W}$ of the point-wise stabilizer G_W is also Cohen-Macaulay.*

The above proposition is very similar in style to Proposition 3.7.8, and in fact a unified proof can be given for both (see [142]).

We would like to determine the structure of all the invariant rings of finite irreducible reflection groups (which appear in Table 3.3 on page 106). What remains to be done for that purpose? The invariants of the symmetric groups S_{n+2} with $p \mid (n+2)$ were shown to be not Cohen-Macaulay by Kemper [138] if $p \geq 5$. So only the cases $p = 2, 3$ remain open. For the symplectic groups we obtain complete intersections by Carlisle and Kropholler [41] (see Benson [18, Section 8.3]). Here the invariant rings are in general not hypersurfaces. Finally, there are the orthogonal and unitary groups and their relatives. For these, the invariant rings are not known to date. There seems to be some evidence suggesting that these invariant rings are all complete intersections.

4 Invariant Theory of Reductive Groups

4.1 Computing Invariants of Linearly Reductive Groups

Throughout this section G will be a linearly reductive group over an algebraically closed field K and V will be an n-dimensional rational representation. We will present an algorithm for computing generators of the invariant ring $K[V]^G$ (see Derksen [53]). This algorithm is actually quite simple and it is easy to implement. The essential step is just one Gröbner basis computation. We will also need the Reynolds operator. For now, the Reynolds operator $\mathcal{R}$ is just a black box which has the required properties (see Definition 2.2.2). In Section 4.5 we will study how to compute the Reynolds operator for several examples of linearly reductive groups.

4.1.1 The Heart of the Algorithm

We will discuss the theoretical results on which the algorithm is based. Suppose that G is a linearly reductive group and V is an n-dimensional rational representation of G. As in the proof of Hilbert's finiteness theorem (Theorem 2.2.10) we take $I \subset K[V]$ to be the ideal generated by all homogeneous invariants of degree > 0. As the proof of Theorem 2.2.10 shows, if I is generated by homogeneous invariants $f_1, \ldots, f_r$, then $f_1, \ldots, f_r$ will generate the invariant ring $K[V]^G$ as a K-algebra. If $f_1, \ldots, f_r \in K[V]$ are *any* homogeneous generators of I (not necessarily invariant), then by the following proposition we can obtain generators of $K[V]^G$ using the Reynolds operator $\mathcal{R}$ (see Definition 2.2.2).

Proposition 4.1.1. *If $I = (f_1, \ldots, f_r)$ with $f_1, \ldots, f_r \in K[V]$ homogeneous, then $I = (\mathcal{R}(f_1), \ldots, \mathcal{R}(f_r))$ and $K[V]^G = K[\mathcal{R}(f_1), \ldots, \mathcal{R}(f_r)]$.*

Proof. Suppose that $K[V]^G$ is generated by homogeneous elements $h_1, \ldots, h_s$. Let $\mathfrak{m}$ be the homogeneous maximal ideal of $K[V]$. Then the images of $h_1, \ldots, h_s$ modulo $\mathfrak{m}I$ span the vector space $I/\mathfrak{m}I$. Notice that $\mathcal{R}(I) \subseteq I$ and $\mathcal{R}(\mathfrak{m}I) \subseteq \mathfrak{m}I$, because I and $\mathfrak{m}I$ are G-stable (see Corollary 2.2.7). Now $\mathcal{R}$ induces a map $I/\mathfrak{m}I \to I/\mathfrak{m}I$ which is the identity because $I/\mathfrak{m}I$ is spanned by the images of invariants. Notice that the images of f_i and $\mathcal{R}(f_i)$ in $I/\mathfrak{m}I$ are the same. It follows that $I/\mathfrak{m}I$ is spanned by the images of $\mathcal{R}(f_1), \ldots, \mathcal{R}(f_r)$. By the graded Nakayama Lemma (Lemma 3.5.1),

$I = (\mathcal{R}(f_1), \ldots, \mathcal{R}(f_r))$. From the proof of Theorem 2.2.10 it follows that $K[V]^G = K[\mathcal{R}(f_1), \ldots, \mathcal{R}(f_r)]$. $\qquad\square$

Remark 4.1.2. If $f_1, \ldots, f_r \in K[V]$ is a *minimal* set of homogeneous ideal generators of I, then the images of $f_1, \ldots, f_r$ in $I/\mathfrak{m}I$ will form a vector space basis. So $\mathcal{R}(f_1), \ldots, \mathcal{R}(f_r)$ are homogeneous invariants whose images form a basis of $I/\mathfrak{m}I$. It follows that $\{\mathcal{R}(f_1), \ldots, \mathcal{R}(f_r)\}$ is a *minimal* set of generators of $K[V]^G$. $\qquad\triangleleft$

Let $\psi : G \times V \to V \times V$ be the map defined by $\psi(\sigma, v) = (v, \sigma \cdot v)$. We will identify $K[V \times V]$ with the polynomial ring $K[x, y]$ where x and y are abbreviations for $x_1, \ldots, x_n$ and $y_1, \ldots, y_n$. Let B be the image of ψ and let $\overline{B}$ be its Zariski closure. Suppose that $(w, 0) \in \overline{B}$. If $f \in K[V]^G$ is a homogeneous invariant of degree > 0, then $f(x) - f(y)$ vanishes on B, because $f(v) - f(\sigma \cdot v) = 0$ for all $v \in V$ and $\sigma \in G$. So $f(x) - f(y)$ also vanishes on $\overline{B}$, in particular $f(w) - f(0) = f(w) = 0$. It follows that $w \in \mathcal{N}_V$, where $\mathcal{N}_V$ is Hilbert's nullcone (see Definition 2.4.1). Conversely, if $w \in \mathcal{N}_V$, then $\{w\} \times G \cdot w \subset B$, so $(w, 0) \in \overline{B}$ because 0 lies in the Zariski closure of the orbit $G \cdot w$ (see Lemma 2.4.2). We have proven

$$\overline{B} \cap (V \times \{0\}) = \mathcal{N}_V \times \{0\}. \tag{4.1.1}$$

Equation (4.1.1) is also true if G is geometrically reductive and not linearly reductive. Let $\mathfrak{b} \subseteq K[V \times V]$ be the vanishing ideal of $\overline{B}$. The left-hand side of Equation (4.1.1) is the zero set of the ideal $\mathfrak{b} + (y_1, y_2, \ldots, y_n)$ and the right-hand side is the zero set of the ideal $I + (y_1, \ldots, y_n)$ (where $I \subset K[x] \subset K[x, y]$ is as before). The following theorem shows that in fact both ideals are the same under the assumption that G is linearly reductive.

Theorem 4.1.3. *If $f_1(x, y), f_2(x, y), \ldots, f_r(x, y) \in K[x, y]$ are homogeneous and generate the ideal $\mathfrak{b}$, then $f_1(x, 0), f_2(x, 0), \ldots, f_r(x, 0)$ generate I.*

Proof. An equivalent statement of the theorem is that

$$\mathfrak{b} + (y_1, \ldots, y_n) = I + (y_1, \ldots, y_n).$$

"$\supseteq$": If $f \in K[V]^G$ is a homogeneous invariant of positive degree, then $f(x) = (f(x) - f(y)) + f(y)$. Now $f(y) \in (y_1, \ldots, y_n)$ and $f(x) - f(y) \in \mathfrak{b}$ because $f(x) - f(y)$ vanishes on $\overline{B}$: $(f(x) - f(y))(v, \sigma \cdot v) = f(v) - f(\sigma \cdot v) = 0$ for all $v \in V$ and $\sigma \in G$. Since I is generated by homogeneous invariants of degree > 0, we have shown this inclusion.

"$\subseteq$": Suppose that $f(x) \in (y_1, \ldots, y_n) + \mathfrak{b}$. We can write

$$f(x) = \sum_i c_i(x) f_i(y) + b(x, y), \tag{4.1.2}$$

where $c_i(x) \in K[x]$, $f_i(y) \in (y_1, \ldots, y_n) \subset K[y]$ homogeneous for all i and $b(x, y) \in \mathfrak{b}$. We will view $V \times V$ as a G-variety where G acts as usual on the second factor and trivially on the first. The corresponding Reynolds operator

$$\mathcal{R}_y : K[x,y] \to K[x,y]^G = K[y]^G[x]$$

is a $K[x]$-module homomorphism (see Corollary 2.2.7) and the restriction to $K[y]$ is the Reynolds operator $K[y] \to K[y]^G$. Let us apply $\mathcal{R}_y$ to Equation (4.1.2):

$$f(x) = \mathcal{R}_y(f(x)) = \sum_i c_i(x)\mathcal{R}_y(f_i(y)) + \mathcal{R}_y(b(x,y)). \qquad (4.1.3)$$

Let $\Delta : V \hookrightarrow V \times V$ be the diagonal morphism $v \mapsto (v,v)$. The corresponding algebra homomorphism $\Delta^* : K[x,y] \to K[x]$ is defined by $p(x,y) \mapsto p(x,x)$. We apply Δ^* to Equation (4.1.3):

$$f(x) = \sum_i c_i(x)\mathcal{R}(f_i(x)).$$

In fact, $\Delta^*(\mathcal{R}_y(b(x,y))) = 0$, because $\mathcal{R}_y(b(x,y)) \in \mathfrak{b}$ ($\mathfrak{b}$ is G-stable, see Corollary 2.2.7) and $\Delta(V) \subset \overline{B}$. Since $\mathcal{R}(f_i(x))$ is a homogeneous invariant of degree > 0, we conclude $f \in I$. $\qquad\square$

Remark 4.1.4. Theorem 4.1.3 can be generalized in the following way. Let X be an arbitrary affine G-variety, and let again $\overline{B}$ be the Zariski closure of the image of the map $\psi : G \times X \to X \times X$ defined by $(\sigma, x) \mapsto (x, \sigma \cdot x)$. Let $\mathfrak{b} \subset K[X \times X] \cong K[X] \otimes K[X]$ be the vanishing ideal of $\overline{B}$. A proof, similar to the one of Theorem 4.1.3, shows that if $J \subset K[X]$ is any G-stable ideal, then

$$(\mathfrak{b} + K[X] \otimes J) \cap (K[X] \otimes K)$$

is the ideal in $K[X]$ generated by $J \cap K[X]^G$. In a more algebraic setting, this generalization is carried out in Theorem 4.2.6. In geometric terms this means that if $W \subset X$ is any G-stable subset, then

$$\overline{p_1((X \times W) \cap \overline{B})} = \pi^{-1}(\pi(W)),$$

where $p_1 : X \times X \twoheadrightarrow X$ is the projection onto the first factor, and $\pi : X \to X /\!/ G$ is the categorical quotient (see Section 2.3.1). This geometric statement is also true if G is only geometrically reductive. $\qquad\triangleleft$

Example 4.1.5. Let $K = \overline{\mathbb{F}}_2$, the algebraic closure of the field of 2 elements. Let G be the group of order 2, generated by σ. The group G acts on $V := K^4$ as follows

$$\sigma(x_1, x_2, x_3, x_4) = (x_2, x_1, x_4, x_3).$$

The invariant ring is generated by $x_1 + x_2$, $x_3 + x_4$, $x_1 x_2$, $x_3 x_4$, $x_1 x_3 + x_2 x_4$. The image B of $\psi : G \times V \to V \times V$ is the union of the diagonal $\{(v,v) \mid v \in V\}$ and $\{(v, \sigma \cdot v) \mid v \in V\}$. It is easy to check that $f := (x_1 + y_2)(x_3 + y_3)$ vanishes on B. Now $f(x, 0) = x_1 x_3$ does not lie in the ideal

$$I = (x_1 + x_2, x_3 + x_4, x_1 x_2, x_3 x_4, x_1 x_3 + x_2 x_4).$$

This shows that Theorem 4.1.3 is not always true if G is only geometrically reductive. Notice also that $x_1 x_3 + x_2 x_4$ lies in the ideal generated by $x_1 + x_2$ and $x_3 + x_4$ because $x_1 x_3 + x_2 x_4 = x_3(x_1 + x_2) + x_2(x_3 + x_4)$. So in fact I is generated by the invariants $x_1 + x_2, x_3 + x_4, x_1 x_2, x_3 x_4$. However, these invariants do not generate the invariant ring. The proof of Theorem 2.2.10 does not generalize to geometrically reductive groups.   ◁

4.1.2 The Input: the Group and the Representation

We will concentrate for the moment on the input of the algorithm. We will need a convenient way to tell the algorithm what group and which representation we are dealing with. We will represent an algebraic group G by its affine coordinate ring and the representation V is given by a polynomial map $\rho : G \to \mathrm{GL}(V) \subset \mathrm{End}(V)$.

Let us start with a linearly reductive group G. We will view G as an affine algebraic group. Since G is an affine variety, there exists a closed embedding $G \hookrightarrow K^l$ for some positive integer l. We will view G as a Zariski closed subset of K^l. This gives us a surjective ring homomorphism $K[z_1, z_2, \ldots, z_l] \twoheadrightarrow K[G]$, whose kernel will be denoted by $I(G)$. Suppose that V is an n-dimensional rational representation. By choosing a basis of V, we will identify $V \cong K^n$ and $\mathrm{End}(V) \cong \mathrm{Mat}_{n,n}(K)$, where $\mathrm{Mat}_{m,n}(K)$ is the set of $m \times n$ matrices with entries in K. The G-action on V defines a group homomorphism $\rho : G \to \mathrm{GL}(V) \subset \mathrm{End}(V) \cong \mathrm{Mat}_{n,n}(K)$ which is also a morphism of affine varieties. We can choose $a_{i,j} \in K[z_1, \ldots, z_l]$, $1 \le i, j \le n$ (which are unique modulo $I(G)$), such that

$$\rho(g) = \begin{pmatrix} a_{1,1}(g) & a_{1,2}(g) & \cdots & a_{1,n}(g) \\ a_{2,1}(g) & a_{2,2}(g) & \cdots & a_{2,n}(g) \\ \vdots & \vdots & \ddots & \vdots \\ a_{n,1}(g) & a_{n,2}(g) & \cdots & a_{n,n}(g) \end{pmatrix}.$$

So the G-action is represented by the matrix $A = (a_{i,j})_{i,j=1}^n \in \mathrm{Mat}_{n,n}(K[z_1, \ldots, z_l])$.

The input of the algorithm for computing invariants will be the matrix A together with a set of generators of the ideal $I(G) \subset K[z_1, \ldots, z_l]$. We could have chosen a different embedding $G \hookrightarrow K^l$. Sometimes there is no canonical choice for the embedding. Since were are going to do Gröbner basis computations, it is advisable to take l, the number of variables, small. We also would like the generators of $I(G)$ to have small degree, and $I(G)$ should have a "small" Gröbner basis.

Example 4.1.6. Let $T = (K^*)^r$ be the r-dimensional torus. We have a closed embedding $i : T \hookrightarrow K^{r+1}$ defined by

$$(t_1, t_2, \ldots, t_r) \in T \mapsto (t_1, t_2, \ldots, t_r, (t_1 t_2 \cdots t_r)^{-1}) \in K^{r+1}.$$

The image is given by the equation $z_1 z_2 \cdots z_{r+1} = 1$. So $K[T] \cong K[z_1,\ldots,z_{r+1}]/I(T)$ where $I(T)$ is the ideal generated by the polynomial $z_1 z_2 \cdots z_{r+1} - 1$.

Suppose that V is an n-dimensional representation of T. We can always choose a basis of V such that the action of T is diagonal. The action is given by

$$\sigma \cdot (x_1,\ldots,x_n) = (\chi_1(\sigma)x_1,\ldots,\chi_n(\sigma)x_n),$$

where $\chi_1,\ldots,\chi_n$ are one-dimensional **characters** of T, i.e., $\chi_i : T \to \mathbb{G}_m = K^*$ is a homomorphism of algebraic groups for all i. We can write $\chi_i(t_1,\ldots,t_r) = t_1^{b_{i,1}} t_2^{b_{i,2}} \cdots t_r^{b_{i,r}}$ with $b_{i,j} \in \mathbb{Z}$ for all i,j. The r-tuple $(b_{i,1},\ldots,b_{i,r})$ is called a **weight**. Define $c_{i,r+1} := \max\{0,-b_{i,1},-b_{i,2},\ldots,-b_{i,r}\}$ for all i and $c_{i,j} = c_{i,r+1} + b_{i,j}$ for all i,j with $j \leq r$. Let

$$A = \begin{pmatrix} \underline{z}^{\underline{c}_1} & 0 & \cdots & 0 \\ 0 & \underline{z}^{\underline{c}_2} & \cdots & 0 \\ \vdots & \vdots & \ddots & \vdots \\ 0 & 0 & \cdots & \underline{z}^{\underline{c}_n} \end{pmatrix},$$

where $\underline{z}^{\underline{c}_i}$ is an abbreviation for $\prod_{j=1}^{r+1} z_j^{c_{i,j}}$. Then A represents the T-action on V.

$\triangleleft$

Example 4.1.7. The group SL_q has a natural embedding into $\mathrm{Mat}_{q,q}(K) \cong K^{q^2}$, namely it is the set of all matrices $B \in \mathrm{Mat}_{q,q}(K)$ such that $\det(B) = 1$. Therefore, the coordinate ring of SL_q is $K[\{z_{i,j}\}_{i,j=1}^q]/I(\mathrm{SL}_q)$ where $I(\mathrm{SL}_q)$ is generated by the polynomial $\det((z_{i,j})_{i,j=1}^q) - 1$. For convenience we use here a double indexing of the z-variables instead of a single indexing. The natural representation of SL_q is just given by the matrix

$$A = \begin{pmatrix} z_{1,1} & z_{1,2} & \cdots & z_{1,q} \\ z_{2,1} & z_{2,2} & \cdots & z_{2,q} \\ \vdots & \vdots & \ddots & \vdots \\ z_{q,1} & z_{q,2} & \cdots & z_{q,q} \end{pmatrix}.$$

Let us consider the action of SL_2 on the binary forms of degree 2 (see Example 2.1.2). As a basis of V_2 we can take x^2, xy and y^2. On this basis, the action is represented by the matrix

$$\begin{pmatrix} z_{1,1}^2 & z_{1,1}z_{1,2} & z_{1,2}^2 \\ 2z_{1,1}z_{2,1} & z_{2,1}z_{1,2} + z_{1,1}z_{2,2} & 2z_{1,2}z_{2,2} \\ z_{2,1}^2 & z_{2,1}z_{2,2} & z_{2,2}^2 \end{pmatrix}.$$

$\triangleleft$

Example 4.1.8. Let G be the symmetric group S_3. As an affine variety it is just a set of 6 points and we can identify G with $\{1,2,3,4,5,6\} \subset K$ (assume that $\mathrm{char}(K) = 0$ or $\mathrm{char}(K) > 6$). We let $1,2,3,4,5,6 \in K$ correspond to the elements e, (1 2), (1 3), (2 3), (1 2 3), (1 3 2) (the elements of S_3 are represented by their cycle structure). So G is embedded in K and $K[G] \cong K[z]/I(G)$ where $I(G)$ is the ideal generated by $(z-1)(z-2)\cdots(z-6)$. Let $V \cong K^3$ be the 3-dimensional representation, where S_3 acts by permuting the coordinates. The action of the elements of S_3 is given by the permutation matrices

$$\begin{pmatrix} 1 & 0 & 0 \\ 0 & 1 & 0 \\ 0 & 0 & 1 \end{pmatrix}, \begin{pmatrix} 0 & 1 & 0 \\ 1 & 0 & 0 \\ 0 & 0 & 1 \end{pmatrix}, \begin{pmatrix} 0 & 0 & 1 \\ 0 & 1 & 0 \\ 1 & 0 & 0 \end{pmatrix}, \begin{pmatrix} 1 & 0 & 0 \\ 0 & 0 & 1 \\ 0 & 1 & 0 \end{pmatrix}, \begin{pmatrix} 0 & 0 & 1 \\ 1 & 0 & 0 \\ 0 & 1 & 0 \end{pmatrix}, \begin{pmatrix} 0 & 1 & 0 \\ 0 & 0 & 1 \\ 1 & 0 & 0 \end{pmatrix}.$$

By interpolation we can find a matrix $A \in \mathrm{Mat}_{3,3}(K[z])$ such that $A(1), A(2), \ldots, A(6)$ are exactly the permutation matrices above:

$$A = \begin{pmatrix} a_{1,1} & a_{1,2} & a_{1,3} \\ a_{2,1} & a_{2,2} & a_{2,3} \\ a_{3,1} & a_{3,2} & a_{3,3} \end{pmatrix}$$

with

$$a_{1,1} = \frac{(z-2)(z-3)(z-5)(z-6)(3z-2)}{40},$$
$$a_{1,2} = \frac{(z-1)(z-3)(z-4)(z-5)(3z-16)}{60},$$
$$a_{1,3} = \frac{(z-1)(z-2)(z-4)(z-6)(-3z+13)}{24},$$
$$a_{2,1} = \frac{(z-1)(z-3)(z-4)(z-6)(3z-11)}{40},$$
$$a_{2,2} = \frac{(z-2)(z-4)(z-5)(z-6)(-11z+13)}{120},$$
$$a_{2,3} = \frac{(z-1)(z-2)(z-3)(z-5)(11z-64)}{120},$$
$$a_{3,1} = \frac{(z-1)(z-2)(z-4)(z-5)(-3z+19)}{40},$$
$$a_{3,2} = \frac{(z-1)(z-2)(z-3)(z-6)^2}{24},$$
$$a_{3,3} = \frac{(z-3)(z-4)(z-5)(z-6)(4z-3)}{120}.$$

Instead of identifying S_3 with $\{1, 2, \ldots, 6\} \subset K$ we could have identified S_3 with the sixth roots of unity in K, or with all pairs $(z, w) \in K^2$ with $z^2 = w^3 = 1$. The last identification may be the most natural one and generalizes to all S_n: For each i, let ζ_i be a primitive i-th root of unity. We define a map

$$\{(z_2, z_3, \ldots, z_n) \mid z_2^2 = z_3^3 = \cdots = z_n^n = 1\} \to S_n$$

by

$$(\zeta_2^{k_2}, \zeta_3^{k_3}, \ldots, \zeta_n^{k_n}) \mapsto (1\ 2)^{k_2}(1\ 2\ 3)^{k_3}\cdots(1\ 2\cdots n)^{k_n} \tag{4.1.4}$$

for all integers $k_2, \ldots, k_n$. Suppose that

$$(1\ 2)^{k_2}(1\ 2\ 3)^{k_3}\cdots(1\ 2\cdots n)^{k_n} = (1\ 2)^{l_2}(1\ 2\ 3)^{l_3}\cdots(1\ 2\cdots n)^{l_n},$$

then $(1\ 2\cdots n)^{k_n - l_n}$ lies in S_{n-1}, so $k_n \equiv l_n \bmod n$. We have

$$(1\ 2)^{k_2}(1\ 2\ 3)^{k_3}\cdots(1\ 2\cdots n-1)^{k_{n-1}} = (1\ 2)^{l_2}(1\ 2\ 3)^{l_3}\cdots(1\ 2\cdots n-1)^{l_{n-1}},$$

and repeating the argument gives $k_{n-1} \equiv l_{n-1} \bmod n-1, \ldots, k_2 \equiv l_2 \bmod 2$. This shows that (4.1.4) is injective, and comparing cardinalities shows that it is a bijection onto S_n. The identification (4.1.4) embeds S_n into K^{n-1} and $K[S_n] \cong K[z_2, z_3, \ldots, z_n]/I(S_n)$ where $I(S_n) = (z_2^2 - 1, z_3^3 - 1, \ldots, z_n^n - 1)$. $\triangleleft$

4.1.3 The Algorithm

In view of Theorem 4.1.3 and Proposition 4.1.1, the computation of generators
of the invariant ring boils down to finding generators of the ideal $\mathfrak{b}$. As before,
$\mathfrak{b}$ is the vanishing ideal of the Zariski closure of the image B of the map
$\psi : G \times V \to V \times V$, defined by $(\sigma, v) \mapsto (v, \sigma \cdot v)$.

There is a general procedure for computing the vanishing ideal of the image of a morphism of affine varieties using Gröbner bases (see Section 1.2.1):
If $\varphi : X \to Y$ is a morphism of affine varieties, one can consider the graph
$\Gamma \subset X \times Y$, defined by $\Gamma = \{(v, \varphi(v)) \mid v \in X\}$. Then $\varphi(X)$ is the projection
of $\Gamma \subseteq X \times Y$ onto Y. Let $I(\Gamma) \subseteq K[X \times Y]$ be the vanishing ideal of Γ.
An element $f \in K[Y]$ vanishes on $\varphi(X)$ if and only if it vanishes on Γ as a
function in $K[X \times Y]$. In other words, $I(\Gamma) \cap K[Y]$ is the vanishing ideal of
$\varphi(X)$ (and of $\overline{\varphi(X)}$). The intersection $I(\Gamma) \cap K[Y]$ can be computed using
Gröbner basis techniques (see Algorithm 1.2).

In our case, we should consider the subset of $(G \times V) \times (V \times V)$ defined by
$\{(\sigma, v, v, \sigma \cdot v) \mid \sigma \in G, v \in V\}$. One of the V-factors is redundant. Since we
want to minimize the number of variables, we consider instead $\Gamma \subseteq G \times V \times V$
defined by $\{(\sigma, v, \sigma \cdot v) \mid \sigma \in G, v \in V\}$. As in the previous section, we will
view G as a Zariski closed subset of K^l. Now $\Gamma \subseteq G \times V \times V \subseteq K^l \times V \times V$
and B is the projection of Γ onto $V \times V$. The coordinate ring of $K^l \times V \times V$ is
the polynomial ring $K[z, x, y] = K[z_1, \ldots, z_l, x_1, \ldots, x_n, y_1, \ldots, y_n]$. Let $I(\Gamma)$
be the ideal of Γ. We have $\mathfrak{b} = I(\Gamma) \cap K[x, y]$. In other words: We get the
ideal $\mathfrak{b}$ from $I(\Gamma)$ by eliminating the variables $z_1, \ldots, z_l$.

Generators of the ideal $I(\Gamma)$ can be given explicitly. The graph Γ is a
Zariski closed subset of $K^l \times V \times V$. First of all, we will need generators
$h_1, \ldots, h_t \in K[z]$ of the ideal $I(G) \subseteq K[z]$, to define $G \subseteq K^l$. The equation

$$\begin{pmatrix} y_1 \\ y_2 \\ \vdots \\ y_n \end{pmatrix} = {}^t A \begin{pmatrix} x_1 \\ x_2 \\ \vdots \\ x_n \end{pmatrix}$$

(with ${}^t A$ the transpose of the matrix A from Section 4.1.2) gives us the
relations $y_i = \sum_j a_{i,j} x_j$. The ideal $I(\Gamma)$ is generated by all

$$h_1, h_2, \ldots, h_t \quad \text{and} \quad \left\{ y_i - \sum_{j=1}^n a_{j,i} x_j \,\middle|\, i = 1, 2, \ldots, n \right\}.$$

Choose a monomial ordering ">" on $K[z, x, y]$ such that $z_i \gg x_j$ and
$z_i \gg y_j$ for $i = 1, \ldots, l$, $j = 1, \ldots, n$ (see on page 9), i.e., z_i is larger than
any monomial in $x_1, \ldots, x_n, y_1, \ldots, y_n$. An example of such an ordering is the
lexicographic ordering with $z_1 > z_2 > \cdots > z_l > x_1 > \cdots > y_n$. Suppose
that $\mathcal{G}$ is a Gröbner basis for $I(\Gamma)$. If $f \in \mathfrak{b} = I(\Gamma) \cap K[x, y]$, then the
leading monomial $\mathrm{LM}(f)$ is divisible by $\mathrm{LM}(g)$ for some $g \in \mathcal{G}$. It follows

that $\mathrm{LM}(g) \in K[x,y]$, and by our choice of the monomial ordering, every monomial appearing in g lies in $K[x,y]$. This shows that $\mathcal{G} \cap K[x,y]$ is a Gröbner basis for $\mathfrak{b} = I(\Gamma) \cap K[x,y]$.

Algorithm 4.1.9. Suppose that G is a linearly reductive algebraic group acting rationally on an n-dimensional vector space. As in Section 4.1.2, we will view G as an affine variety in K^l given by the vanishing ideal $I(G) \subseteq K[z_1,\ldots,z_l]$. The representation of G on V is given by a matrix $A = (a_{i,j})_{i,j=1}^n$ with $a_{i,j} \in K[z_1,\ldots,z_l]$. The following steps will give generators of the invariant ring $K[V]^G$:

(1) Input: ideal generators $h_1,\ldots,h_t \in K[z_1,\ldots,z_l]$ such that $h_1 = h_2 = \cdots = h_t = 0$ defines an affine variety isomorphic to G; a matrix $A = (a_{i,j})_{i,j=1}^n$ with $a_{i,j} \in K[z_1,\ldots,z_l]$ for all i,j, corresponding to the representation of G on V.

(2) Choose a monomial ordering ">" on $K[z_1,\ldots,z_l,x_1,\ldots,x_n,y_1,\ldots,y_n]$ such that $z_i \gg x_j$ and $z_i \gg y_j$ for $i = 1,\ldots,l$, $j = 1,\ldots,n$ (see on page 9). Compute a Gröbner basis $\mathcal{G}$ of the ideal

$$\left(h_1,\ldots,h_t, \left\{ y_i - \sum_{j=1}^n a_{j,i} x_j \;\middle|\; i = 1,\ldots,n \right\} \right).$$

(3) $\mathcal{B} = \mathcal{G} \cap K[x_1,\ldots,x_n,y_1,\ldots,y_n]$ ($\mathcal{B}$ is a Gröbner basis of $\mathfrak{b}$).
(4) $\mathcal{I} = \{f(x,0) \mid f \in \mathcal{B}\}$ ($\mathcal{I}$ is a generating set for I, see Theorem 4.1.3).
(5) Output: $\{\mathcal{R}(f) \mid f \in \mathcal{I}\}$ (generators of the invariant ring, see Proposition 4.1.1).

Example 4.1.10. We give a simple illustrative example. Let $G = C_2$ the cyclic group of order 2, and assume $\mathrm{char}(K) \neq 2$. We can identify G as an affine variety with the set $\{-1,1\} \subset K$. So $G \subset K$ has the vanishing ideal $I(G) = (z^2 - 1) \subset K[z]$. We consider the representation of G on $V = K^2$ which interchanges the two coordinates. This representation is given for example by the matrix

$$A = \begin{pmatrix} \frac{z+1}{2} & \frac{1-z}{2} \\ \frac{1-z}{2} & \frac{z+1}{2} \end{pmatrix}.$$

For $z = 1$ we get the identity matrix and for $z = -1$ we get the permutation matrix

$$\begin{pmatrix} 0 & 1 \\ 1 & 0 \end{pmatrix}.$$

We follow the steps of the algorithm:

(1) Input: G is given by the equation $z^2 - 1 \in K[z]$. The representation is given by the matrix

$$A = \begin{pmatrix} \frac{z+1}{2} & \frac{1-z}{2} \\ \frac{1-z}{2} & \frac{z+1}{2} \end{pmatrix}.$$

(2) Let ">" be the lexicographic ordering on $K[z, x_1, x_2, y_1, y_2]$ with $z > x_1 > x_2 > y_1 > y_2$. We compute a Gröbner basis $\mathcal{G}$ of the ideal

$$(z^2 - 1, y_1 - \tfrac{z+1}{2}x_1 - \tfrac{1-z}{2}x_2, y_2 - \tfrac{1-z}{2}x_1 - \tfrac{z+1}{2}x_2).$$

A reduced Gröbner basis is:

$$\mathcal{G} = \{z^2 - 1, zx_2 - zy_2 + x_2 - y_2, zy_1 - zy_2 + 2x_2 - y_1 - y_2, x_1 + x_2 - y_1 - y_2,$$
$$x_2^2 + y_1 y_2 - y_1 x_2 - x_2 y_2\}. \quad (4.1.5)$$

(3)

$$\mathcal{B} = \mathcal{G} \cap K[x_1, x_2, y_1, y_2] = \{x_1 + x_2 - y_1 - y_2, x_2^2 + y_1 y_2 - y_1 x_2 - x_2 y_2\}.$$

(4) We substitute $y_1 = y_2 = 0$ in $\mathcal{B}$:

$$\mathcal{I} = \{x_1 + x_2, x_2^2\}.$$

(5) We apply the Reynolds operator: $\mathcal{R}(x_1 + x_2) = x_1 + x_2$, because $x_1 + x_2$ is already invariant; $\mathcal{R}(x_2^2) = (x_2^2 + x_1^2)/2$. So the output is: $x_1 + x_2, (x_1^2 + x_2^2)/2$. These are generators of the invariant ring $K[x_1, x_2]^{C_2}$.

◁

Example 4.1.11. We take the multiplicative group $\mathbb{G}_m$. As in Example 4.1.6, we embed $\mathbb{G}_m$ into K^2 and $\mathbb{G}_m$ is defined by the equation $z_1 z_2 = 1$. The group acts diagonally on $V \cong K^4$ with weights $-5, -3, 2, 4$. We use the algorithm to compute $K[x_1, x_2, x_3, x_4]^{\mathbb{G}_m}$.

(1) The group $\mathbb{G}_m \subset K^2$ is defined as an affine variety by $z_1 z_2 - 1$. The representation is given by the matrix

$$A = \begin{pmatrix} z_2^5 & 0 & 0 & 0 \\ 0 & z_2^3 & 0 & 0 \\ 0 & 0 & z_1^2 & 0 \\ 0 & 0 & 0 & z_1^4 \end{pmatrix}.$$

(2) We choose a lexicographic ordering ">" on $K[z_1, z_2, x_1, x_2, x_3, x_4, y_1, y_2, y_3, y_4]$ with $z_1 > z_2 > x_1 > x_2 > x_3 > x_4 > y_1 > y_2 > y_3 > y_4$.

$$\mathcal{G} = \{x_3^2 y_4 - x_4 y_3^2, x_2^2 x_4^2 y_3 - x_3 y_2^2 y_4^2, x_2^2 x_3 x_4 - y_2^2 y_3 y_4, x_2^2 x_3^3 - y_2^2 y_3^3,$$

$$x_2^4 x_4^3 - y_2^4 y_4^3, x_1 y_2^3 y_4 - x_2^3 x_4 y_1, x_1 y_2^3 y_3^2 - x_2^3 x_3^2 y_1, x_1 x_4 y_2 y_3 - x_2 x_3 y_1 y_4,$$

$$x_1 x_3 y_2 - x_2 y_1 y_3, x_1 x_2 x_4^2 - y_1 y_2 y_4^2, x_1 x_2 x_3^2 x_4 - y_1 y_2 y_3^2 y_4, x_1 x_2 x_3^4 - y_1 y_2 y_3^4,$$

$$x_1^2 y_2^4 y_3 - x_2^4 x_3 y_1^2, x_1^2 x_4 y_2^2 - x_2^2 y_1^2 y_4, x_1^2 x_4^3 y_3 - x_3 y_1^2 y_4^3, x_1^2 x_3 x_4^2 - y_1 y_3 y_4^2,$$

$$x_1^2 x_3^3 x_4 - y_1^2 y_3^3 y_4, x_1^2 x_3^5 - y_1^2 y_3^5, x_1^3 y_2^5 - x_2^5 y_1^3, x_1^3 x_4^4 y_2 - x_2 y_1^3 y_4^3, x_1^4 x_4^5 - y_1^4 y_4^5,$$

$$z_2 y_2 y_4 - x_2 x_4, z_2 y_2 y_3^2 - x_2 x_3^2, z_2 y_1 y_3 y_4 - x_1 x_3 x_4, z_2 y_1 y_3^3 - x_1 x_3^3,$$

$$z_2 y_1^3 y_4^4 - x_1^3 x_4^4, z_2 x_3 y_1 y_4^2 - x_1 x_4^2 y_3, z_2 x_2 y_1^2 y_4^2 - x_1^2 x_4^2 y_2, z_2 x_2 x_4 y_3 - x_3 y_2 y_4,$$

$$z_2 x_2 x_3 - y_2 y_3, z_2 x_2^2 y_1 - x_1 y_2^2, z_2 x_2^3 x_4^2 - y_2^3 y_4^2, z_2 x_1 y_2^2 y_3 - x_2^2 x_3 y_1,$$

$$z_2 x_1 x_4 - y_1 y_4, z_2 x_1 x_3^2 - y_1 y_3^2, z_2 x_1^2 y_2^3 - x_2^3 y_1^2, z_2^2 y_3 - x_3, z_2^2 y_1^2 y_4^4 - x_1^2 x_4^3,$$

$$z_2^2 x_3 y_4 - x_4 y_3, z_2^2 x_2 y_1 y_4 - x_1 x_4 y_2, z_2^2 x_2^2 x_4 - y_2^2 y_4, z_2^2 x_1 y_2 - x_2 y_1,$$

$$z_2^3 y_1 y_4^2 - x_1 x_4^2, z_2^3 x_2 - y_2, z_2^3 x_1 x_3 - y_1 y_3, z_2^4 y_4 - x_4, z_2^5 x_1 - y_1, z_1 y_2 - z_2^2 x_2,$$

$$z_1 y_1 - z_2^4 x_1, z_1 x_4 - z_2^3 y_4, z_1 x_3 - z_2 y_3, z_1 z_2 - 1\}. \quad (4.1.6)$$

(3) Now we take the intersection $\mathcal{B} = \mathcal{G} \cap K[x_1, x_2, x_3, x_4, y_1, y_2, y_3, y_4]$:

$$\mathcal{B} = \{x_3^2 y_4 - x_4 y_3^2, x_2^2 x_4^2 y_3 - x_3 y_2^2 y_4^2, x_2^2 x_3 x_4 - y_2^2 y_3 y_4, x_2^2 x_3^3 - y_2^2 y_3^3,$$

$$x_2^4 x_4^3 - y_2^4 y_4^3, x_1 y_2^3 y_4 - x_2^3 x_4 y_1, x_1 y_2^3 y_3^2 - x_2^3 x_3^2 y_1, x_1 x_4 y_2 y_3 - x_2 x_3 y_1 y_4,$$

$$x_1 x_3 y_2 - x_2 y_1 y_3, x_1 x_2 x_4^2 - y_1 y_2 y_4^2, x_1 x_2 x_3^2 x_4 - y_1 y_2 y_3^2 y_4, x_1 x_2 x_3^4 - y_1 y_2 y_3^4,$$

$$x_1^2 y_2^4 y_3 - x_2^4 x_3 y_1^2, x_1^2 x_4 y_2^2 - x_2^2 y_1^2 y_4, x_1^2 x_4^3 y_3 - x_3 y_1^2 y_4^3, x_1^2 x_3 x_4^2 - y_1 y_3 y_4^2,$$

$$x_1^2 x_3^3 x_4 - y_1^2 y_3^3 y_4, x_1^2 x_3^5 - y_1^2 y_3^5, x_1^3 y_2^5 - x_2^5 y_1^3, x_1^3 x_4^4 y_2 - x_2 y_1^3 y_4^3,$$

$$x_1^4 x_4^5 - y_1^4 y_4^5\}. \quad (4.1.7)$$

(4) We substitute $y_1 = y_2 = y_3 = y_4 = 0$ to obtain:

$$\mathcal{I} = \{x_2^2 x_3 x_4, x_2^2 x_3^3, x_2^4 x_4^3, x_1 x_2 x_4^2, x_1 x_2 x_3^2 x_4, x_1 x_2 x_3^4,$$

$$x_1^2 x_3 x_4^2, x_1^2 x_3^3 x_4, x_1^2 x_3^5, x_1^4 x_4^5\}.$$

(5) The elements of $\mathcal{I}$ are already invariant, so we do not have to apply the Reynolds operator here (this is typical for invariants of a diagonal action). The generators of $K[x_1, x_2, x_3, x_4]^{\mathbb{G}_m}$ are therefore:

$$x_2^2 x_3 x_4, x_2^2 x_3^3, x_2^4 x_4^3, x_1 x_2 x_4^2, x_1 x_2 x_3^2 x_4,$$

$$x_1 x_2 x_3^4, x_1^2 x_3 x_4^2, x_1^2 x_3^3 x_4, x_1^2 x_3^5, x_1^4 x_4^5.$$

In the torus case Algorithm 4.1.9 does the same as Algorithm 1.4.5 in Sturmfels [239]. $\triangleleft$

Example 4.1.12. Let us do a simple example with a nice group. We take $G = \mathrm{SL}_2$ acting on the binary forms of degree 2.

(1) Input: The group is given by $z_{1,1}z_{2,2} - z_{1,2}z_{2,1} - 1 \in K[z_{1,1}, z_{1,2}, z_{2,1}, z_{2,2}]$ and the representation is given by the matrix

$$\rho = \begin{pmatrix} z_{1,1}^2 & z_{1,1}z_{1,2} & z_{1,2}^2 \\ 2z_{1,1}z_{2,1} & z_{1,1}z_{2,2} + z_{1,2}z_{2,1} & 2z_{1,2}z_{2,2} \\ z_{2,1}^2 & z_{2,1}z_{2,2} & z_{2,2}^2 \end{pmatrix}.$$

(2) We choose the lexicographic ordering with

$$z_{1,1} > z_{2,1} > z_{1,2} > z_{2,2} > x_1 > x_2 > x_3 > y_1 > y_2 > y_3$$

on $K[z_{1,1}, z_{2,1}, z_{1,2}, z_{2,2}, x_1, x_2, x_3, y_1, y_2, y_3]$. We have

$$I(\Gamma) = (z_{1,1}z_{2,2} - z_{2,1}z_{1,2} - 1, y_1 - z_{1,1}^2 x_1 - 2z_{1,1}z_{2,1}x_2 - z_{2,1}^2 x_3,$$
$$y_2 - z_{1,1}z_{1,2}x_1 - (z_{1,1}z_{2,2} + z_{2,1}z_{1,2})x_2 - z_{2,1}z_{2,2}x_3,$$
$$y_3 - z_{1,2}^2 x_1 - 2z_{1,2}z_{2,2}x_2 - z_{2,2}^2 x_3). \quad (4.1.8)$$

A reduced Gröbner basis is

$$\mathcal{G} = \{z_{1,1}z_{2,2} - z_{2,1}z_{1,2} - 1, 2z_{1,1}x_1 - 2z_{2,2}y_1 + z_{2,1}y_2 + z_{2,1}x_2,$$
$$x_2 z_{1,1} - z_{1,1}y_2 + 2z_{2,1}x_3 + 2z_{1,2}y_1, -z_{1,2}y_2 + 2z_{1,1}y_3 - z_{1,2}x_2 - 2z_{2,2}x_3,$$
$$2x_1 z_{2,1}z_{1,2} + 2x_1 - 2z_{2,2}^2 y_1 + z_{2,2}z_{2,1}y_2 + z_{2,2}z_{2,1}x_2,$$
$$- y_2 + 2z_{2,1}z_{2,2}x_3 + 2z_{1,2}z_{2,2}y_1 - z_{1,2}z_{2,1}y_2 + x_2 z_{2,1}z_{1,2} + x_2,$$
$$- z_{2,2}y_2 + 2z_{1,2}x_1 + x_2 z_{2,2} + 2z_{2,1}y_3, -y_3 + z_{1,2}^2 x_1 + z_{1,2}z_{2,2}x_2 + z_{2,2}^2 x_3,$$
$$- 4y_3 x_3 + 4x_3 z_{1,2}z_{2,2}x_2 + 4z_{2,2}^2 x_3^2 + 4z_{1,2}^2 y_3 y_1 - z_{1,2}^2 y_2^2 + z_{1,2}^2 x_2^2,$$
$$- 4y_3 y_1 + y_2^2 + 4x_1 x_3 - x_2^2\}. \quad (4.1.9)$$

(3) The intersection $\mathcal{B} = \mathcal{G} \cap K[x_1, x_2, x_3, y_1, y_2, y_3]$ is equal to

$$\mathcal{B} = \{-4y_3 y_1 + y_2^2 + 4x_1 x_3 - x_2^2\}.$$

(4) Substituting $y_1 = y_2 = y_3 = 0$, we obtain

$$\mathcal{I} = \{4x_1 x_3 - x_2^2\}.$$

(5) We do not have to apply the Reynolds operator, because $4x_1 x_3 - x_2^2$ is already invariant. In fact, it is the discriminant up to a sign. So we conclude

$$K[x_1, x_2, x_3]^{\mathrm{SL}_2} = K[x_2^2 - 4x_1 x_3].$$

Compare this to Example 2.1.2 ◁

In Example 5.9.1 we will consider a situation where the set $\mathcal{I}$ of generators of I does not consist entirely of invariants. We will replace the application of the Reynolds operator (step 5) by a different approach, which comes down to solving a linear system in order to determine invariant generators of the ideal I.

4.2 Improvements and Generalizations

Algorithm 4.1.9 works in a general setting. It is quite fast for tori. In that case, every polynomial in the Gröbner basis of $I(\Gamma)$, and even every polynomial appearing in the Buchberger algorithm is a difference between two monomials. This makes the Gröbner basis computation quite efficient. For examples other than tori, the speed varies. Sometimes the Gröbner basis computation can be quite time consuming. If more is known about the group or the representation, this might be used to improve the algorithm. In this section we discuss an adaptation of Algorithm 4.1.9 which might speed up the computation. We will also see how the algorithm generalizes to the computation of the *module of covariants*, the *ring of covariants* and to rings of invariants of graded rings.

4.2.1 Localization of the Invariant Ring

We do not necessarily need generators of the ideal $\mathfrak{b} \subset K[x,y]$ for the algorithm to work. Instead of $\mathfrak{b}$ we might use another ideal with sufficiently nice properties. Let $\mathfrak{c}$ be the ideal generated by all $f(x) - f(y)$, with $f \in K[V]^G$. The following corollary follows from Theorem 4.1.3.

Corollary 4.2.1. *Suppose that* $f_1(x,y), f_2(x,y), \ldots, f_r(x,y)$ *are homogeneous and generate a homogeneous ideal* $\mathfrak{a} \subset K[x_1, \ldots, x_n, y_1, \ldots, y_n]$ *with* $\mathfrak{c} \subseteq \mathfrak{a} \subseteq \mathfrak{b}$. *Then* $I = (f_1(x,0), \cdots, f_r(x,0))$ *where* $I \subset K[V]$ *is the ideal generated by all homogeneous invariants of positive degree.*

Proof. If $\mathfrak{a} = \mathfrak{b}$, then the corollary follows from Theorem 4.1.3 and if $\mathfrak{a} = \mathfrak{c}$, then the statement is trivial. It follows that the corollary is true for all $\mathfrak{a}$ with $\mathfrak{c} \subseteq \mathfrak{a} \subseteq \mathfrak{b}$. $\qquad\qquad\square$

The corollary gives us more flexibility. Such an ideal $\mathfrak{a}$ might be easier to compute than the ideal $\mathfrak{b}$. Also, a Gröbner basis of such an ideal $\mathfrak{a}$ might be smaller than the Gröbner basis of $\mathfrak{b}$.

Sometimes we know the invariant ring $K[V]_h^G$ for some localization $K[V]_h$, but we do not know the invariant ring $K[V]^G$ itself. In that case, the following proposition gives an ideal $\mathfrak{a} \subset K[V \times V]$ which satisfies the conditions in Corollary 4.2.1.

Proposition 4.2.2. *Let* $h \in K[V]^G$ *be homogeneous and suppose that the localization* $K[V]_h^G$ *is generated by* $h, 1/h$ *and* $f_1, \ldots, f_r \in K[V]^G$. *Let* $\mathfrak{a}' \subset K[z, x_1, \ldots, x_n, y_1, \ldots, y_n]$ *be the ideal*

$$(zh(x) - 1, h(x) - h(y), f_1(x) - f_1(y), f_2(x) - f_2(y), \ldots, f_r(x) - f_r(y)).$$

Then $\mathfrak{a} = \mathfrak{a}' \cap K[x,y]$ *satisfies* $\mathfrak{c} \subseteq \mathfrak{a} \subseteq \mathfrak{b}$.

Proof. Suppose that $f \in K[V]^G$ is homogeneous. For some positive power k, $h^k f$ lies in the ring generated by $h, f_1, \ldots, f_r$. This implies that $h^k(x)f(x) - h^k(y)f(y) \in \mathfrak{a}'$. Modulo $\mathfrak{a}'$, $z^k(h^k(x)f(x) - h^k(y)f(y))$ is equal to $f(x) - f(y)$, so $f(x) - f(y) \in \mathfrak{a}'$. This shows that $\mathfrak{c} \subseteq \mathfrak{a}$.

Clearly $\mathfrak{a}'$ vanishes on all $(1/h(v), v, v) \in K \times V \times V$ with $h(v) \neq 0$. This shows that $\mathfrak{a}$ vanishes on the diagonal $\Delta(V) = \{(v, v) \mid v \in V\} \subset V \times V$. Let

$$I(\Delta(V)) = (x_1 - y_1, \ldots, x_n - y_n) \subset K[V \times V]$$

be the vanishing ideal of $\Delta(V)$. The ideals $\mathfrak{a}'$ and $\mathfrak{a}$ are G-stable (with G acting trivially on z and on the y_i), so $\mathfrak{a} \subseteq \sigma \cdot I(\Delta(V))$ for all $\sigma \in G$. We have $\mathfrak{a} \subseteq \mathfrak{b}$ because the intersection of all $\sigma \cdot I(\Delta(V))$ is equal to $\mathfrak{b}$. $\qquad\square$

Notice that again we can compute the ideal $\mathfrak{a}$ by elimination (see Algorithm 1.2.1). Namely, choose a monomial ordering ">" on $K[z, x_1, \ldots, x_n, y_1, \ldots, y_n]$ such that $z \gg x_i$ and $z \gg y_i$ for $i = 1, \ldots, n$. If $\mathcal{G}$ is a Gröbner basis of $\mathfrak{a}' \subset K[z, x, y]$, then $\mathcal{G} \cap K[x, y]$ is a Gröbner basis of $\mathfrak{a}$.

Often, it is much easier to find a localization of the invariant ring $K[V]_h^G$ ($h \in K[V]^G$ homogeneous), than to find the invariant ring itself.

Suppose V is a representation of G containing a $v \in V$ such that $G \cdot v$ is a closed orbit, and the stabilizer of v is trivial. It is a consequence of Luna's Slice Theorem (see Luna [163]) that there exists an affine variety U and an étale G-equivariant map

$$\gamma : G \times U \to V$$

such that the image of γ contains v (G acts on the left on G and trivially on U). In other words, the action is locally trivial in an étale neighborhood of v. For some groups, called **special groups**, we can actually choose a Zariski open neighborhood, i.e., we may assume that γ is an open immersion. Examples of such special groups are SL_q and GL_q. Now $K[G \times U]$ is just a localization $K[V]_h$ with respect to some semi-invariant h and $K[G \times U]^G$ is just equal to $K[U]$.

We will discuss this more explicitly in the case $G = \mathrm{SL}_q$ and we will show how we can use Proposition 4.2.2 to improve Algorithm 4.1.9.

Proposition 4.2.3. *Let $W \cong K^q$ be the natural representation of SL_q and suppose that V is a representation of $G = \mathrm{SL}_q$ (extendible to a GL_q-representation). Suppose that*

$$\varphi_1, \varphi_2, \ldots, \varphi_q : V \to W$$

are SL_q-equivariant morphisms. We can view $\varphi = (\varphi_1, \ldots, \varphi_q)$ as a $q \times q$ matrix with entries in $K[V]$. We assume that $h := \det(\varphi) \neq 0$. For $f \in K[V]$, define $\varphi \cdot f \in K[V]_h^G$ by $(\varphi \cdot f)(v) = f(\varphi(v)^{-1} \cdot v)$, where $\varphi(v)^{-1} \in \mathrm{GL}_q(K)$ is applied to v by the action extended from G to GL_q. Then the ring $K[V]_h^G$ is generated by

$$1/h, h, \varphi \cdot x_1, \varphi \cdot x_2, \ldots, \varphi \cdot x_n,$$

where $x_1, \ldots, x_n \in K[V]$ form a basis of V^*.

Proof. We have $\varphi(\sigma \cdot v) = \sigma\varphi(v)$ for all $v \in V$ and $\sigma \in G$. This shows that h is SL_q-invariant. If $f \in K[V]$, then $\varphi \cdot f$ is SL_q invariant: For $\sigma \in \mathrm{SL}_q$ we have

$$\sigma \cdot f(\varphi(v)^{-1} \cdot v) = f(\varphi^{-1}(\sigma^{-1} \cdot v)\sigma^{-1} \cdot v) = f(\varphi^{-1}(v)\sigma\sigma^{-1} \cdot v) = f(\varphi^{-1}(v) \cdot v).$$

Note that φ^{-1} is a matrix with coefficients in $K[V]_h$. If $f \in K[V]^G$ is a GL_q-semi-invariant, then

$$h^k f(x_1, \ldots, x_n) = \varphi \cdot f(x_1, \ldots, x_n) = f(\varphi \cdot x_1, \ldots, \varphi \cdot x_n)$$

for some integer k, so f lies in the ring generated by $h, 1/h, \varphi \cdot x_1, \ldots, \varphi \cdot x_n$. This completes the proof because $K[V]^{\mathrm{SL}_q}$ is spanned by GL_q-semi-invariants (see Corollary 2.2.12). $\qquad\square$

An SL_q-equivariant morphism $\varphi_i : V \to W$ as above is called a covariant with values in W (see Definition 4.2.9)

Example 4.2.4. Let us consider SL_2 and $V = V_1 \oplus V_2$. We identify $K[V_1] \cong K[x_1, x_2]$, $K[V_2] \cong K[x_3, x_4, x_5]$, and $K[V] \cong K[x_1, x_2, x_3, x_4, x_5]$. The action on the functions $x_1, \ldots, x_5$ is given by

$$\begin{pmatrix} a & b \\ c & d \end{pmatrix} \cdot \begin{pmatrix} x_1 \\ x_2 \\ x_3 \\ x_4 \\ x_5 \end{pmatrix} = \begin{pmatrix} (ad - bc)^{-1}(dx_1 - bx_2) \\ (ad - bc)^{-1}(ax_2 - cx_1) \\ (ad - bc)^{-2}(d^2 x_3 - bdx_4 + b^2 x_5) \\ (ad - bc)^{-2}(-2cdx_3 + (bc + ad)x_4 - 2abx_5) \\ (ad - bc)^{-2}(c^2 x_3 - acx_4 + a^2 x_5) \end{pmatrix}. \quad (4.2.1)$$

We take

$$\varphi_1 = \begin{pmatrix} x_1 \\ x_2 \end{pmatrix}, \quad \varphi_2 = \begin{pmatrix} 2x_2 x_3 - x_1 x_4 \\ x_2 x_4 - 2x_1 x_5 \end{pmatrix}.$$

The matrix φ is given by

$$\varphi = \begin{pmatrix} x_1 & 2x_2 x_3 - x_1 x_4 \\ x_2 & x_2 x_4 - 2x_1 x_5 \end{pmatrix}.$$

Thus $h = \det(\varphi) = 2x_1 x_2 x_4 - 2x_1^2 x_5 - 2x_2^2 x_3$. We use $a = x_1$, $b = 2x_2 x_3 - x_1 x_4$, $c = x_2$ and $d = x_2 x_4 - 2x_1 x_5$ in (4.2.1) to obtain

$$\varphi \cdot x_1 = 1, \ \varphi \cdot x_2 = 0, \ \varphi \cdot x_3 = (x_4^2 - 4x_3 x_5)/h, \ \varphi \cdot x_4 = 0, \ \varphi \cdot x_5 = -1/4h.$$

We take the ideal

$$\mathfrak{a}' = (zx_1 x_2 x_4 - zx_1^2 x_5 - zx_2^2 x_3 - 1, x_1 x_2 x_4 - x_1^2 x_5 - x_2^2 x_3 - y_1 y_2 y_4 + y_1^2 y_5 + y_2^2 y_3,$$

$$x_4^2 - 4x_3 x_5 - y_4^2 + 4y_3 y_5). \quad (4.2.2)$$

We use the lexicographic ordering with $z > x_1 > x_2 > \cdots > y_5$ and compute a Gröbner basis:

$$\mathcal{G} = \{zy_1y_2y_4 - zy_1^2y_5 - zy_2^2y_3 - 1, -4x_3x_1x_2x_4 + 4x_2^2x_3^2 + 4x_3y_1y_2y_4 -$$
$$4x_3y_1^2y_5 - 4x_3y_2^2y_3 + x_1^2x_4^2 - x_1^2y_4^2 + 4x_1^2y_3y_5, x_1x_2x_4 - x_1^2x_5 - x_2^2x_3 - y_1y_2y_4 +$$
$$y_1^2y_5 + y_2^2y_3, x_4^2 - 4x_3x_5 - y_4^2 + 4y_3y_5\}. \quad (4.2.3)$$

A Gröbner basis $\mathcal{A}$ of $\mathfrak{a} = \mathfrak{a}' \cap K[x, y]$ is obtained by intersecting $\mathcal{G}$ with $K[x, y]$:

$$\mathcal{A} = \{-4x_3x_1x_2x_4 + 4x_2^2x_3^2 + 4x_3y_1y_2y_4 - 4x_3y_1^2y_5 - 4x_3y_2^2y_3 + x_1^2x_4^2 - x_1^2y_4^2 +$$
$$4x_1^2y_3y_5, x_1x_2x_4 - x_1^2x_5 - x_2^2x_3 - y_1y_2y_4 + y_1^2y_5 + y_2^2y_3,$$
$$x_4^2 - 4x_3x_5 - y_4^2 + 4y_3y_5\}. \quad (4.2.4)$$

We substitute $y_1 = y_2 = y_3 = 0$ to get generators of the ideal I:

$$I = (-4x_3x_1x_2x_4 + 4x_2^2x_3^2 + x_1^2x_4^2, x_1x_2x_4 - x_1^2x_5 - x_2^2x_3, x_4^2 - 4x_3x_5).$$

The first generator lies in the ideal generated by the latter two. Therefore

$$I = (x_1x_2x_4 - x_1^2x_5 - x_2^2x_3, x_4^2 - 4x_3x_5)$$

and since these two generators are already SL_2-invariant, we do not have to apply the Reynolds operator and conclude

$$K[V]^{\mathrm{SL}_2} = K[x_1x_2x_4 - x_1^2x_5 - x_2^2x_3, x_4^2 - 4x_3x_5].$$

$$\lhd$$

The following proposition shows that under mild conditions, a morphism $\varphi : V \to W^q$ as in Proposition 4.2.3 can be found.

Proposition 4.2.5. *Suppose that V is a representation of SL_q, $v \in V$ has trivial stabilizer and the orbit $\mathrm{SL}_q \cdot v$ is closed. Then there exist SL_q-equivariant morphisms $\varphi_1, \ldots, \varphi_q : V \to W$ such that $\det(\varphi) \neq 0$ where $\varphi = (\varphi_1, \ldots, \varphi_q)$. Here, as before, W denotes the natural representation of SL_q.*

Proof. We will identify SL_q with the orbit $\mathrm{SL}_q \cdot v$ in V. The inclusion $\mathrm{SL}_q \cdot v \hookrightarrow V$ induces a surjective SL_q-equivariant ring homomorphism $K[V] \twoheadrightarrow K[\mathrm{SL}_q]$ whose kernel we denote by J. Choose a basis of W. There is a SL_q-equivariant embedding $\psi = (\psi_1, \ldots, \psi_q) : \mathrm{SL}_q \to W^q$ which identifies SL_q with the linear maps $K^q \to W$ with determinant equal to 1 (with respect to the chosen bases). In particular we have $\det(\psi) \neq 0$. For each i, $\psi_i : \mathrm{SL}_q \cdot v \to W$ is dominant because $\mathrm{SL}_q \cdot v$ maps to the dense orbit of W. The morphism $\psi_i : \mathrm{SL}_q \cdot v \to W$ corresponds to an injective ring homomorphism $\psi_i^* : K[W] \cong S(W^*) \to K[\mathrm{SL}_q] \cong K[V]/J$. Choose a subrepresentation $W_i \subset K[V]$ isomorphic to W^* such that $W_i + J = \psi_i^*(W^*)$. The inclusion $W^* \cong W_i \subset K[V]$ defines an SL_q-equivariant morphism $\varphi_i : V \to W$ which extends ψ_i. In particular $\det(\varphi) + J = \det(\psi) \neq 0$, so $\det(\varphi) \neq 0$. $\qquad\square$

4.2.2 Generalization to Arbitrary Graded Rings

Let us prove the results of Section 4.1.1 in the most general (but less geometric) setting. Let R be an arbitrary commutative algebra of finite type over K. Suppose that G is a linearly reductive group acting rationally on R by automorphisms, i.e., there exists a map $\widehat{\mu} : R \to R \otimes K[G]$ such that if

$$\widehat{\mu}(f) = \sum_i f_i \otimes g_i,$$

then $\sigma \cdot f = \sum_i f_i g_i(\sigma)$ (see Definition A.1.7). We define a homomorphism $\Delta^* : R \otimes R \to R$ by $\Delta^*(f \otimes h) = fh$.

Theorem 4.2.6. *Suppose that $\mathfrak{a} \subset R \otimes R$ is an ideal with the following properties:*

(1) $\mathfrak{a}$ is G-stable (G acts trivially on the first factor of $R \otimes R$, and as usual on the second);
(2) for every $f \in R^G$ we have $f \otimes 1 - 1 \otimes f \in \mathfrak{a}$;
(3) $\Delta^*(\mathfrak{a}) = \{0\}$, i.e., $\mathfrak{a}$ lies in the kernel of Δ^*.

Assume that J is a G-stable ideal of R. Then

$$(R \otimes J + \mathfrak{a}) \cap R \otimes 1 = J' \otimes 1,$$

where J' is the ideal in R generated by $J \cap R^G$.

Proof. "$\supseteq$": Clearly, if $f \in J \cap R^G$, then

$$f \otimes 1 = (f \otimes 1 - 1 \otimes f) + 1 \otimes f$$

and $f \otimes 1 - 1 \otimes f \in \mathfrak{a}$ and $1 \otimes f \in R \otimes J$.
    "$\subseteq$": Suppose that $f \otimes 1$ lies in $R \otimes J + \mathfrak{a}$, say

$$f \otimes 1 = \sum_i f_i \otimes h_i + a \tag{4.2.5}$$

with $f_i \in R$, $h_i \in J$ for all i and $a \in \mathfrak{a}$. We apply $\mathrm{id} \otimes \mathcal{R} : R \otimes R \to R \otimes R^G$ to (4.2.5), where $\mathcal{R}$ is the Reynolds operator and id is the identity:

$$f \otimes 1 = \sum_i f_i \otimes \mathcal{R}(h_i) + (\mathrm{id} \otimes \mathcal{R})(a) \tag{4.2.6}$$

Let us apply Δ^* to (4.2.6):

$$f = \sum_i f_i \mathcal{R}(h_i).$$

Notice that $\Delta^*((\mathrm{id} \otimes \mathcal{R})(a)) = 0$ because $\mathfrak{a}$ is G-stable, so $(\mathrm{id} \otimes \mathcal{R})(a) \in \mathfrak{a}$. Now for every i, $\mathcal{R}(h_i) \in R^G \cap J$ because J is G-stable. This proves that $f \in J'$. $\square$

Suppose that $R = \bigoplus_{d=0}^{\infty} R_d$ is graded and $R_0 = K$. We assume that G acts homogeneously, i.e., on each component R_d.

Proposition 4.2.7. *Let I be the ideal generated by all homogeneous invariants of positive degree. If $I = (f_1, \ldots, f_r)$ with f_i homogeneous for all i, then $R^G = R[\mathcal{R}(f_1), \ldots, \mathcal{R}(f_r)]$.*

Proof. This proof is exactly the same as the proof of Proposition 4.1.1. $\square$

Using the construction of Lemma A.1.9, we can find a representation V of G and a G-equivariant surjective ring homomorphism $\varphi : K[V] \twoheadrightarrow R$. In fact, we may assume that $K[V]$ is graded, i.e., $K[V] = \bigoplus_{d=0}^{\infty} K[V]_d$, φ is surjective, and $V^* \subset K[V]$ is spanned by homogeneous functions, say spanned by $x_1, \ldots, x_n \in V^*$ (possibly not all of the same degree). Let $I_R = (u_1, \ldots, u_p) \subset K[x_1, \ldots, x_n]$ be the kernel. This is a homogeneous ideal.

We write $K[G] \cong K[z_1, \ldots, z_l]/I(G)$ where $I(G) = (h_1, \ldots, h_t)$. As usual, the action of G on V is given by

$$\rho(g) = \begin{pmatrix} a_{1,1}(g) & a_{1,2}(g) & \cdots & a_{1,n}(g) \\ a_{2,1}(g) & a_{2,2}(g) & \cdots & a_{2,n}(g) \\ \vdots & \vdots & \ddots & \vdots \\ a_{n,1}(g) & a_{n,2}(g) & \cdots & a_{n,n}(g) \end{pmatrix}, \qquad g \in G,$$

with $a_{i,j} \in K[z_1, \ldots, z_l]$ for all i, j.

The following algorithm is an obvious generalization of Algorithm 4.1.9.

Algorithm 4.2.8. (1) Input: ideal generators $h_1, \ldots, h_t \in K[z_1, \ldots, z_l]$ such that the ideal $(h_1, \ldots, h_t)$ defines the group G; a matrix $A = (a_{i,j})_{i,j=1}^{n}$ with $a_{i,j} \in K[z_1, \ldots, z_l]$ defining the representation of G on V; $u_1, \ldots, u_p \in K[x_1, \ldots, x_n]$ defining the homogeneous ideal I_R.

(2) Choose a monomial ordering "$>$" on $K[z_1, \ldots, z_l, x_1, \ldots, x_n, y_1, \ldots, y_n]$ such that $z_i \gg x_j$ and $z_i \gg y_j$ for $i = 1, \ldots, l$, $j = 1, \ldots, n$ (see on page 9). Compute a Gröbner basis $\mathcal{G}$ of the ideal generated by

$$h_1(z), \ldots, h_t(z), u_1(x), \ldots, u_p(x), u_1(y), \ldots, u_p(y), \quad \text{and}$$

$$\{y_i - \sum_{j=1}^{n} a_{j,i}(z)x_j \mid i = 1, \ldots, n\}.$$

(3) $\mathcal{B} = \mathcal{G} \cap K[x_1, \ldots, x_n, y_1, \ldots, y_n]$.

(4) $\mathcal{I} = \{f(x, 0) \mid f \in \mathcal{B}\}$.

(5) Output: $\{\mathcal{R}(f) \mid f \in \mathcal{I}\}$. These are modulo I_R generators of $(K[V]/I_R)^G$.

Proof. We prove the correctness of the algorithm. Let $X \subseteq V$ be the zero set of $u_1, \ldots, u_p$, so $R \cong K[X]$. The ideal in Step (2) is equal to

$$I(\Gamma) \subseteq K[z_1, \ldots, z_l, x_1, \ldots, x_n, y_1, \ldots, y_n],$$

where $\Gamma = \{(\sigma, x, \sigma \cdot x) \mid \sigma \in G, \ x \in X\}$. Let us write $\mathfrak{b} = I(\Gamma) \cap K[x,y]$. We have

$$R \otimes R \cong K[x,y]/(u_1(x), \ldots, u_p(x), u_1(y), \ldots, u_p(y)).$$

Let $\mathfrak{a} \in R \otimes R$ be the image of the ideal $\mathfrak{b}$. We prove that $\mathfrak{a}$ has the desired properties of Theorem 4.2.6. First of all, if $f \in R^G$, then there exists an $\widetilde{f} \in K[V]^G$ whose image modulo I_R is equal to f. It is easy to see that $\widetilde{f}(x) - \widetilde{f}(y) \in \mathfrak{b}$, so $f \otimes 1 - 1 \otimes f \in \mathfrak{a}$. Also note that if $\mathfrak{m}_e \subset K[z_1, \ldots, z_l]$ is the maximal ideal corresponding to the identity element, then

$$\mathfrak{m}_e + I(\Gamma) =$$
$$\mathfrak{m}_e + (u_1(x), \ldots, u_p(x), u_1(y), \ldots, u_p(y), x_1 - y_1, x_2 - y_2, \ldots, x_n - y_n).$$

This shows that

$$\mathfrak{b} \subseteq (u_1(x), \ldots, u_p(x), u_1(y), \ldots, u_p(y), x_1 - y_1, x_2 - y_2, \ldots, x_n - y_n)$$

and $\mathfrak{a}$ lies in the kernel of Δ^*. In Step (3) we obtain a Gröbner basis $\mathcal{B}$ of $\mathfrak{b}$. The correctness of the algorithm now follows from Theorem 4.2.6 and Proposition 4.2.7. $\qquad\square$

4.2.3 Covariants

Invariants and semi-invariants are special cases of so-called covariants. In this section we will study covariants, and how to compute them.

Definition 4.2.9. *Suppose that X is a G-variety and W is a G-module. A* **covariant** *(sometimes also called an* **equivariant***) of X with values in W is a G-equivariant morphism $\varphi : X \to W$. For fixed X and W, we will denote the set of covariants with values in W by $\mathrm{Mor}(X, W)^G$.*

If $W = K$ is the trivial representation, then $\mathrm{Mor}(X, K)^G = K[X]^G$ is the ring of invariants. In general $\mathrm{Mor}(X, K)^G$ carries the structure of a vector space. Also, if $f \in K[X]^G$, and $\varphi : X \to W$ is a covariant, then $f\varphi$ (scalar multiplication on W) is also a covariant. In this way $\mathrm{Mor}(X, W)^G$ is a $K[X]^G$-module. We will call $\mathrm{Mor}(X, W)^G$ the **module of covariants** with values in W.

Theorem 4.2.10. *For a linearly reductive group G and an affine G-variety, the module of covariants is finitely generated.*

Proof. See Popov and Vinberg [194, Theorem 3.24]). $\qquad\square$

A covariant $\varphi : X \to W$ naturally induces a ring homomorphism $\varphi^* : K[W] \to K[X]$. Since $K[W]$ is the symmetric algebra on W^*, we obtain $\varphi^* : S(W^*) \to K[X]$. This homomorphism is uniquely determined by its restriction to the linear part W^* and every linear map $\varphi^* : W^* \to K[X]$ induces a unique $\varphi^* : S(W^*) \to K[X]$. In this way, we can identify $\mathrm{Mor}(X, W)^G$ with the space $\mathrm{Hom}(W^*, K[X])^G$ of G-equivariant linear maps from W^* to $K[X]$. In other words

$$\mathrm{Mor}(X, W)^G \cong (W \otimes K[X])^G. \tag{4.2.7}$$

Let us write $K[W^*] \cong K[y_1, \ldots, y_m]$. Then $K[X \times W^*] \cong K[X][y_1, \ldots, y_m]$. Let $\mathfrak{m}_y$ be the ideal in $K[X \times W^*]$ generated by $y_1, \ldots, y_m$. We have a grading on $K[X \times W^*]$ given by the degree in the y-variables. The degree 0 part is equal to $K[X \times W^*]_0 = K[X]$, and the degree 1 part is equal to $K[X] \otimes W$. So we have an isomorphism of $K[X]$-modules

$$K[X \times W^*]/\mathfrak{m}_y^2 \cong K[X] \oplus K[X] \otimes W.$$

Taking invariants, we get

$$(K[X \times W^*]/\mathfrak{m}_y^2)^G \cong K[X]^G \oplus (K[X] \otimes W)^G,$$

and, as remarked before, $(K[X] \otimes W)^G$ is the module of covariants with values in W. Suppose $f_1, \ldots, f_r, u_1, \ldots, u_s$ are homogeneous generators of $(K[X \times W^*]/\mathfrak{m}_y^2)^G$. We assume that $f_1, \ldots, f_r$ have degree 0, and $u_1, \ldots, u_s$ have degree 1. Then $f_1, \ldots, f_r$ must be generators of $K[X]^G$. The polynomials $u_1, \ldots, u_s$ are $K[X]^G$-module generators of the module of covariants $(K[X] \otimes W)^G$.

Suppose V is a representation of G. The previous discussion shows that in order to compute the module of *covariants* with values in W, we need to compute the ring of *invariants* of $K[V \times W^*]/\mathfrak{m}_y^2$. This can be done by using Algorithm 4.2.8.

Let X be an affine G-variety with G linearly reductive. Let $B \subset G$ be a Borel subgroup with $B = T \ltimes U$ where T is a maximal torus and U is the maximal unipotent subgroup in B. The **ring of covariants** is defined as $K[X]^U$. The relation to modules of covariants is the following. Suppose that V_λ is an irreducible representation with highest weight λ (see Theorem A.5.1). Let $v_\lambda \in V_\lambda$ be the highest weight vector. A covariant $\varphi \in \mathrm{Hom}(V_\lambda, K[X])^G$ is uniquely determined by $\varphi(v_\lambda) \in K[X]^U$ because the orbit $G \cdot v_\lambda$ spans the vector space V_λ.

The ring of covariants is finitely generated (see Khadzhiev [149] and Grosshans [100]). Let U act on G by multiplication on the right. The quotient G/U is a quasi-affine variety. We will study $K[G]^U = K[G/U]$. Let V_λ be an irreducible representation with λ a dominant weight (see Theorem A.5.1). From (4.2.7) with $X = G/U$ follows that

$$(K[G]^U \otimes V_\lambda)^G = \mathrm{Mor}(G/U, V_\lambda)^G. \tag{4.2.8}$$

A G-equivariant morphism $\varphi : G/U \to V_\lambda$ is determined by $\varphi(eU) \in V_\lambda^U$, where $e \in G$ is the unit element. The space V_λ^U is 1-dimensional and spanned by a highest weight vector v_λ. It follows from (4.2.8) that the module of covariants $(K[G]^U \otimes V_\lambda)^G$ is 1-dimensional. In other words, every irreducible representation appears exactly once in $K[G]^U$. We can write

$$K[G]^U = K[G/U] = \bigoplus_{\lambda \in X(T)_+} V_\lambda,$$

where $X(T)_+$ is the set of dominant weights. The multiplicative structure of $K[G]^U$ can be understood as well. For dominant weights λ, μ, the λ-weightspace of V_λ and the μ-weightspace of V_μ are one dimensional. Now $V_\lambda \otimes V_\mu$ has maximal weight $\lambda + \mu$, and the $\lambda + \mu$-weight space of $V_\lambda \otimes V_\mu$ is one-dimensional. This shows that $V_\lambda \otimes V_\mu$ contains a unique copy of $V_{\lambda+\mu}$. Now there exists a unique non-zero projection (up to scalar multiplication)

$$V_\lambda \otimes V_\mu \to V_{\lambda+\mu},$$

and this projection defines the multiplication in $K[G]^U$.

Example 4.2.11. If $G = \mathrm{SL}_2$ and U is the set of all

$$\begin{pmatrix} 1 & a \\ 0 & 1 \end{pmatrix}, \quad a \in K.$$

The subgroup U is a maximal unipotent subgroup of SL_2. The coordinate ring of SL_2 is $K[z_{1,1}, z_{1,2}, z_{2,1}, z_{2,2}]/(z_{1,1}z_{2,2} - z_{1,2}z_{2,1} - 1)$. We let U act on SL_2 by right multiplication. It is easy to check that $K[\mathrm{SL}_2]^U = K[z_{1,1}, z_{1,2}]$ which is SL_2-equivariantly isomorphic to $K[V_1]$ where V_1 is the space of binary forms of degree 1. Notice that $K[V_1] \cong \bigoplus_{d \geq 0} V_d$ as an SL_2-representation, where V_d is the space of binary forms of degree d. ◁

Let V be a rational representation of G. We consider $X_1 = G \times V$ and $X_2 = G \times V$ with an action of $G \times U$ on both X_1 and X_2 as follows

$$(\tau, \gamma) \cdot (\sigma, v) = (\tau\sigma\gamma^{-1}, \gamma v), \quad \tau \in G, \ \gamma \in U, \ (\sigma, v) \in X_1.$$

$$(\tau, \gamma) \cdot (\sigma, v) = (\tau\sigma\gamma^{-1}, \tau v), \quad \tau \in G, \ \gamma \in U, \ (\sigma, v) \in X_2.$$

The morphism $\psi : X_1 \to X_2$ defined by

$$(\sigma, v) \mapsto (\sigma, \sigma \cdot v)$$

defines a $G \times U$-equivariant isomorphism between X_1 and X_2.

Since $K[X_1]^{G \times U} = K[V]^U$ and $K[X_2]^{G \times U} = (K[G]^U \otimes K[V])^G$, we conclude

$$(K[G]^U \otimes K[V])^G \cong K[V]^U \tag{4.2.9}$$

(see also Popov and Vinberg [194, Lemma 3.10]). In fact, $K[G]^U$ can be graded in a natural way and using Algorithm 4.2.8 one can compute $K[V]^U$.

Example 4.2.12. Consider $G = \mathrm{SL}_2$ with maximal unipotent subgroup U (see Example 4.2.11). If V is any rational representation of SL_2, then it follows from (4.2.9) that the ring of covariants is equal to

$$K[V_1 \oplus V]^{\mathrm{SL}_2} \cong (K[\mathrm{SL}_2]^U \otimes K[V])^{\mathrm{SL}_2} \cong K[V]^U.$$

◁

4.3 Invariants of Tori

In this section we introduce a new algorithm for computing invariants of tori. Although Algorithm 4.1.9 is efficient for most examples, the algorithm here is simpler and more efficient. Where Algorithm 4.1.9 (and equivalently Algorithm 1.4.5 of Sturmfels [239]) uses a Gröbner basis computation, the algorithm here relies only on divisibility tests of two monomials. The computation of torus invariants is equivalent to an integer programming problem (see Sturmfels [239, Section 1.4]) and has therefore a large scope of applications. For example, the computation of SAGBI bases involves the solution of a great number of integer programming problems (see Robbiano and Sweedler [203]).

Suppose that $T = (K^*)^r$ is a torus acting diagonally on an n-dimensional vector space V. We can identify $K[V] \cong K[x_1, \ldots, x_n]$. If $\omega = (\omega^{(1)}, \ldots, \omega^{(r)}) \in \mathbb{Z}^r$ is a weight, then we write t^ω instead of $t_1^{\omega^{(1)}} \cdots t_r^{\omega^{(r)}}$. For $i = 1, \ldots, n$ let ω_i be the weight with which T acts on the variable x_i, i.e.,

$$t \cdot x_i = t^{\omega_i} x_i.$$

If $m = x_1^{a_1} x_2^{a_2} \cdots x_n^{a_n}$, then T acts on m with weight $a_1\omega_1 + \cdots + a_n\omega_n$. We also say that m *has* this weight. The idea of the algorithm is to choose a suitable finite set $\mathcal{C}$ of weights, and then produce sets S_ω, $\omega \in \mathcal{C}$ of monomials of weight ω. These sets grow during the course of the algorithm, until upon termination we have that S_0 generates $K[V]^G$ (this will be guaranteed by the choice of $\mathcal{C}$). We now present the algorithm.

Algorithm 4.3.1. This algorithm computes a minimal set of generating invariants of a torus.

(1) Input: Weights $\omega_1, \ldots, \omega_n \in \mathbb{Z}^r$ defining an action of $T = (K^*)^r$ on $K[x_1, \ldots, x_n]$.

(2) If T is one-dimensional (i.e., $r = 1$), then let $\mathcal{C} \subset \mathbb{Z}^r$ be the set of integral points in the convex hull of $\omega_1, \ldots, \omega_n$. If $r > 1$, then let $\mathcal{C} \subset \mathbb{Z}^r$ be the set of integral points in the convex hull of $2r\omega_1, \ldots, 2r\omega_n, -2r\omega_1, \ldots, -2r\omega_n$.

(3) Define $S_\omega := \emptyset$ for all $\omega \in \mathcal{C}$. Then put $S_{\omega_i} := S_{\omega_i} \cup \{x_i\}$ for $i = 1, 2, \ldots, n$. Finally put $U_\omega := S_\omega$ for all $\omega \in \mathcal{C}$.

(4) If $U_\omega = \emptyset$ for all $\omega \in \mathcal{C}$ then terminate the algorithm with output S_0. Otherwise, choose $\omega \in \mathcal{C}$ such that $U_\omega \neq \emptyset$ and choose $m \in U_\omega$.

(5) For i from 1 to n perform steps (6)–(7).

(6) Put $u := m x_i$ and let $\nu = \omega \cdot \omega_i$ be its weight.

(7) If $\nu \in \mathcal{C}$ and u is not divisible by any element of S_ν, then put $S_\nu := S_\nu \cup \{u\}$, $U_\nu := U_\nu \cup \{u\}$.

(8) Set $U_\omega := U_\omega \setminus \{m\}$. Go to Step 4.

Proof of Algorithm 4.3.1. We first show that the algorithm terminates. Note that a monomial m lies in the ideal generated by S_ω if and only if m is divisible by one of the elements of S_ω. Ideals are finitely generated. This means that after a certain number of steps in the algorithm, the sets S_ω, $\omega \in \mathcal{C}$ will not increase anymore. If the sets S_ω do not increase, then the sets U_ω will not increase either. At each step in the loop the sum of the cardinalities of U_ω, $\omega \in \mathcal{C}$ decreases by 1. After a finite number of steps, $U_\omega = \emptyset$ for all $\omega \in \mathcal{C}$ and the algorithm terminates.

We now prove that after termination of the algorithm, S_0 is is a generating set of invariants. Suppose that m is an invariant monomial of degree d and m is not divisible by a non-constant invariant monomial of smaller degree. We will show that m lies in S_0 after termination of the algorithm. By Lemma 4.3.2 below we can write

$$m = x_{i_1} x_{i_2} \cdots x_{i_d}$$

with $i_1, \ldots, i_d \in \{1, 2, \ldots, n\}$ such that for every j, the weight ν_j of

$$m_j := x_{i_1} x_{i_2} \cdots x_{i_j} \tag{4.3.1}$$

lies in $\mathcal{C}$. The monomial $m_1 = x_{i_1}$ lies in S_{ν_1} right from the start. We will show that as the algorithm runs, m_2 is added to S_{ν_2}, m_3 is added to S_{ν_3}, etc., until finally $m = m_d$ is added to $S_{\nu_d} = S_0$. Suppose at a certain step, m_j is added to S_{ν_j}. Then m_j is also added to U_{ν_j} and after a few more steps $u = m_{j+1} = m_j x_{j+1}$ will be tested in Step 7. Suppose that m_{j+1} is divisible by another monomial of $S_{\nu_{j+1}}$, say $m_{j+1} = vw$ with $w \in S_{\nu_{j+1}}$ and v an invariant monomial. If v is constant then m_{j+1} was already added to $S_{\nu_{j+1}}$. Otherwise, m_{j+1} is divisible by the non-constant invariant monomial v and therefore m is divisible by v which is in contradiction with our assumptions on m. If m_{j+1} is not divisible by any monomial of $S_{\nu_{j+1}}$ then m_{j+1} is added to $S_{\nu_{j+1}}$. In this way we see that after the algorithm terminates, we have $m_j \in S_{\nu_j}$ for $j = 1, 2, \ldots, d$. In particular $m = m_d \in S_{\nu_d} = S_0$. This shows that S_0 is a set of generators of $K[V]^T$. In fact, since every monomial in S_0 is not divisible by any other monomial in S_0, S_0 consists of a *minimal* set of generators of $K[V]^T$. $\qquad\square$

Lemma 4.3.2. *Suppose that $\omega_1, \ldots, \omega_d \in \mathbb{R}^r$ such that $\sum_{i=1}^d \omega_i = 0$. If $r = 1$, let $\mathcal{C}$ be the convex hull of $\omega_1, \ldots, \omega_d$. If $r > 1$, let $\mathcal{C}$ be the convex hull of*

$$2r\omega_1, \ldots, 2r\omega_d, -2r\omega_1, \ldots, -2r\omega_d.$$

Then there exists a permutation σ of $\{1, \ldots, d\}$, such that for every $j \leq d$ we have

$$\omega_{\sigma(1)} + \cdots + \omega_{\sigma(j)} \in \mathcal{C}.$$

Proof. We first do the case $r = 1$. We can take $\sigma(1) = 1$. For $i > 1$ we can define $\sigma(i)$ as follows. If $\omega_1 + \cdots + \omega_{i-1} \geq 0$ then we can choose $\sigma(i) \in \{1, 2, \ldots, d\} \setminus \{\sigma(1), \ldots, \sigma(i-1)\}$ such that $\omega_{\sigma(i)} \leq 0$. If $\omega_1 + \cdots + \omega_{i-1} < 0$, then we can choose $\sigma(i) \in \{1, \ldots, d\} \setminus \{\sigma(1), \ldots, \sigma(i-1)\}$ such that $\omega_{\sigma(i)} > 0$. It is now clear that

$$\omega_{\sigma(1)} + \cdots + \omega_{\sigma(j)} \in \mathcal{C}$$

for all j.

Let us now treat the case $r > 1$. We can define a norm $\|\cdot\|$ on $\mathbb{R}^r$ by

$$\|\omega\| = \inf\{|\lambda|^{-1} \mid 0 \neq \lambda \in \mathbb{R},$$
$$\lambda\omega \text{ lies in the convex hull of } \omega_1, \ldots, \omega_n, -\omega_1, \ldots, -\omega_n\}.$$

Then we have $\|\omega_i\| \leq 1$ for all i. Moreover, $\|\omega\| \leq 2r$ if and only if $\omega \in C$. A theorem of Bárány and Grinberg (see Beck and Sós [14, Corollary 4.9]) tells us that $\|\omega_i\| \leq 1$ and $\sum_{i=1}^n \omega_i = 0$ imply the existence of a permutation σ such that

$$\|\omega_{\sigma(1)} + \cdots + \omega_{\sigma(j)}\| \leq 2r$$

for all j. So we have

$$\omega_{\sigma(1)} + \cdots + \omega_{\sigma(j)} \in C$$

for all j. $\qquad\square$

Remark 4.3.3. Note that in the norm $\|\cdot\|$ in the above proof is constructed in such a way that $\|\omega_i\| \leq 1$ and the set $\{\omega \in \mathbb{R}^r \mid \|\omega\| \leq 2r\}$ is as small as possible. This means that the set C in Algorithm 4.3.1 is the smallest possible set that allows the application of the theorem of Bárány and Grinberg for the proof of correctness. $\qquad\triangleleft$

Example 4.3.4. Suppose that a one-dimensional torus $T = \mathbb{G}_m$ acts diagonally on a 4-dimensional vector space V. Then $K[V] \cong K[x_1, x_2, x_3, x_4]$. Suppose that the weights of the monomial x_1, x_2, x_3, x_4 are $-3, -1, 1, 2$ respectively. Then $C = \{-3, -2, -1, 0, 1, 2\}$. The course of the algorithm can be summarized in the following table.

	S_{-3}	S_{-2}	S_{-1}	S_0	S_1	S_2
1	x_1		x_2		x_3	x_4
2		$x_1 x_3$ x_2^2	$x_1 x_4$	$x_2 x_3$	$x_2 x_4$	x_3^2
3	x_2^3	$x_1 x_2 x_4$	$x_1 x_3^2$	$x_2^2 x_4$ $x_1 x_3 x_4$	$x_1 x_4^2$	
4		$x_1^2 x_4^2$		$x_1 x_2 x_4^2$ $x_1 x_3^3$		
5				$x_1^2 x_4^3$		

At the start, we take $S_{-3} = \{x_1\}$, $S_{-1} = \{x_2\}$, $S_1 = \{x_3\}$, $S_2 = \{x_4\}$. Then we look at monomials of degree 2, i.e., we multiply elements of the S_i's with x_1, x_2, x_3 and x_4. Monomials like x_1^2, $x_1 x_2$, x_2^2, x_4^2, $x_3 x_4$ have weights which do not lie in C. The other monomials $x_1 x_3$, $x_1 x_4$, x_2^2, $x_2 x_3$, $x_2 x_4$, x_3^2 are added to the appropriate S_i's and U_i's. If we take now for example $x_2^2 \in U_{-2}$ and multiply it with x_3, then $x_3 x_2^2$ is divisible by $x_2 \in S_{-1}$. Therefore $x_3 x_2^2$ will not be added to S_{-1}. The algorithm continues in this way and after termination we get $S_0 = \{x_2 x_3, x_2^2 x_4, x_1 x_3 x_4, x_1 x_2 x_4^2, x_1 x_3^3, x_1^2 x_2^3\}$. We conclude

$$K[x_1, x_2, x_3, x_4]^T = K[x_2x_3, x_2^2x_4, x_1x_3x_4, x_1x_2x_4^2, x_1x_2^3, x_1^2x_2^3].$$

Remark 4.3.5. The algorithm can also be used for invariants of any finite abelian linearly reductive groups G. One may again assume that G acts diagonally. For $\mathcal{C}$ one has to take the set of *all* irreducible characters of G.

4.4 Invariants of SL_n and GL_n

In this section we briefly discuss some classical results on the invariant theory of SL_n and GL_n. The theorems of this section can all be found in §9.3 and §9.4 of Popov and Vinberg [194]. Our base field K is assumed to be of characteristic 0. Suppose that V is an n-dimensional vector space. Let V^* be the dual space of V and let $\langle \cdot, \cdot \rangle : V \times V^* \to K$ be the canonical pairing. We will study invariants of the representation $V^r \oplus (V^*)^s$. For each $i \leq r$ and each $j \leq s$ we have an invariant

$$V^r \oplus (V^*)^s \ni (v_1, \ldots, v_r, w_1, \ldots, w_s) \mapsto \langle v_i, w_j \rangle,$$

which we symbolically denote by $\langle i, j \rangle$.

Theorem 4.4.1 (First Fundamental Theorem for GL_n). *The invariant ring*

$$K[V^r \oplus (V^*)^s]^{\mathrm{GL}_n}$$

is generated by all $\langle i, j \rangle$.

Remark 4.4.2. Suppose that W and Z are finite dimensional vector spaces. Let

$$\pi : \mathrm{Hom}(W, V) \times \mathrm{Hom}(V, Z) \to \mathrm{Hom}(W, Z)$$

be the composition map. Let $Y \subset \mathrm{Hom}(W, Z)$ be the image of π. It is easy to see that Y is the set of all $A \in \mathrm{Hom}(W, Z)$ of rank at most $\min(n, \dim(W), \dim(Z))$ with $n = \dim(V)$ (in particular, Y is Zariski-closed). Note that $\mathrm{Hom}(W, V)$ is isomorphic to V^r and $\mathrm{Hom}(V, Z)$ is isomorphic to $(V^*)^s$ as representations of $\mathrm{GL}(V)$, where $r := \dim(W)$ and $s := \dim(Z)$. Now Theorem 4.4.1 shows that $\pi : \mathrm{Hom}(W, V) \times \mathrm{Hom}(V, Z) \to Y$ is the categorical quotient with respect to the action of $\mathrm{GL}(V)$.   ◁

In the previous remark, the ideal $I(Y) \subset K[\mathrm{Hom}(W, Z)]$ of Y is generated by all $(n + 1) \times (n + 1)$ minors. In the notation of Theorem 4.4.1 this means that we have the following description of the relations between the generating invariants (see also Popov and Vinberg [194, §9.4]).

Theorem 4.4.3 (Second Fundamental Theorem for GL_n). *All polynomial relations between the invariants $\langle i, j \rangle$ are generated by*

$$\det \begin{pmatrix} \langle i_1, j_1 \rangle & \langle i_1, j_2 \rangle & \cdots & \langle i_1, j_{n+1} \rangle \\ \langle i_2, j_1 \rangle & \langle i_2, j_2 \rangle & & \langle i_2, j_{n+1} \rangle \\ \vdots & & \ddots & \vdots \\ \langle i_{n+1}, j_1 \rangle & \langle i_{n+1}, j_2 \rangle & \cdots & \langle i_{n+1}, j_{n+1} \rangle \end{pmatrix}$$

with $1 \le i_1 < i_2 < \cdots < i_{n+1} \le r$ and $1 \le j_1 < j_2 < \cdots < j_{n+1} \le s$.

Let us describe invariants for $\mathrm{SL}(V)$. Besides the $\mathrm{GL}(V)$ invariants which we already found, we also have invariant determinants. If $1 \le i_1 < i_2 < \cdots < i_n \le r$, we have a **bracket invariant**

$$V^r \oplus (V^*)^s \ni (v_1, \ldots, v_r, w_1, \ldots, w_s) \mapsto \det(v_{i_1}\ v_{i_2}\ \cdots\ v_{i_n}),$$

which will be denoted by $[i_1 i_2 \cdots i_n]$. Similarly for $1 \le j_1 < j_2 < \cdots < j_n \le s$ we have an invariant

$$V^r \oplus (V^*)^s \ni (v_1, \ldots, v_r, w_1, \ldots, w_s) \mapsto \det(w_{j_1}\ w_{j_2}\ \cdots\ w_{j_n}),$$

which will be denoted by $|j_1 j_2 \ldots, j_n|$.

Theorem 4.4.4 (First Fundamental Theorem for SL_n). *The invariant ring*

$$K[V^r \oplus (V^*)^s]^{\mathrm{SL}_n}$$

is generated by all $\langle i, j \rangle$ *($1 \le i \le r$, $1 \le j \le s$), all $[i_1 i_2 \cdots i_n]$ ($1 \le i_1 < i_2 < \cdots < i_n \le r$) and all $|j_1 j_2 \cdots j_n|$ ($1 \le j_1 < j_2 < \cdots < j_n \le s$).*

For a description of the relations between all these invariants, see for example Popov and Vinberg [194, §9.4]. Let us just explain the case where $s = 0$. In that case the invariant ring $K[V^r]^{\mathrm{SL}_n}$ is generated by all $[i_1 i_2 \cdots i_n]$ with $1 \le i_1 < i_2 < \cdots < i_n \le r$.

Theorem 4.4.5 (Second Fundamental Theorem for SL_n). *The ideal of relations between the generating invariants in $K[V^r]^{\mathrm{SL}_n}$ is generated by*

$$\sum_{k=1}^{n+1} (-1)^{k-1} [i_1 i_2 \cdots i_{n-1} j_k][j_1 j_2 \cdots \widehat{j_k} \cdots j_{n+1}]$$

with $1 \le i_1 < i_2 < \cdots < i_{n-1} \le r$ and $1 \le j_1 < j_2 < \cdots < j_{n+1} \le r$.

The relations given in Theorem 4.4.5 are called **Grassmann-Plücker relations**. Similar descriptions for generating invariants and their relations exist for other classical groups as well (see Popov and Vinberg [194, §9.4]).

Example 4.4.6. If V is 2-dimensional, then $K[V^4]^{\mathrm{SL}_2}$ is generated by [1 2], [1 3], [1 4], [2 3], [2 4] and [3 4]. There is one Plücker relation, namely

$$[1\ 2][3\ 4] - [1\ 3][2\ 4] + [1\ 4][2\ 3].$$

We refer to Sturmfels [239, Chapter 3] for more details. The invariant ring $K[V^r]^{\mathrm{SL}_n}$ is generated by all brackets, and sometimes it is called the bracket ring. To present the invariant ring $K[V^r]^{\mathrm{SL}_n}$, one can start with the polynomial ring on all brackets, and then divide out the ideal $I_{r,n}$ of all relations as in Theorem 4.4.5. In Sturmfels [239, Chapter 3] more general generators of $I_{r,n}$ are described. In fact, a Gröbner basis of $I_{r,n}$ is given. The so-called straightening algorithm of Young can be related to finding the normal form with respect to this Gröbner basis (see Sturmfels [239], Sturmfels and White [240], Young [263] and Hodge and Pedoe [114]).

4.4.1 Binary Forms

Let us now study the invariants for SL_2. The base field K will be algebraically closed and of characteristic 0. We will give an introduction to the so called symbolic method. Denote the space of binary forms of degree d by V_d, so

$$V_d = \{a_0 x^d + a_1 x^{d-1} y + \cdots + a_d y^d \mid a_0, \ldots, a_d \in K\}.$$

The action of SL_2 on V_d is given by

$$\begin{pmatrix} a & b \\ c & d \end{pmatrix} \cdot p(x, y) = p(ax + cy, bx + dy)$$

for every $p(x, y) \in V_d$. The coordinate ring $K[V_d]$ can be identified with $K[a_0, \ldots, a_d]$. Let us put $V = V_1$. Notice that we have a surjective map $\pi : V^d \to V_d = S^d(V)$ defined by

$$(f_1, f_2, \ldots, f_d) \mapsto f_1 f_2 \cdots f_d.$$

Let $T = \mathbb{G}_m{}^{d-1}$, the $(d-1)$-dimensional torus. We can define an action of T on V^d by

$$(\sigma_1, \sigma_2, \ldots, \sigma_{d-1}) \cdot (f_1, \ldots, f_d) = (\sigma_1 f_1, \sigma_1^{-1} \sigma_2 f_2, \ldots, \sigma_{d-2}^{-1} \sigma_{d-1} f_{d-1}, \sigma_{d-1}^{-1} f_d).$$

The symmetric group S_d also acts on V^d by permuting the factors. Combining the two actions gives an action of the semidirect product $S_d \ltimes T$ on V^d. For a non-zero $f \in V_d$, the fiber $\pi^{-1}(f)$ is exactly the set of all possible factorizations of f into d linear forms, so the fiber $\pi^{-1}(f)$ is exactly one $S_d \ltimes T$-orbit. It follows that the dual homomorphism $\pi^* : K[V_d] \hookrightarrow K[V^d]$ induces an isomorphism $K[V_d] \cong K[V^d]^{S_d \ltimes T}$. Moreover, we have

$$K[V_d]^{\mathrm{SL}_2} \cong (K[V^d]^{S_d \ltimes T})^{\mathrm{SL}_2} = (K[V^d]^{\mathrm{SL}_2})^{S_d \ltimes T} = ((K[V^d]^{\mathrm{SL}_2})^T)^{S_d}.$$

This is very nice because we have a good description of $K[V^d]^{\mathrm{SL}_2}$: it is generated by all brackets $[i\ j]$ with $1 \leq i < j \leq d$ and the relations are all Plücker relations

$$[i\ j][k\ l] - [i\ k][j\ l] + [i\ l][j\ k]$$

with $i < j < k < l$ by Theorem 4.4.4 and Theorem 4.4.5. We would like to find all $S_d \ltimes T$-invariant bracket-polynomials. Let us first describe the T-invariant bracket polynomials. Every bracket monomial is a semi-invariant for T. It is easy to see that a bracket monomial is invariant if and only if every integer between 1 and d appears the same number of times. We will call such a monomial *regular*. A bracket polynomial is called regular if all of its monomials are regular.

Example 4.4.7. For $d = 3$ we have that

$$[1\ 2][1\ 3][2\ 3]$$

is T-invariant, because 1,2 and 3 all appear twice.   ◁

From the previous considerations it follows that $K[V^d]^{SL_2 \times T}$ is the set of all regular bracket polynomials. Notice that if $i > j$, we can also define $[i\ j]$ by $[i\ j] = -[j\ i]$. The action of S_d on $K[V^d]^{SL_2}$ is given by $\sigma([i\ j]) = [\sigma(i)\ \sigma(j)]$. We call a bracket polynomial symmetric if it is invariant under S_d. The invariant ring $K[V_d]^{SL_2}$ can be identified with all *symmetric* regular bracket polynomials.

Example 4.4.8. For $d = 3$ the bracket monomial

$$[1\ 2][1\ 3][2\ 3]$$

is *not* symmetric, because if we interchange 1 and 2, we get

$$[2\ 1][2\ 3][1\ 3] = -[1\ 2][1\ 3][2\ 3].$$

However,
$$([1\ 2][1\ 3][2\ 3])^2$$

is symmetric and regular. This polynomial corresponds to the discriminant in $K[V_3]^{SL_2}$. In fact, for any d we have a symmetric bracket monomial

$$([1\ 2][1\ 3] \cdots [1\ d][2\ 3][2\ 4] \cdots [2\ d] \cdots [d-1\ d])^2,$$

which corresponds to the discriminant in $K[V_d]^{SL_2}$.   ◁

To obtain invariants for the binary forms of degree d, one can just take a regular bracket monomial and symmetrize it over the symmetric group S_d. However, the invariant obtained in this way can be zero.

Example 4.4.9. For $d = 4$, we can take $[1\ 2]^2[3\ 4]^2$ and symmetrize it to

$$[1\ 2]^2[3\ 4]^2 + [1\ 3]^2[2\ 4]^2 + [1\ 4]^2[2\ 3]^2,$$

which is a non-zero invariant. Another invariant can be obtained by symmetrizing $[1\ 2]^2[3\ 4]^2[1\ 3][2\ 4]$. Both invariants together generate $K[V_4]^{SL_2}$.
◁

The method explained above generalizes to simultaneous invariants of several binary forms. The ring of covariants on binary forms of degree d is isomorphic to $K[V_d \oplus V]^{\mathrm{SL}_2}$. We have

$$K[V_d \oplus V]^{\mathrm{SL}_2} \cong (K[V^d \oplus V]^{\mathrm{SL}_2})^{S_d \ltimes T}.$$

To describe covariants, we add another symbol, say $\mathbf{u}$, to our alphabet $\{1, 2, \ldots, d\}$ corresponding to the extra copy of V. Covariants are now polynomials in the brackets $[i\ j]$, $1 \le i < j \le d$ and the brackets $[i\ \mathbf{u}]$ with $1 \le i \le r$. Again, a bracket monomial is called regular if every integer between 1 and d appears the same number of times, and a bracket polynomial is called regular if all its bracket monomials are regular. The set of covariants are exactly all symmetric regular bracket polynomials.

Example 4.4.10. The **Hessian** of a binary form f of degree d is defined by

$$H(f) := \frac{\partial^2 f}{(\partial x)^2} \frac{\partial^2 f}{(\partial y)^2} - \left(\frac{\partial^2 f}{\partial x \partial y} \right)^2.$$

The corresponding bracket polynomial is the symmetrization of

$$([1\ 2][3\ \mathbf{u}][4\ \mathbf{u}] \cdots [d\ \mathbf{u}])^2.$$

$\lhd$

4.5 The Reynolds Operator

In this section we will study the Reynolds operator.

Example 4.5.1. For finite groups G we have seen that the Reynolds operator $\mathcal{R} : K[X] \to K[X]^G$ is just averaging over the group (see Example 2.2.3 and 3.1.2).

$$\mathcal{R}(f) = \frac{1}{|G|} \sum_{\sigma \in G} \sigma \cdot f.$$

$\lhd$

Example 4.5.2. Suppose that $G = T$ is a torus. If $\dim T = 1$ the Reynolds operator was studied in Example 2.2.4. This easily generalizes to an r-dimensional torus T. Let X be an affine T-variety. The coordinate ring $K[T]$ can be identified with the ring of Laurent polynomials in r variables $K[z_1, \ldots, z_r, z_1^{-1}, \ldots, z_r^{-1}]$. The action $\mu : T \times X \to X$ corresponds to a ring homomorphism $\mu^* : K[X] \to K[T] \otimes K[X]$. If $f \in K[X]$, then $\mu^*(f)$ is a Laurent polynomial in $z_1, \ldots, z_r$ with coefficients in $K[X]$. Now $\mathcal{R}(f)$ is the coefficient of $z_1^0 z_2^0 \cdots z_r^0$ in $\mu^*(f)$.

$\lhd$

Suppose that G is an arbitrary linearly reductive group. Let G° be the connected component of the identity and let $[G^\circ, G^\circ]$ be the commutator subgroup of G°. Then G/G° is a finite group, $G^\circ/[G^\circ, G^\circ]$ is a torus and $[G^\circ, G^\circ]$ is semi-simple. Because of the following lemma, computing the Reynolds operator for G can be reduced to computing the Reynolds operator for the connected semi-simple group $[G^\circ, G^\circ]$.

Lemma 4.5.3. *Suppose that G has a normal subgroup N and that X is an affine G-variety. Let $\mathcal{R}_G : K[X] \to K[X]^G$ be the Reynolds operator with respect to G, $\mathcal{R}_N : K[X] \to K[X]^N$ the Reynolds operator with respect to N and $\mathcal{R}_{G/N} : K[X]^N \to K[X]^G$ be the Reynolds operator with respect to G/N. Then we have $\mathcal{R}_G = \mathcal{R}_{G/N}\mathcal{R}_N$.*

Proof. This is clear from the definition of the Reynolds operator (Definition 2.2.2). $\qquad\qquad\square$

4.5.1 The Dual Space $K[G]^*$

Let $K[G]^*$ be the dual space to $K[G]$. It is very natural to study $K[G]^*$ because it contains all important operators, such as the Reynolds operator and the Casimir operator. We will give a brief overview of the main properties of $K[G]^*$ in this section. For details and proofs, see Appendix A.2. First of all, to every $\sigma \in G$ we can associate a linear map $\epsilon_\sigma : K[G] \to K$ defined by

$$f \mapsto f(\sigma).$$

An important object we need is the **Lie algebra** $\mathfrak{g}$ of G. We define the Lie algebra as the set of all point derivations at the identity element $e \in G$ (see Definition A.2.4). In particular $\mathfrak{g}$ is contained in $K[G]^*$.

The space $K[G]^*$ has the structure of an associative algebra. Let $m : G \times G \to G$ be the multiplication map and let $m^* : K[G] \to K[G] \otimes K[G]$ be the dual homomorphism. For every $\delta, \gamma \in K[G]^*$, the **convolution** $\delta * \gamma$ is defined as the composition

$$(\delta \otimes \gamma) \circ m^* : K[G] \to K[G] \otimes K[G] \to K.$$

One can show that $*$ defines an associative K-bilinear multiplication (see Proposition A.2.2). The unit element of $K[G]^*$ is $\epsilon := \epsilon_e$ where $e \in G$ is the identity.

Remark 4.5.4. For any $\delta, \gamma \in \mathfrak{g}$, we have

$$[\delta, \gamma] = \delta * \gamma - \gamma * \delta \in \mathfrak{g},$$

and this defines the Lie algebra structure on $\mathfrak{g}$ (see Proposition A.2.5).

Example 4.5.5. Let $G = \mathrm{GL}_n$ and we denote its Lie algebra by $\mathfrak{gl}_n$. For every $n \times n$ matrix A we can define $\partial_A \in \mathfrak{gl}_n$, the derivative in direction A at the identity matrix I by

$$\partial_A f := \frac{d}{dt} f(I + tA)\Big|_{t=0}$$

for all $f \in K[\mathrm{GL}_n]$. Let $z_{i,j}$, $1 \le i,j \le n$, be the coordinate functions and let Z be the matrix $(z_{i,j})_{i,j=1,\dots,n}$. We write $\partial_A Z$ for the matrix $(\partial_A(z_{i,j}))_{i,j=1,\dots,n}$. We have

$$\partial_A Z = \frac{d}{dt} I + tA\Big|_{t=0} = A.$$

By definition

$$(\partial_A * \partial_B)f = \frac{d}{dt}\frac{d}{ds} f((I + tA)(I + sB))\Big|_{t=s=0}$$

and

$$[\partial_A, \partial_B]f = \frac{d}{dt}\frac{d}{ds}\left(f((I + tA)(I + sB)) - f((I + sB)(I + tA))\right)\Big|_{t=s=0}.$$

Applying this to Z yields

$$[\partial_A, \partial_B]Z = \frac{d}{dt}\frac{d}{ds}\left((I + tA)(I + sB) - (I + sB)(I + tA)\right)\Big|_{t=s=0} =$$
$$AB - BA = [A, B] = \partial_{[A,B]}Z.$$

This shows that $\partial_{[A,B]} = [\partial_A, \partial_B]$. So the Lie algebra $\mathfrak{gl}_n$ is canonically isomorphic to the Lie algebra $\mathrm{Mat}_{n,n}(K)$. A basis of $\mathfrak{gl}_n$ is given by all $\partial_{i,j}$ with $1 \le i,j \le n$ where $\partial_{i,j}$ is defined by $\partial_{i,j} = \partial_{E_{i,j}}$ and $E_{i,j}$ is the matrix with a 1 at entry (i,j) and 0's everywhere else.   ◁

Remark 4.5.6. Suppose that $H \subset G$ is a Zariski closed subgroup of G. The Lie algebras of G and H are $\mathfrak{g}$ and $\mathfrak{h}$ respectively. Let I be the kernel of the surjective ring homomorphism $K[G] \to K[H]$. Clearly every point derivation of H at e can be seen as a point derivation on G. On the other hand, $\delta \in \mathfrak{g}$ is a point derivation on H if and only if $\delta(I) = \{0\}$. If I is generated by $f_1, \dots, f_r \in K[G]$, then this is equivalent to $\delta(f_1) = \delta(f_2) = \cdots = \delta(f_r) = 0$. ◁

Example 4.5.7. The subgroup $\mathrm{SL}_n \subset \mathrm{GL}_n$ is determined by $\det(Z) = 1$. We can view the Lie algebra $\mathfrak{sl}_n$ as a subalgebra of $\mathfrak{gl}_n$. Let $\partial_A \in \mathfrak{gl}_n$. By Remark 4.5.6, $\partial_A \in \mathfrak{sl}_n$ if and only if

$$0 = \partial_A(\det(Z) - 1) = \frac{d}{dt}\det(I + tA)\Big|_{t=0} =$$
$$\frac{d}{dt}(1 + t\,\mathrm{Tr}(A) + \cdots)\Big|_{t=0} = \mathrm{Tr}(A),$$

where $\mathrm{Tr}(A)$ is the trace of A. Thus $\mathfrak{sl}_n$ is the set of all $\partial_A \in \mathfrak{gl}_n$ with $\mathrm{Tr}(A) = 0$. A basis of $\mathfrak{sl}_n$ is given by all $\partial_{i,j}$ with $i \neq j$ and $\partial_{i,i} - \partial_{i+1,i+1}$ for $i = 1, \ldots, n - 1$. $\lhd$

Example 4.5.8. Let $G = \mathrm{O}_n$ be the orthogonal group and let $\mathfrak{o}_n$ be its Lie algebra. The group $\mathrm{O}_n \subset \mathrm{GL}_n$ is determined by the relation ${}^t Z Z = I$ where ${}^t Z$ is the transposed matrix of Z. We can view $\mathfrak{o}_n$ as a subspace of $\mathfrak{gl}_n$. We have

$$\partial_A({}^t Z Z - I) = \frac{d}{dt}(I + {}^t At)(I + At)\bigg|_{t=0} = A + {}^t A.$$

From Remark 4.5.6 it follows that $\mathfrak{o}_n \subset \mathfrak{gl}_n$ is the set of all ∂_A with $A + {}^t A = 0$. A basis of $\mathfrak{o}_n$ is given by all $\partial_{i,j} - \partial_{j,i}$ with $1 \leq i < j \leq n$. $\lhd$

Suppose that X is an affine G-variety. G acts rationally on $K[X]$, so this action extends to an action of $K[G]^*$ on $K[X]$. In particular, we have $\epsilon_\sigma \cdot f = \sigma \cdot f$ for all $\sigma \in G$ and $f \in K[X]$. If $\delta \in \mathfrak{g}$, then δ acts on $K[X]$ as a derivation.

Proposition 4.5.9. *Let* $\mathcal{R}_G : K[G] \to K$ *be the Reynolds operator. If* X *is an affine G-variety, then the Reynolds operator* $\mathcal{R}_X : K[X] \to K[X]^G$ *is given by* $f \mapsto \mathcal{R}_G \cdot f$.

Proof. We define $\mathcal{R}_X : K[X] \to K[X]^G$ by $f \mapsto \mathcal{R}_G \cdot f$ and prove that it satisfies the properties of a Reynolds operator as in Definition 2.2.2. Let $\mu^* : K[X] \to K[X] \otimes K[G]$ be as in Example A.2.12. (a) If $f \in K[X]^G$, then $\mu^*(f) = f \otimes 1$ and $\mathcal{R}_X(f) = f\mathcal{R}_G(1) = f$. (b) Since $\mathcal{R}_G : K[G] \to K$ is the Reynolds operator, we have $\mathcal{R}_G(\sigma \cdot g) = \mathcal{R}_G(g)$ for all $\sigma \in G$ and all $g \in K[G]$. In other words, $\mathcal{R}_G * \epsilon_\sigma = \mathcal{R}_G$. This shows that

$$\mathcal{R}_G \cdot (\sigma \cdot f) = \mathcal{R}_G \cdot (\epsilon_\sigma \cdot f) = (\mathcal{R}_G * \epsilon_\sigma) \cdot f = \mathcal{R}_G \cdot f.$$

$\square$

4.5.2 The Reynolds Operator for Semi-simple Groups

Suppose that G is connected and semi-simple with Lie algebra $\mathfrak{g}$, i.e., $[\mathfrak{g}, \mathfrak{g}] = \mathfrak{g}$. For $\delta \in \mathfrak{g}$, we define $\mathrm{ad}(\delta) : \mathfrak{g} \to \mathfrak{g}$ by $\gamma \mapsto [\delta, \gamma]$. The **Killing form** is a symmetric bilinear form $\kappa : \mathfrak{g} \times \mathfrak{g} \to K$ defined by

$$\kappa(\delta, \gamma) = \mathrm{Tr}(\mathrm{ad}(\delta)\,\mathrm{ad}(\gamma)),$$

where Tr is the trace. Because $\mathfrak{g}$ is semi-simple, it follows from Humphreys [117, 5.1] that the Killing form is non-degenerate.

Definition 4.5.10. *Let* $\delta_1, \ldots, \delta_m$ *be a basis of $\mathfrak{g}$ and let* $\gamma_1, \ldots, \gamma_m$ *be the dual basis of $\mathfrak{g}$ with respect to the bilinear form κ. The* **Casimir operator** c *is defined by*

$$c = \sum_{i=1}^{m} \delta_i * \gamma_i \in K[G]^*.$$

The Casimir operator does not depend on the choice of the basis $\delta_1, \ldots, \delta_m$. In fact, one can define the Casimir operator basis-independently as follows. The Killing form defines an element $\kappa \in S^2(\mathfrak{g})$. Because κ gives an isomorphism between $\mathfrak{g}$ and $\mathfrak{g}^*$, it also induces an isomorphism between $S^2(\mathfrak{g})$ and $S^2(\mathfrak{g}^*)$. So κ determines an element in $S^2(\mathfrak{g})$ which is the Casimir operator. Here $S^2(\mathfrak{g})$ is identified with the subspace of $K[G]^*$ spanned by all $\delta * \gamma + \gamma * \delta$ with $\delta, \gamma \in \mathfrak{g}$.

Lemma 4.5.11. *If $\sigma \in G$, then $\epsilon_\sigma * c = c * \epsilon_\sigma$. This means that for every rational representation V we have*

$$\sigma \cdot (c \cdot v) = c \cdot (\sigma \cdot v), \qquad v \in V.$$

Proof. Let $\delta_1, \ldots, \delta_m$ be a basis of $\mathfrak{g}$. If $\gamma \in \mathfrak{g}$, then the matrix of $\mathrm{ad}(\mathrm{Ad}(\sigma)\gamma)$ with respect to the basis $\mathrm{Ad}(\sigma)\delta_1, \ldots, \mathrm{Ad}(\sigma)\delta_m$ is the same as the matrix of $\mathrm{ad}(\gamma)$ with respect to $\delta_1, \ldots, \delta_m$. This shows that $\kappa(\mathrm{Ad}(\sigma)\gamma, \mathrm{Ad}(\sigma)\varphi) = \kappa(\gamma, \varphi)$ for all $\gamma, \varphi \in \mathfrak{g}$. Let $\gamma_1, \ldots, \gamma_m$ be the dual basis to $\delta_1, \ldots, \delta_m$. Then $\mathrm{Ad}(\sigma)\gamma_1, \ldots, \mathrm{Ad}(\sigma)\gamma_m$ is the dual basis of $\mathrm{Ad}(\sigma)\delta_1, \ldots, \mathrm{Ad}(\sigma)\delta_m$ and

$$c = \sum_{i=1}^m (\mathrm{Ad}(\sigma)\delta_i) * (\mathrm{Ad}(\sigma)\gamma_i) = \mathrm{Ad}(\sigma)\Big(\sum_{i=1}^m \delta_i * \gamma_i\Big) = \mathrm{Ad}(\sigma)c.$$

$\square$

Remark 4.5.12. As remarked in the proof, the Killing form κ is invariant under the group action of G. Since κ is non-degenerate, it defines an isomorphism between $\mathfrak{g}$ and $\mathfrak{g}^*$, the dual space. If $(\cdot, \cdot)$ is any other G-invariant bilinear form on $\mathfrak{g}$, then this determines an linear map $\mathfrak{g} \to \mathfrak{g}^* \cong \mathfrak{g}$ which commutes with the action of G. By Schur's Lemma, this must be a multiple of the identity. Consequently, $(\cdot, \cdot)$ is a scalar multiple of κ.   ◁

Example 4.5.13. Let $G = \mathrm{SL}_n$ with Lie algebra $\mathfrak{sl}_n$. We can define a G-invariant bilinear form $(\cdot, \cdot)$ on $\mathfrak{sl}_n$ by

$$(\partial_A, \partial_B) = \mathrm{Tr}(AB),$$

where Tr is the trace. By Remark 4.5.12 this must be a multiple of the Killing Form. It is not hard to check that $\kappa(\partial_{1,2}, \partial_{2,1}) = 2n$ and $(\partial_{1,2}, \partial_{2,1}) = 1$ so we conclude that

$$\kappa(\cdot, \cdot) = 2n(\cdot, \cdot).$$

A basis of $\mathfrak{sl}_n$ is given by $\partial_{i,j}$ for all $i \neq j$ and $\partial_{i,i} - \partial_{i+1,i+1}$ for $i = 0, \ldots, n-1$. A dual basis with respect to $(\cdot, \cdot)$ is $\partial_{j,i}$ for all $i \neq j$ and

$$\frac{n-i}{n}(\partial_{1,1} + \cdots + \partial_{i,i}) - \frac{i}{n}(\partial_{i+1,i+1} + \cdots + \partial_{n,n})$$

for $i = 1, \ldots, n-1$. We divide by $2n$ to get the dual basis with respect to κ. Now the Casimir operator is defined by

$$\frac{1}{2n}\Big(\sum_{i\neq j}\partial_{i,j}*\partial_{j,i}+\sum_{i=1}^{n-1}\Big(\frac{n-i}{n}\sum_{j=1}^{i}\partial_{j,j}*(\partial_{i,i}-\partial_{i+1,i+1})-$$

$$\frac{i}{n}\sum_{j=i+1}^{n-1}\partial_{j,j}*(\partial_{i,i}-\partial_{i+1,i+1}))\Big)=$$

$$=\frac{1}{2n}\sum_{i,j}\partial_{i,j}*\partial_{j,i}-\frac{1}{2n^2}\sum_{i,j}\partial_{i,i}*\partial_{j,j}. \qquad (4.5.1)$$

For our applications, we only need to know the Casimir operator up to a constant multiple, so we might as well work with the operator

$$\sum_{i,j}\partial_{i,j}*\partial_{j,i}-\frac{1}{n}\sum_{i,j}\partial_{i,i}*\partial_{j,j}.$$

◁

Example 4.5.14. Let $G=\mathrm{O}_n$ be the orthogonal group and let $\mathfrak{o}_n$ be its Lie algebra. A basis of $\mathfrak{o}_n$ is given by $\partial_{i,j}-\partial_{j,i}$ with $i<j$. The Killing form κ is a multiple of the bilinear form

$$(\partial_A,\partial_B)=\mathrm{Tr}(AB)$$

restricted to $\mathfrak{o}_n$. A basis of $\mathfrak{o}_n$ is given by $\partial_{i,j}-\partial_{j,i}$ with $i<j$. The dual basis is given by $\frac{1}{2}(\partial_{i,j}-\partial_{j,i})$ with $i<j$. The Casimir operator up to a scalar multiple is equal to

$$\sum_{i<j}(\partial_{i,j}-\partial_{j,i})*(\partial_{i,j}-\partial_{j,i})=\sum_{i\neq j}(\partial_{i,j}*\partial_{i,j}+\partial_{i,j}*\partial_{j,i}).$$

◁

Remark 4.5.15. For every rational representation V of G, the Casimir operator $c\in K[G]^*$ commutes with the action of G on V. If V is an irreducible representation, then by Schur's Lemma, the only linear maps commuting with the action are scalar multiples of the identity map. This shows that the Casimir operator acts as a scalar on every irreducible representation. ◁

Lemma 4.5.16. *Suppose that V is an irreducible rational representation of G. Then c acts as the zero map $V\to V$ if and only if V is the trivial representation.*

Proof. This follows from Humphreys [117, §22.3]. For this we need some representation theory. Let ρ be the sum of the fundamental weights (which is equal to half the sum of the positive roots). If V is a representation with highest weight λ, then c acts as the scalar $(\lambda,\rho+\lambda)$ which is equal to 0 if and only if $\lambda=0$. $\qquad\square$

Suppose that V is a rational representation of G. Let $V = V^G \oplus C$ where C is the unique G-stable complement. The action of the Reynolds operator $\mathcal{R} : K[G] \to K$ on V is the projection of V onto V^G.

Let us write $c^i = c * c * \cdots * c$ (i times) and $c^0 = \epsilon_e$. If $P(t) = \sum_{i=0}^{l} a_i t^i \in K[t]$ is a polynomial, we define $P(c) := \sum_{i=0}^{l} a_i c^i \in K[G]^*$.

Proposition 4.5.17. *Suppose $v \in V$. Let P be the monic polynomial of smallest degree such that $P(c) \cdot v = 0$.*

(a) If $P(0) \neq 0$, then $\mathcal{R} \cdot v = 0$;

(b) If $P(0) = 0$, write $P(t) = tQ(t)$. Then we have $\mathcal{R} \cdot v = Q(0)^{-1} Q(c) \cdot v$.

Proof. Write $v = v_0 + v_1$ with $v_0 \in V^G$ and $v_1 \in C$. Then we have $\mathcal{R} \cdot v = v_0$. Moreover, $P(c) \cdot v = (P(c) \cdot v_0) + (P(c) \cdot v_1)$. Notice that $P(c) \cdot v_0 \in V^G$ and $P(c) \cdot v_1 \in C$. Because $P(c) \cdot v = 0$, we have $P(c) \cdot v_0 = P(c) \cdot v_1 = 0$. By Lemma 4.5.16, $c \cdot v_0 = 0$, so $P(0)v_0 = P(c) \cdot v_0 = 0$. If $P(0) \neq 0$, then $v_0 = 0$.

If $P(0) = 0$, then we have $P(c) \cdot v_1 = c(Q(c) \cdot v_1) = 0$. Because c is injective on C by Lemma 4.5.16, we have $Q(c) \cdot v_1 = 0$. We have $Q(c) \cdot v_0 = Q(0)v_0$ and $Q(c) \cdot v = Q(c)v_0 + Q(c)v_1 = Q(0)v_0$. Because of the minimality of the degree of P, we know that $Q(0) \neq 0$. $\qquad\square$

Algorithm 4.5.18. Suppose that V is a rational representation of a connected semi-simple group G. The input of the algorithm is an element v in V, and the output will be $\mathcal{R} \cdot v$ where $\mathcal{R} : K[G] \to K$ is the Reynolds operator (if $V = K[X]$, then $\mathcal{R} \cdot v = \mathcal{R}_X v$ where $\mathcal{R}_X : K[X] \to K[X]^G$ is the Reynolds operator on $K[X]$).

(1) Input: $v \in V$
(2) $l := 0$; $v_0 := v$
(3) If there is a linear combination $\sum_{i=0}^{l} a_i v_i = 0$ with $a_l \neq 0$, then
(4)   If $a_0 \neq 0$, then
(5)   Output: 0
(6)   else
(7)   Output: $a_1^{-1} \sum_{i=1}^{l} a_i v_{i-1}$
(8) $l := l + 1$
(9) compute $v_l = c \cdot v_{l-1}$.
(10) goto step 3

In Step (7) we have that $a_1 \neq 0$. This follows from the proof of Proposition 4.5.17 (b), because $a_0 = P(0)$ and $a_1 = Q(0)$.

Let us explain more precisely how this works in the case where G acts on $K[V]$ with V an n-dimensional rational representation of G. Suppose we would like to compute $\mathcal{R}(f)$ for some $f \in K[V]$.

As in the algorithm we compute $f_0, f_1, f_2, \ldots$ defined by $f_0 = f$ and $f_i := c \cdot f_{i-1}$ for all $i > 0$. How to determine whether $f_0, f_1, \ldots, f_l$ are linearly dependent? Of course, for every l we can use straightforward linear algebra to test whether $f_0, \ldots, f_l$ are linearly dependent. We can do it slightly more efficient here.

For every l we define W_l as the vector space spanned by $f_0, f_1, \ldots, f_l$ (and $W_{-1} = \{0\}$). The coordinate ring $K[V]$ is isomorphic to $K[x_1, \ldots, x_n]$. Choose a total ordering ">" on the monomials. We will construct $g_0, g_1, \ldots$ with the following properties:

(a) W_l is spanned by $g_0, \ldots, g_l$ for all l;
(b) If $g_l \neq 0$, then $\mathrm{LM}(g_l) \neq \mathrm{LM}(g_i)$ for all $i < l$;
(c) The leading coefficient $\mathrm{LC}(g_l)$ equals 1.

The construction is inductively. We define $g_0 = f_0$. Assume that $g_0, \ldots, g_{l-1}$ are constructed. Then we can construct g_l as follows. Start with $g_l := f_l$ and subtract multiples of the g_i's with $i < l$ until property (b) is satisfied. Then divide by a constant to get $\mathrm{LC}(g_l) = 1$ which is property (c). Clearly property (a) is satisfied. Algorithm 4.5.18 can be refined to

Algorithm 4.5.19. Let V be a rational representation of a connected semi-simple group G. The input of the algorithm is an element $f \in K[V]$ and the output is $\mathcal{R}f$ where $\mathcal{R}$ is the Reynolds operator.

(1) Input: f
(2) $f_0 := f$; $l := 0$;
(3) $g_l := f_l$; $b_{l,l} := 1$; $b_{l,i} := 0$ for $i = 0, \ldots, l-1$.
(4) $a := \mathrm{LC}(g_l)$
(5) If there is an $i < l$ such that $\mathrm{LM}(g_l) = \mathrm{LM}(g_i)$, then
(6) $g_l := g_l - a g_i$
(7) $b_{l,j} := b_{l,j} - a b_{i,j}$ for $j = 0, \ldots, i$
(8)   goto step 4
(9) If $g_l = 0$, then
(10)   if $b_{l,0} \neq 0$, then
(11)   Output: 0
(12)   else
(13)   Output: $(\sum_{j=1}^{l} b_{l,j} f_{j-1})/b_{l,1}$
(14) $g_l := g_l / a$
(15) $b_{l,j} := b_{l,j} / a$ for $j = 0, \ldots, l$
(16) $l := l + 1$
(17) $f_l := c \cdot f_{l-1}$
(18) goto step 3

We first note that throughout the algorithm, we will have

$$g_l = \sum_{i=0}^{l} b_{l,i} f_i. \qquad (4.5.2)$$

This explains Step 3, 6 and 7. The loop covering the Steps (4)–(8) is finite because $\mathrm{LM}(g_l)$ decreases each time because of Step 6. Once $g_l = 0$, then

$f_0, f_1, \ldots, f_l$ are linearly dependent, because of (4.5.2) and the algorithm terminates.

Example 4.5.20. Let $G = \mathrm{SL}_2$ act on V_2. Here V_2 is the set of binary forms of degree 2, i.e.,

$$V_2 := \{a_0 x^2 + a_1 xy + a_2 y^2\}.$$

A basis of $\mathfrak{sl}_2$ is $\mathbf{x} = \partial_{2,1}$, $\mathbf{y} = \partial_{1,2}$ and $\mathbf{h} = \partial_{1,1} - \partial_{2,2}$. The action of the Lie algebra of $\mathfrak{sl}_2$ on V_2 is given by

$$\mathbf{x} \cdot v = y \frac{\partial}{\partial x} v,$$

$$\mathbf{y} \cdot v = x \frac{\partial}{\partial y} v,$$

$$\mathbf{h} \cdot v = (x \frac{\partial}{\partial x} - y \frac{\partial}{\partial y}) v.$$

The coordinate ring $K[V_2]$ can be identified with $K[a_0, a_1, a_2]$. The action of $\mathfrak{sl}_2$ on $K[a_0, a_1, a_2]$ is given by

$$\mathbf{x} \cdot f = (-2a_0 \frac{\partial}{\partial a_1} - a_1 \frac{\partial}{\partial a_2}) f,$$

$$\mathbf{y} \cdot f = (-a_1 \frac{\partial}{\partial a_0} - 2a_2 \frac{\partial}{\partial a_1}) f,$$

$$\mathbf{h} \cdot f = (-2a_0 \frac{\partial}{\partial a_0} + 2a_2 \frac{\partial}{\partial a_2}) f,$$

where $f \in K[a_0, a_1, a_2]$. A dual basis of $\mathbf{x}, \mathbf{y}, \mathbf{h}$ is $\mathbf{y}/4, \mathbf{x}/4, \mathbf{h}/8$. The Casimir operator is therefore defined by

$$c = \frac{\mathbf{x} * \mathbf{y}}{4} + \frac{\mathbf{y} * \mathbf{x}}{4} + \frac{\mathbf{h} * \mathbf{h}}{8}.$$

A straightforward calculation shows that c acts by the differential operator

$$a_0 a_1 \frac{\partial}{\partial a_0} \frac{\partial}{\partial a_1} + a_1 a_2 \frac{\partial}{\partial a_1} \frac{\partial}{\partial a_2} + 2a_0 a_2 \frac{\partial^2}{(\partial a_1)^2} + a_0 \frac{\partial}{\partial a_0} + a_1 \frac{\partial}{\partial a_1} + a_2 \frac{\partial}{\partial a_2} +$$

$$+ \frac{1}{2} \left(a_0^2 \frac{\partial^2}{(\partial a_0)^2} + a_2^2 \frac{\partial^2}{(\partial a_2)^2} + (a_1^2 - 2a_0 a_2) \frac{\partial}{\partial a_0} \frac{\partial}{\partial a_2} \right). \quad (4.5.3)$$

Let us compute $\mathcal{R}(a_1^2)$ where $\mathcal{R} : K[V_2] \to K[V_2]^{\mathrm{SL}_2}$ is the Reynolds operator. We apply c:

$$c \cdot a_1^2 = 2a_1^2 + 4a_0 a_2, \qquad c \cdot (2a_1^2 + 4a_0 a_2) = 6a_1^2 + 12a_0 a_2 = 3(2a_1^2 + 4a_0 a_2).$$

We have found the relation $c^2 \cdot a_1^2 - 3c \cdot a_1 = 0$. Therefore,

$$\mathcal{R}(a_1^2) = \frac{(c-3)\cdot a_1^2}{-3} = \frac{2a_1^2 + 4a_0a_2 - 3a_1^2}{-3} = \frac{1}{3}\left(a_1^2 - 4a_0a_2\right).$$

Let us compute $\mathcal{R}(a_1^4)$ following Algorithm 4.5.19. We choose the graded reverse lexicographic ordering on the monomials. We define $f_0 = g_0 = a_1^4$. Now

$$f_1 := c \cdot f_0 = c \cdot a_1^4 = 4a_1^4 + 24a_0a_1^2a_2.$$

The leading term of f_1 is $4a_1^4$, and the leading term of g_0 is a_1^4. The leading term of $f_1 - 4g_0$ is $24a_0a_1^2a_2$ so we define $g_1 = (f_1 - 4g_0)/24$. We have $g_1 = \frac{1}{24}f_1 - \frac{1}{6}f_0$. We apply c again:

$$f_2 := c \cdot f_1 = 28a_1^4 + 264a_0a_1^2a_2 + 96a_0^2a_2^2.$$

We take f_2, subtract 28 times g_0 and 264 times g_1. The leading term of $f_2 - 28g_0 - 264g_1$ is $96a_0^2a_2^2$. We define $g_2 := (f_2 - 28g_0 - 264g_1)/96$. We have $g_2 = \frac{1}{96}f_2 - \frac{7}{24}f_0 - \frac{11}{4}(\frac{1}{24}f_1 - \frac{1}{6}f_0) = \frac{1}{96}f_2 - \frac{11}{96}f_1 + \frac{1}{6}f_0$. We apply c again:

$$f_3 := c \cdot f_2 = 244a_1^4 + 2712a_0a_1^2a_2 + 1248a_0^2a_2^2.$$

Thus

$$f_3 = 244g_0 + 2712g_1 + 1248g_2 =$$
$$244f_0 + 2712(\frac{1}{24}f_1 - \frac{1}{6}f_0) + 1248(\frac{1}{96}f_2 - \frac{11}{96}f_1 + \frac{1}{6}f_0) = 13f_2 - 30f_1.$$

We have found the relation

$$f_3 - 13f_2 + 30f_1 = (c^3 - 13c^2 + 30c)\cdot a_1^4 = 0.$$

Therefore $\mathcal{R}(a_1^4)$ is equal to

$$\frac{(c^2 - 13c + 30\epsilon)\cdot f_0}{30} = \frac{f_2 - 13f_1 + 30f_0}{30} =$$
$$\frac{1}{5}a_1^4 - \frac{8}{5}a_0a_1^2a_2 + \frac{16}{5}a_0^2a_2^2 = \frac{1}{5}(a_1^2 - 4a_0a_2)^2.$$

$\triangleleft$

4.5.3 Cayley's Omega Process

For GL_n and SL_n there is an alternative method for computing the Reynolds operator. This is the so-called Ω process, which was already known in the 19th century.

Let us consider the coordinate ring of GL_n. We denote the coordinate function corresponding to the (i,j) entry by $z_{i,j}$. Let Z be the matrix

$$\begin{pmatrix} z_{1,1} & z_{1,2} & \cdots & z_{1,n} \\ z_{2,1} & z_{2,2} & & z_{2,n} \\ \vdots & & \ddots & \vdots \\ z_{n,1} & z_{n,2} & \cdots & z_{n,n} \end{pmatrix} \in K\,[\mathrm{GL}_n]\,.$$

We have

$$K[\mathrm{GL}_n] = K[\{z_{i,j} \mid 1 \le i, j \le n\}, \det(Z)^{-1}],$$

where $\det(Z)$ is the determinant of Z. Let us write $\frac{\partial}{\partial Z}$ for the matrix

$$\begin{pmatrix} \dfrac{\partial}{\partial z_{1,1}} & \dfrac{\partial}{\partial z_{1,2}} & \cdots & \dfrac{\partial}{\partial z_{1,n}} \\[2mm] \dfrac{\partial}{\partial z_{2,1}} & \dfrac{\partial}{\partial z_{2,2}} & & \dfrac{\partial}{\partial z_{2,n}} \\[2mm] \vdots & & \ddots & \vdots \\[2mm] \dfrac{\partial}{\partial z_{n,1}} & \dfrac{\partial}{\partial z_{n,2}} & \cdots & \dfrac{\partial}{\partial z_{n,n}} \end{pmatrix}.$$

We define a differential operator $\Omega : K[\mathrm{GL}_n] \to K[\mathrm{GL}_n]$ by $\Omega = \det(\frac{\partial}{\partial Z})$. Let $m : \mathrm{GL}_n \times \mathrm{GL}_n \to \mathrm{GL}_n$ be the group multiplication and let $m^* : K[\mathrm{GL}_n] \to K[\mathrm{GL}_n] \otimes K[\mathrm{GL}_n]$ be the dual ring homomorphism. For $f \in K[\mathrm{GL}_n]$ we denote multiplication by f as a map $K[\mathrm{GL}_n] \to K[\mathrm{GL}_n]$ also by f.

Lemma 4.5.21. *We have*

$$(\det(Z)^{-1} \otimes \Omega) \circ m^* = m^* \circ \Omega = (\Omega \otimes \det(Z)^{-1}) \circ m^*.$$

Proof. Let $\sigma, \tau \in \mathrm{GL}_n$ and $f \in K[\mathrm{GL}_n]$. We get

$$\big(\mathrm{id} \otimes \frac{\partial}{\partial z_{i,j}}\big)(m^*(f))(\sigma, \tau) = \frac{\partial}{\partial \tau_{i,j}} f(\sigma\tau) = \sum_{i,j} \sigma_{k,i}\Big(\big(\frac{\partial}{\partial z_{k,j}} f\big)(\sigma\tau)\Big) =$$

$$= \sum_{k=1}^{n} \sigma_{k,i} m^*\big(\frac{\partial}{\partial z_{i,j}} f\big)(\sigma, \tau) = \sum_{k=1}^{n} (z_{k,i} \otimes \mathrm{id}) \circ m^*\big(\frac{\partial}{\partial z_{k,j}} f\big)(\sigma, \tau). \quad (4.5.4)$$

Or in matrix notation we have

$$\big(\mathrm{id} \otimes \frac{\partial}{\partial Z}\big)(m^*(f)) = ({}^t Z \otimes \mathrm{id}) \circ m^*\big(\frac{\partial}{\partial Z} f\big).$$

Taking determinants gives us

$$(\mathrm{id} \otimes \Omega) \circ m^* = (\det(Z) \otimes \mathrm{id}) \circ m^* \circ \Omega.$$

This shows the first equality of the Lemma and the other equality follows by symmetry. $\qquad\square$

Remark 4.5.22. If $\delta \in K[\mathrm{GL}_n]^*$, then δ is a linear function on $K[\mathrm{GL}_n]$. The value of δ on $f \in K[\mathrm{GL}_n]$ is denoted by δf. But GL_n acts on $K[\mathrm{GL}_n]$, and this induces a $K[\mathrm{GL}_n]^*$-module structure on $K[\mathrm{GL}_n]$:

$$(\delta, f) \in K[\mathrm{GL}_n]^* \times K[\mathrm{GL}_n] \mapsto \delta \cdot f \in K[\mathrm{GL}_n].$$

The two distinct notations δf and $\delta \cdot f$ should not be confused.   ◁

Remark 4.5.23. The operator Ω cannot be seen as the action of an element in $K[\mathrm{GL}_n]^*$. However, the operator $\widetilde{\Omega} := \det(Z)\Omega$ can be seen as an action of an element $\omega \in K[\mathrm{GL}_n]^*$. Notice that

$$(\mathrm{id} \otimes \widetilde{\Omega}) \circ m^* = (\det(Z) \otimes \det(Z)) \circ (\det(Z)^{-1} \otimes \Omega) \circ m^* =$$
$$= (\det(Z) \otimes \det(Z)) \circ m^* \circ \Omega = m^* \circ (\det(Z)\Omega) = m^* \circ \widetilde{\Omega}. \quad (4.5.5)$$

Let us define $\omega \in K[G]^*$ by $\omega f = (\widetilde{\Omega}f)(e)$ for all $f \in K[G]$. We have

$$\widetilde{\Omega}f = ((\mathrm{id} \otimes \epsilon) \circ m^* \circ \widetilde{\Omega})f = ((\mathrm{id} \otimes \epsilon) \circ (\mathrm{id} \otimes \widetilde{\Omega}) \circ m^*)(f)$$
$$= ((\mathrm{id} \otimes \omega) \circ m^*)(f) = \omega \cdot f.$$

◁

Remark 4.5.24. Suppose that $f \in K[\mathrm{GL}_n]$ and $\sigma, \tau \in \mathrm{GL}_n$. From Lemma 4.5.21 it follows that

$$(\det(\sigma)^{-1}\Omega(\sigma^{-1} \cdot f))(\tau) = ((\det(Z)^{-1} \otimes \Omega) \circ m^*(f))(\sigma, \tau) =$$
$$(m^* \circ \Omega(f))(\sigma, \tau) = (\sigma^{-1} \cdot \Omega(f))(\tau).$$

So we get

$$\det(\sigma)\Omega(\sigma \cdot f) = \sigma \cdot \Omega(f).$$

◁

Lemma 4.5.25. *For any positive integer p, the element $c_{p,n} := \Omega^p(\det(Z)^p) \in K[\mathrm{GL}_n]$ is a non-zero constant.*

Proof. Write $\det(Z)^p = \sum_i a_i m_i(Z_{1,1}, \ldots, Z_{n,n})$ where all m_i are different monomials, all of degree pn. Clearly we have $\Omega^p = \sum_i a_i m_i(\frac{\partial}{\partial z_{1,1}}, \ldots, \frac{\partial}{\partial z_{n,n}})$. Notice that

$$m_i\left(\frac{\partial}{\partial z_{1,1}}, \ldots, \frac{\partial}{\partial z_{n,n}}\right)m_j(Z_{1,1}, \ldots, Z_{n,n})$$

is equal to 0 if $i \neq j$ and is a positive constant if $i = j$. It follows that

$$\Omega^p(\det(Z)^p) = \sum_i a_i^2 m_i\left(\frac{\partial}{\partial z_{1,1}}, \ldots, \frac{\partial}{\partial z_{n,n}}\right)m_i(Z_{1,1}, \ldots, Z_{n,n})$$

is a non-zero constant.   □

The Reynolds operator $\mathcal{R} : K[\mathrm{GL}_n] \to K$ can be expressed in Ω as follows.

Proposition 4.5.26. *Suppose that $f \in K[\{z_{i,j} \mid 1 \le i, j \le n\}]$ is homogeneous. If $\deg(f) = np$, then*

$$\mathcal{R}\left(\frac{f}{\det(Z)^p}\right) = \frac{\Omega^p(f)}{c_{p,n}}$$

with $c_{p,n} = \Omega^p(\det(Z)^p)$ and if $\deg(f) \ne np$, then $\mathcal{R}(f/\det(Z)^p) = 0$.

Proof. We have $K[\mathrm{GL}_n]^{\mathrm{GL}_n} = K$. Let $V_{p,q} \subset K[\mathrm{GL}_n]$ be the set of all $f/\det(Z)^p$ with f homogeneous of degree q. If $q \ne np$, then $K \cap V = \{0\}$ and $V^{\mathrm{GL}_n} = 0$. It follows that $\mathcal{R}(g) = 0$ for all $g \in V_{p,q}$.

Let $V = V_{p,pn}$. Define a map $\mathcal{R}_V : V \to V^{\mathrm{GL}_n} \cong K$ by

$$\mathcal{R}_V\left(\frac{f}{\det(Z)^p}\right) = \frac{\Omega^p f}{c_{p,n}},$$

and we will show that $\mathcal{R}_V$ is the restriction of the Reynolds operator. First of all $\mathcal{R}_V$ is GL_n-invariant because of Lemma 4.5.21. Also, the restriction of $\mathcal{R}_V$ to K is the identity because

$$\mathcal{R}_V(1) = \mathcal{R}_V\left(\frac{\det(Z)^p}{\det(Z)^p}\right) = \frac{\Omega^p(\det(Z)^p)}{c_{p,n}} = 1.$$

If W is the kernel of $\mathcal{R}_V$, then $\mathcal{R}_V$ is the projection of $V = W \oplus K$ onto K. From Remark 2.2.6 it follows that the restriction of $\mathcal{R}$ to V is equal to $\mathcal{R}_V$. $\square$

Proposition 4.5.27. *Let $\mathcal{R}_{\mathrm{SL}_n} : K[\mathrm{GL}_n] \to K[\mathrm{GL}_n]^{\mathrm{SL}_n}$ be the Reynolds operator, where SL_n acts on GL_n by left multiplication. Suppose that $f \in K[\{z_{i,j} \mid 1 \le i, j \le n\}]$ is homogeneous. If $\deg(f) = nr$, then*

$$\mathcal{R}_{\mathrm{SL}_n}\left(\frac{f}{\det(Z)^p}\right) = \det(Z)^{r-p}\frac{\Omega^r f}{c_{r,n}}.$$

If $\deg(f)$ is not divisible by n, then $\mathcal{R}_{\mathrm{SL}_n}(f/\det(Z)^p) = 0$ for every integer p.

Proof. Notice that $K[\mathrm{GL}_n]^{\mathrm{SL}_n} = K[\det(Z), \det(Z)^{-1}]$. If $V_{p,q}$ is the vector space of all $f/\det(Z)^p$ with $\deg(f) = q$. If q is not divisible by n, then clearly $V_{p,q}$ cannot contain any invariants and $\mathcal{R}(f/\det(Z)^p) \in V_{p,q}^{\mathrm{SL}_n} = \{0\}$.

If $q = nr$, then a similar argument as in the proof of Proposition 4.5.26 shows that the restriction of $\mathcal{R}_{\mathrm{SL}_n}$ to $V_{p,nr}$ is given by

$$\frac{f}{\det(Z)^p} \mapsto \det(Z)^{r-p}\frac{\Omega^r f}{c_{r,n}}.$$

$\square$

Remark 4.5.28. The coordinate ring of SL_n is $K[\{z_{i,j} \mid 1 \le i,j \le n\}]/I$ where I is the principal ideal generated by $\det(Z) - 1$. The Reynolds operator $\mathcal{R}_{\mathrm{SL}_n} : K[\mathrm{SL}_n] \to K$ can be computed as follows. Suppose $g \in K[\mathrm{SL}_n]$ and suppose that it is represented by $f \in K[\{z_{i,j} \mid 1 \le i,j \le n\}]$. Then $\mathcal{R}_{\mathrm{SL}_n}(g) = \mathcal{R}_{\mathrm{SL}_n}(f) + I$. If we assume that f is homogeneous, then by Proposition 4.5.27

$$\mathcal{R}_{\mathrm{SL}_n}(g) = \frac{\Omega^r f}{c_{r,n}} + I$$

if $\deg(f) = rn$ and $\mathcal{R}_{\mathrm{SL}_n}(g) = 0$ if the degree of f is not divisible by n.   ◁

Remark 4.5.29. In Proposition 4.5.26 and Remark 4.5.28 we have described the Reynolds operators $\mathcal{R}_{\mathrm{GL}_n} : K[\mathrm{GL}_n] \to K$ and $\mathcal{R}_{\mathrm{SL}_n} : K[\mathrm{SL}_n] \to K$. Suppose that V is a rational representation of G where $G = \mathrm{GL}_n$ or $G = \mathrm{SL}_n$ and let $\mathcal{R}_V : K[V] \to K[V]^G$ be the Reynolds operator. For every $f \in K[V]$, we have $\mathcal{R}_V(f) = \mathcal{R}_G \cdot f$. Let $\mu^* : K[V] \to K[V] \otimes K[G]$ be as in Example A.2.12. If we write $\mu^*(f) = \sum_i f_i \otimes g_i$, then $\mathcal{R}_G \cdot f = \sum_i f_i \mathcal{R}_G(g_i)$ (see Proposition 4.5.9). Since we have described $\mathcal{R}_G$ in terms of Ω, we also can describe $\mathcal{R}_V$ in terms of Ω.   ◁

Example 4.5.30. Let SL_2 act on the binary forms

$$V_2 = \{a_0 x^2 + a_1 xy + a_2 y^2\}$$

and on the coordinate ring $K[V_2] \cong K[a_0, a_1, a_2]$. Let

$$\begin{pmatrix} z_{1,1} & z_{1,2} \\ z_{2,1} & z_{2,2} \end{pmatrix} \in \mathrm{SL}_2$$

act on $a_0 x^2 + a_1 xy + a_2 y^2$ by

$$a_0(z_{1,1}x + z_{2,1}y)^2 + a_1(z_{1,1}x + z_{2,1}y)(z_{1,2}x + z_{2,2}y) + a_2(z_{1,2}x + z_{2,2}y)^2$$
$$= (a_0 z_{1,1}^2 + a_1 z_{1,1} z_{1,2} + a_2 z_{1,2}^2)x^2 +$$
$$(2a_0 z_{1,1} z_{2,1} + a_1 z_{1,1} z_{2,2} + a_1 z_{2,1} z_{1,2} + 2a_2 z_{1,2} z_{2,2})xy +$$
$$(a_0 z_{2,1}^2 + a_1 z_{2,1} z_{2,2} + a_2 z_{2,2}^2)y^2. \quad (4.5.6)$$

Let $\mu^* : K[V_2] \to K[V_2] \otimes K[\mathrm{SL}_2]$ be as in Example A.2.12, so μ^* is given by:

$$a_0 \mapsto a_0 z_{1,1}^2 + a_1 z_{1,1} z_{1,2} + a_2 z_{1,2}^2,$$

$$a_1 \mapsto 2a_0 z_{1,1} z_{2,1} + a_1 z_{1,1} z_{2,2} + a_1 z_{2,1} z_{1,2} + 2a_2 z_{1,2} z_{2,2},$$

$$a_2 \mapsto a_0 z_{2,1}^2 + a_1 z_{2,1} z_{2,2} + a_2 z_{2,2}^2.$$

We will compute $\mathcal{R}(a_1^2)$. We have

$$\mu^*(a_1^2) = (2a_0 z_{1,1} z_{2,1} + a_1 z_{1,1} z_{2,2} + a_1 z_{2,1} z_{1,2} + 2a_2 z_{1,2} z_{2,2})^2 =$$

$$4z_{1,1}^2 z_{2,1}^2 a_0^2 + (4z_{1,1}^2 z_{2,1} z_{2,2} + 4z_{1,1} z_{2,1}^2 z_{1,2})a_0 a_1 + 8z_{1,1} z_{2,1} z_{1,2} z_{2,2} a_0 a_2 +$$

$$+ (z_{1,1}^2 z_{2,2}^2 + 2z_{1,1} z_{2,2} z_{2,1} z_{1,2} + z_{2,1}^2 z_{1,2}^2)a_1^2 + (4z_{1,1} z_{1,2} z_{2,2}^2 + 4z_{2,1} z_{1,2}^2 z_{2,2})a_1 a_2$$

$$+ 4z_{1,2}^2 z_{2,2}^2 a_0^2. \quad (4.5.7)$$

Since the coefficients of the monomials in a are homogeneous of degree 4 in the z variables, we get

$$\mathcal{R}(a_1^2) = \frac{(\mathrm{id} \otimes \Omega^2)(m^*(a_1^2))}{c_{2,2}}.$$

Applying $\Omega^2 = \frac{\partial^2}{(\partial z_{1,1})^2} \frac{\partial^2}{(\partial z_{2,2})^2} - 2\frac{\partial}{\partial z_{1,1}} \frac{\partial}{\partial z_{2,2}} \frac{\partial}{\partial z_{1,2}} \frac{\partial}{\partial z_{2,1}} + \frac{\partial^2}{(\partial z_{2,1})^2} \frac{\partial^2}{(\partial z_{1,2})^2}$ to (4.5.7) yields

$$4a_1^2 - 16a_0 a_2.$$

We compute $c_{2,2} = \Omega^2(\det(Z)^2) = 12$, so we conclude that

$$\mathcal{R}(a_1^2) = \frac{4a_1^2 - 16a_0 a_2}{12} = \frac{1}{3}(a_1^2 - 4a_0 a_2).$$

$\triangleleft$

4.6 Computing Hilbert Series

4.6.1 A Generalization of Molien's Formula

We assume that $\mathrm{char}(K) = 0$. In Theorem 3.2.2 we have seen that for a finite group, the Hilbert series of the invariant ring can easily be computed in advance (without knowing the generators of the invariant ring) with Molien's Formula. If G is a finite group and V is a finite dimensional representation, then

$$H(K[V]^G, t) = \frac{1}{|G|} \sum_{\sigma \in G} \frac{1}{\det_V(1 - t \cdot \sigma)}. \quad (4.6.1)$$

This idea can be generalized to arbitrary reductive groups. Instead of averaging over a finite group, we will have to average over a reductive group. Let us assume in this section that our base field K is the complex numbers $\mathbb{C}$. Averaging over G does not make much sense, because an infinite reductive group is not compact. However, G always contains a maximal compact subgroup C (see Anhang II of Kraft [152] or the references there). We can choose a Haar measure $d\mu$ on C and normalize it such that $\int_C d\mu = 1$. Let V be an finite dimensional rational representation of G. The proper generalization of (4.6.1) is

$$H(\mathbb{C}[V]^G, t) = \int_C \frac{d\mu}{\det_V(1 - t \cdot \sigma)}. \quad (4.6.2)$$

Indeed, for any representation W, let $\Phi_\sigma(W)$ be the trace of the action of $\sigma \in C$ on U. The function $I(W) := \int_C \Phi_\sigma(W) d\mu$ counts the dimension of U^C and $U^C = U^G$ since C is Zariski dense in G. Using this, (4.6.2) can be proven in a similar way as Theorem 3.2.2.

Notice that the Hilbert series $H(\mathbb{C}[V]^G, t)$ converges for $|t| < 1$ because it is a rational function with poles only at $t = 1$. Since C is compact, there exist constants $A > 0$ such that for every $\sigma \in C$ and every eigenvalue λ of σ we have $|\lambda| \leq A$. Since λ^l is an eigenvalue of σ^l, it follows that $|\lambda^l| \leq A$ for all l, so $|\lambda| \leq 1$. It is clear that the integral on the right-hand side of (4.6.2) also is defined for $|t| < 1$.

Assume that G is also connected. Let T be a maximal torus of G, and let D be a maximal compact subgroup of T. We may assume that C contains D. The torus can be identified with $(\mathbb{C}^*)^r$, where r is the rank of G, and D can be identified with the subgroup $(S^1)^r$ of $(\mathbb{C}^*)^r$, where $S^1 \subset \mathbb{C}^*$ is the unit circle. We can choose a Haar measure $d\nu$ on D such that $\int_D d\nu = 1$. Suppose that f is a continuous class function on C. In a compact group, all elements are semi-simple and every C-conjugacy class has a representative in D. An integral like

$$\int_C f(\sigma) d\mu$$

can be viewed as an integral over D, since f is constant on conjugacy classes. More precisely, there exists a weight function $\varphi : D \to \mathbb{R}$ such that for every continuous class function f we have

$$\int_C f(\sigma) d\mu = \int_D \varphi(\sigma) f(\sigma) d\nu$$

(see Weyl [258, 259], Adams [4], and Zhelobenko [267]). Notice that the integrand $\det_V(1 - t \cdot \sigma)^{-1}$ in (4.6.2) only depends on the conjugacy class of σ. We get that

$$H(\mathbb{C}[V]^G, t) = \int_C \frac{d\mu}{\det_V(1 - t \cdot \sigma)} = \int_D \frac{\varphi(\sigma) d\nu}{\det_V(1 - t \cdot \sigma)}. \qquad (4.6.3)$$

The compact torus D acts diagonally on V and its dual space V^* for a convenient choice of bases in V and V^*. So the action of $(z_1, \ldots, z_r) \in D$ on V^* is given by a matrix

$$\begin{pmatrix} m_1(z) & 0 & \cdots & 0 \\ 0 & m_2(z) & & 0 \\ \vdots & & \ddots & \vdots \\ 0 & 0 & \cdots & m_n(z) \end{pmatrix},$$

where $m_1, m_2, \ldots, m_n$ are Laurent monomials in $z_1, \ldots, z_r$. With this notation, we have

$$\det{}_V(1 - t \cdot (z_1, \ldots, z_n)) = (1 - m_1(z)t)(1 - m_2(z)t) \cdots (1 - m_n(z)t),$$

and it follows that

$$H(\mathbb{C}[V]^G, t) = \int_D \frac{\varphi(z)d\nu}{(1 - m_1(z)t)(1 - m_2(z)t) \cdots (1 - m_n(z)t)}. \qquad (4.6.4)$$

In 4.6.3 we will give a similar, but more explicit formula.

Example 4.6.1. Let $G = \mathrm{SL}_2(\mathbb{C})$ with maximal compact subgroup $C = \mathrm{SU}_2(\mathbb{C})$. A maximal torus T of G is the set of diagonal matrices

$$\begin{pmatrix} z & 0 \\ 0 & z^{-1} \end{pmatrix}, \qquad z \in \mathbb{C}^*.$$

A maximal compact subgroup D of T is

$$\begin{pmatrix} z & 0 \\ 0 & z^{-1} \end{pmatrix}, \qquad z \in S^1.$$

We can define a diffeomorphism of the three-dimensional real sphere S^3 to $\mathrm{SU}_2(\mathbb{C})$ by

$$(x, y, v, w) \in S^3 \mapsto \begin{pmatrix} x + yi & v + wi \\ -v + wi & x - yi \end{pmatrix}.$$

It is straightforward to check that

$$d\mu = \frac{dx \wedge dy \wedge dv}{w} = -\frac{dx \wedge dy \wedge dw}{v} = \frac{dx \wedge dv \wedge dw}{y} = -\frac{dy \wedge dv \wedge dw}{x}$$

is a Haar measure on SU_2. A normalized Haar measure is $d\mu/2\pi^2$. Under the diffeomorphism, SU_2-conjugacy classes correspond to the 2-dimensional spheres $x = a$ with $-1 \leq a \leq 1$ and the compact torus D corresponds to $v = w = 0$.

If a function f is constant on conjugacy classes, then

$$\frac{1}{2\pi^2} \int_{S^3} f(x, y, v, w)d\mu = \frac{1}{2\pi^2} \int_{S^3} f(x, y, v, w)\frac{dv \wedge dw \wedge dx}{y} =$$

$$= \frac{1}{2\pi^2} \int_{S^3} f(x, y, v, w)\frac{\sqrt{1 - x^2}\,d\widetilde{v} \wedge d\widetilde{w} \wedge dx}{\widetilde{y}} =$$

$$= \frac{1}{2\pi^2} \int_{-1}^{1} f(x, \sqrt{1 - x^2}, 0, 0)\sqrt{1 - x^2}\,dx \int_{S^2} \frac{d\widetilde{v} \wedge d\widetilde{w}}{\widetilde{y}} =$$

$$= \frac{2}{\pi} \int_{-1}^{1} f(x, \sqrt{1 - x^2}, 0, 0)\sqrt{1 - x^2}\,dx,$$

where $\widetilde{y} = y/\sqrt{1 - x^2}, \widetilde{v} = v/\sqrt{1 - x^2}$ and $\widetilde{w} = w/\sqrt{1 - x^2}$. Notice that $(\widetilde{y}, \widetilde{v}, \widetilde{w})$ lies on the two-dimensional sphere S^2. Using a substitution $x = \cos(u)$, this is equal to

$$\frac{2}{\pi} \int_{\pi}^{0} \sqrt{1 - \cos^2(u)}\, f(\cos(u), \sin(u), 0, 0)(-\sin(u))\, du =$$

$$= \frac{1}{\pi} \int_{0}^{2\pi} \sin^2(u) f(\cos(u), \sin(u), 0, 0)\, du. \quad (4.6.5)$$

Let V_d be the binary forms of degree d. The action of $D \cong S^1$ on V_d is given by the $(d+1) \times (d+1)$-matrix

$$\begin{pmatrix} z^d & 0 & 0 & \cdots & 0 \\ 0 & z^{d-2} & 0 & & 0 \\ 0 & 0 & z^{d-4} & & 0 \\ \vdots & & & \ddots & \vdots \\ 0 & 0 & 0 & \cdots & z^{-d} \end{pmatrix}.$$

We have

$$H(\mathbb{C}[V_d]^{\mathrm{SL}_2}, t) = \frac{1}{\pi} \int_{0}^{2\pi} \frac{\sin^2(u)\, du}{(1 - e^{idu}t)(1 - e^{i(d-2)u}t) \cdots (1 - e^{-idu}t)}.$$

We could also view this integral as a complex contour integral by putting $z = e^{iu}$ so that $du = dz/iz$ and $\sin(u) = (z - z^{-1})/2i$:

$$H(\mathbb{C}[V_d]^{\mathrm{SL}_2}, t) = \frac{-1}{4\pi i} \int_{S^1} \frac{(z - z^{-1})^2\, dz}{z(1 - z^d t)(1 - z^{d-2}t) \cdots (1 - z^{-d}t)} =$$

$$\frac{1}{4\pi i} \int_{S^1} \frac{(1 - z^2)\, dz}{z(1 - z^d t)(1 - z^{d-2}t) \cdots (1 - z^{-d}t)} +$$

$$+ \frac{1}{4\pi i} \int_{S^1} \frac{(1 - z^{-2})\, dz}{z(1 - z^d t)(1 - z^{d-2}t) \cdots (1 - z^{-d}t)}. \quad (4.6.6)$$

By symmetry $z \leftrightarrow z^{-1}$, the latter two integrals are equal to each other. So we obtain

$$H(\mathbb{C}[V_d]^{\mathrm{SL}_2}, t) = \frac{1}{2\pi i} \int_{S^1} \frac{(1 - z^2)\, dz}{z(1 - z^d t)(1 - z^{d-2}t) \cdots (1 - z^{-d}t)}. \quad (4.6.7)$$

As we will see later, an integral like this can be evaluated using the Residue Theorem. An explicit treatment of Hilbert series for binary forms can be found in Springer [232]. $\triangleleft$

4.6.2 Hilbert Series of Invariant Rings of Tori

In this section we will derive a formula for the Hilbert series of the invariant ring of a torus group. We will work over an algebraically closed base field K of characteristic 0.

Let T be an r-dimensional torus group. Suppose that $\rho : T \to \mathrm{GL}(V)$ is a rational representation.

Definition 4.6.2. *The* **character** $\chi^V = \chi_T^V : T \to K^*$ *of T is defined by*

$$\chi^V(\sigma) = \mathrm{Tr}(\rho(\sigma)),$$

where Tr *is the trace.*

Let $X(T) \cong \mathbb{Z}^r$ be the group of one-dimensional characters. Choose generators $z_1, \ldots, z_r$ of $X(T)$. After a convenient basis choice, the action of T on V is diagonal. The action is given by the matrix

$$\rho = \begin{pmatrix} m_1(z) & 0 & \cdots & 0 \\ 0 & m_2(z) & & 0 \\ \vdots & & \ddots & \vdots \\ 0 & 0 & \cdots & m_n(z) \end{pmatrix},$$

where $m_1, \ldots, m_n$ are Laurent monomials in $z_1, \ldots, z_r$. The character of V is the trace of the representation:

$$\chi^V = m_1(z) + m_2(z) + \cdots + m_n(z).$$

Remark 4.6.3. Notice that $\dim V^T$ is the coefficient of $z_1^0 z_2^0 \cdots z_r^0 = 1$ in χ^V. ◁

Definition 4.6.4. *Suppose that $V = \bigoplus_{d=k}^{\infty} V_d$ is a graded vector space and for each d, V_d is a rational representation of T, then we define the T-Hilbert series of V by*

$$H_T(V, z_1, \ldots, z_r, t) = \sum_{d=k}^{\infty} \chi^{V_d} t^d.$$

Remark 4.6.5. Because of Remark 4.6.3 the Hilbert series $H(V^T, t) = \sum_{d=k}^{\infty} \dim(V_d^T) t^i$ of the invariant space V^T is obtained by taking the coefficient of $z_1^0 z_2^0 \cdots z_r^0 = 1$ in $H_T(V, z_1, \ldots, z_r, t)$. ◁

Suppose that V is a rational representation of T and let $\rho : T \to \mathrm{GL}(V^*)$ be the dual representation. We can choose a basis $x_1, \ldots, x_n$ of V^* such that the action of T is diagonal. Suppose that the action on V^* is given by the matrix

$$\begin{pmatrix} m_1(z) & 0 & \cdots & 0 \\ 0 & m_2(z) & & 0 \\ \vdots & & \ddots & \vdots \\ 0 & 0 & \cdots & m_n(z) \end{pmatrix}.$$

Since $K[V]$ is the polynomial ring in $x_1, \ldots, x_n$, we have

$$H_T(K[V], z_1, \ldots, z_r, t) = \frac{1}{(1 - m_1(z)t)(1 - m_2(z)t) \cdots (1 - m_n(z)t)}. \tag{4.6.8}$$

From Remark 4.6.5 we get the following corollary.

Corollary 4.6.6. *With the notation as before, the Hilbert series of $K[V]^T$ is given by the coefficient of $z_1^0 z_2^0 \cdots z_r^0 = 1$ in (4.6.8).*

Example 4.6.7. Let T be a 2-dimensional torus and let z_1, z_2 are two characters such that $K[T] = k[z_1, z_2, z_1^{-1}, z_2^{-1}]$. Suppose that the representation on a vector space V is given by the matrix

$$\begin{pmatrix} z_1^{-1} & 0 & 0 \\ 0 & z_2^{-1} & 0 \\ 0 & 0 & z_1 z_2 \end{pmatrix}.$$

The action on the dual space V^* is then given by the matrix

$$\begin{pmatrix} z_1 & 0 & 0 \\ 0 & z_2 & 0 \\ 0 & 0 & z_1^{-1} z_2^{-1} \end{pmatrix}.$$

From (4.6.8) we get

$$H_T(K[V], z_1, z_2, t) = \frac{1}{(1 - z_1 t)(1 - z_2 t)(1 - z_1^{-1} z_2^{-1} t)}. \tag{4.6.9}$$

The Hilbert series $H(K[V]^T, t)$ of the invariant ring is the coefficient of $z_1^0 z_2^0$ in (4.6.9). If we expand the series in (4.6.9) we get

$$H_T(K[V], z_1, z_2, t) =$$
$$(1 + z_1 t + z_1^2 t^2 + \dots)(1 + z_2 t + z_2^2 t^2 + \dots)(1 + z_1^{-1} z_2^{-1} t + z_1^{-2} z_2^{-2} t^2 + \dots).$$

It is easy to see that the constant coefficient is $1 + t^3 + t^6 + t^9 + \cdots = (1 - t^3)^{-1}$. We conclude $H(K[V]^T, t) = (1 - t^3)^{-1}$. If x_1, x_2, x_3 is the basis of V^*, then it is clear that $K[V]^T = K[x_1 x_2 x_3]$. So we verify that $H(K[V]^T, t) = H(K[x_1 x_2 x_3], t) = (1 - t^3)^{-1}$. Usually it is much harder to compute the constant coefficient. ◁

4.6.3 Hilbert Series of Invariant Rings of Connected Reductive Groups

Let K be again an arbitrary algebraically closed field of characteristic 0. We will assume that G is connected and reductive. We will give an algebraic derivation of a formula for the Hilbert series of the invariant ring. For details on the notation, see Section A.4. We fix a maximal torus $T \subseteq G$, and a Borel subgroup B of G containing T. We define the Weyl group by $W = N_G(T)/Z_G(T)$ where $N_G(T)$ is the normalizer of T in G and $Z_G(T)$ is the centralizer of T in G. The set of roots is denoted by Φ. Let $\alpha_1, \dots, \alpha_r$ be a set of simple roots. We have $\Phi = \Phi_+ \cup \Phi_-$ where Φ_+ and Φ_- are the positive and negative roots, respectively. Let $\lambda_1, \dots, \lambda_r \in X(T) \otimes_{\mathbb{Z}} \mathbb{Q}$ be the fundamental weights, i.e., $\langle \lambda_i, \alpha_j^\vee \rangle = \delta_{i,j}$. For the group of weights $X(T)$ we will use additive notation. The character of T associated to a weight λ will be denoted by z^λ. We define $z_i = z^{\lambda_i}$. Every character of T is a Laurent monomial in $z_1, \ldots, z_r$.

Definition 4.6.8. *For a rational representation V of G we define the character χ_G^V of V by $\chi_G^V = \chi_T^V$, where T is a maximal torus of G (see Definition 4.6.2).*

Let $\rho = \lambda_1 + \lambda_2 + \cdots + \lambda_r$. If $V = \bigoplus_\lambda V_\lambda^{a_\lambda}$ is a rational representation, then the coefficient of $z^{\rho+\lambda}$ in $\sum_{w \in W} \mathrm{sgn}(w) z^{w(\rho)} \chi^V$ is equal to a_λ because of Weyl's Theorem (see Theorem A.5.3). In particular, the coefficient of z^ρ in $\sum_{w \in W} \mathrm{sgn}(w) z^{w(\rho)} \chi^V$ is equal to $a_0 = \dim V^G$.

Suppose that $V = \bigoplus_{d \geq k} V_d$ is a graded vector space and for every d, V_d is a G-module. As in Definition 4.6.4, we define the T-Hilbert series by

$$H_T(V, z_1, \ldots, z_r, t) = \sum_{d=k}^\infty \chi^{V_d} t^d.$$

From Weyl's Theorem (Theorem A.5.3) it follows that the coefficient of z^ρ in

$$\sum_{w \in W} \mathrm{sgn}(w) z^{w(\rho)} H_T(V, z_1, \ldots, z_r, t)$$

is equal to $\sum_{d=k}^\infty \dim V_d^G t^i = H(V^G, t)$.

Suppose that V is a rational representation of G. The action of $T \subset G$ on V^* is diagonal for some choice of basis. Suppose that the action of T on V^* is given by the matrix

$$\begin{pmatrix} m_1(z) & 0 & \cdots & 0 \\ 0 & m_2(z) & & 0 \\ \vdots & & \ddots & \vdots \\ 0 & 0 & \cdots & m_n(z) \end{pmatrix},$$

where $m_1(z), \ldots, m_n(z)$ are Laurent monomials in $z_1, \ldots, z_r$. As in (4.6.8), we have

$$H_T(K[V], z_1, \ldots, z_r, t) = \frac{1}{(1 - m_1(z)t)(1 - m_2(z)t) \cdots (1 - m_n(z)t)}.$$

It follows that $H(K[V]^G, t)$ is the coefficient of z^ρ (as series in $z_1, \ldots, z_r$ with coefficients in $K(t)$) in

$$\frac{\sum_{w \in W} \mathrm{sgn}(w) z^{w(\rho)}}{(1 - m_1(z)t)(1 - m_2(z)t) \cdots (1 - m_n(z)t)}$$

or the coefficient of $z^0 = 1$ in

$$\frac{z^{-\rho}\sum_{w\in W}\mathrm{sgn}(w)z^{w(\rho)}}{(1-m_1(z)t)(1-m_2(z)t)\cdots(1-m_n(z)t)}.$$

If we use the identity $\sum_{w\in W}\mathrm{sgn}(w)z^{w(\rho)} = z^\rho\prod_{\alpha\in\Phi+}(1-z^{-\alpha})$ (Humphreys [117, 24.1 Lemma A, 24.3 Lemma]), then the next Corollary follows. This Corollary is a reformulation of Weyl's integral formula for Hilbert series (see Bröcker and tom Dieck [28]).

Corollary 4.6.9. *The Hilbert series $H(K[V]^G,t)$ is the coefficient of 1 (as series in $z_1,\dots,z_r$ with coefficients in $K(t)$) of*

$$\frac{\prod_{\alpha\in\Phi+}(1-z^{-\alpha})}{(1-m_1(z)t)(1-m_2(z)t)\cdots(1-m_n(z)t)}. \tag{4.6.10}$$

Remark 4.6.10. In (4.6.10), if G is semisimple and we replace z_i by z_i^{-1} for all i, then $m_1(z),\dots,m_n(z)$ will just be permuted. Hence we also obtain that $H(K[V]^G,t)$ is the coefficient of 1 in

$$\frac{\prod_{\alpha\in\Phi+}(1-z^{\alpha})}{(1-m_1(z)t)(1-m_2(z)t)\cdots(1-m_n(z)t)}. \tag{4.6.11}$$

◁

Example 4.6.11. Let $G = \mathrm{SL}_2$ act on the binary forms V_d. In this case $\Phi = \{\alpha_1, -\alpha_1\}$ and $\Phi^+ = \{\alpha_1\}$. We have $\alpha_1 = 2\lambda_1$. The weights appearing in V_d are $d\lambda_1, (d-2)\lambda_1, \dots, -d\lambda_1$. So we get that the Hilbert series $H(K[V_d]^{\mathrm{SL}_2}, t)$ is equal to the coefficient of $z^0 = 1$ in

$$\frac{(1-z^2)}{(1-z^d t)(1-z^{d-2}t)\cdots(1-z^{-d}t)}.$$

This is equivalent to the formula in Example 4.6.1 because $\int_{S^1} z^n\,dz$ is equal to 0 if $n \neq -1$ and equal to $2\pi i$ if $n = -1$.

◁

Example 4.6.12. Suppose that $G = \mathrm{SL}_3$ and let V be the adjoint representation. Let T be the set of diagonal matrices in SL_3. This is a maximal torus. Let λ_1, λ_2 the two fundamental weights and let α_1, α_2 be the simple roots. Then $\alpha_1 = 2\lambda_1 - \lambda_2$ and $\alpha_2 = 2\lambda_2 - \lambda_1$. The set of roots is

$$\Phi = \{\alpha_1, \alpha_2, \alpha_1 + \alpha_2, -\alpha_1, -\alpha_2, -\alpha_1 - \alpha_2\}.$$

The weights appearing in V are all roots α (all with multiplicity 1) and 0 (which has multiplicity 2). So the weights appearing in V are

$$0, 0, 2\lambda_1 - \lambda_2, 2\lambda_2 - \lambda_1, \lambda_1 + \lambda_2, \lambda_2 - 2\lambda_1, \lambda_1 - 2\lambda_2, -\lambda_1 - \lambda_2.$$

Now $H(K[V]^{\mathrm{SL}_3}, t)$ is the coefficient of $z_1^0 z_2^0$ in

$$\frac{(1-z_1^2 z_2^{-1})(1-z_1^{-1} z_2^2)(1-z_1 z_2)}{(1-t)^2(1-z_1^2 z_2^{-1}t)(1-z_1^{-1}z_2^2 t)(1-z_1 z_2 t)(1-z_1^{-2}z_2 t)(1-z_1 z_2^{-2}t)(1-z_1^{-1}z_2^{-1}t)}.$$

◁

In the next section, we will show how the residue theorem can be used to compute the Hilbert series of an invariant ring. In Broer [29] an alternative method can be found.

4.6.4 Hilbert Series and the Residue Theorem

We will first briefly recall the Residue Theorem in complex function theory. This theorem can be applied to compute the Hilbert series of invariant rings.

Suppose that $f(z)$ is a meromorphic function on $\mathbb{C}$. If $a \in \mathbb{C}$, then f can be written as a Laurent series around $z = a$.

$$f(z) = \sum_{i=-d}^{\infty} c_i(z - a)^i.$$

If $d > 0$ and $c_{-d} \neq 0$, then f has a pole at $z = a$ and the pole order is d. The **residue** of f at $z = a$ is denoted by $\mathrm{res}(f, a)$ and defined by

$$\mathrm{res}(f, a) = c_{-1}.$$

If the pole order of f is 1, then the residue can be computed by

$$\mathrm{res}(f, a) = \lim_{z \to a} (z - a) f(z).$$

Suppose D that $\gamma : [0, 1] \to \mathbb{C}$ is a smooth curve. The integral over the curve γ is defined by

$$\int_\gamma f(z)\, dz = \int_0^1 f(\gamma(t)) \gamma'(t)\, dt.$$

Let $\gamma : [0, 1] \to \mathbb{C}$ be defined by $\gamma(t) = e^{2\pi i t}$. Then $\gamma([0, 1])$ is the unit circle. For $n \in \mathbb{Z}$ we have

$$\int_\gamma z^n\, dz = \int_0^1 e^{2\pi n i t}(2\pi i)e^{2\pi i t}\, dt = 2\pi i \int_0^1 e^{2\pi(n+1)it}\, dt,$$

which is 0 if $n \neq -1$ and equal to $2\pi i$ if $n = -1$.

Theorem 4.6.13 (Residue Theorem). *Suppose that D is a connected, simply connected compact region in $\mathbb{C}$ whose border is ∂D, and $\gamma : [0, 1] \to \mathbb{C}$ is a smooth curve such that $\gamma([0, 1]) = \partial D$, $\gamma(0) = \gamma(1)$ and γ circles around D exactly once in counterclockwise direction. Assume that f is a meromorphic function on $\mathbb{C}$ with no poles in ∂D. Then we have*

$$\frac{1}{2\pi i} \int_\gamma f(z)\, dz = \sum_{a \in D} \mathrm{res}(f, a).$$

There are only finitely many points in the compact region D such that f has non-zero residue there.

Example 4.6.14. Let $T = \mathbb{G}_m$ be the one-dimensional torus acting on a 3-dimensional space V by the matrix

$$\rho = \begin{pmatrix} z & 0 & 0 \\ 0 & z & 0 \\ 0 & 0 & z^{-2} \end{pmatrix}.$$

The action of $\mathbb{G}_m$ on V^* is given by

$$\begin{pmatrix} z^{-1} & 0 & 0 \\ 0 & z^{-1} & 0 \\ 0 & 0 & z^2 \end{pmatrix}.$$

So we get

$$H_T(K[V], z, t) = \frac{1}{(1 - z^{-1}t)^2(1 - z^2 t)}. \tag{4.6.12}$$

For the Hilbert series to converge, we need that $|z^{-1}t| < 1$ and $|z^2 t| < 1$. We will assume that $|z| = 1$ and $|t| < 1$. To find the coefficient of z^0, we divide (4.6.12) by $2\pi i z$ and integrate over the unit circle S^1 in $\mathbb{C}$ (counterclockwise). We obtain

$$H(K[V]^{\mathbb{G}_m}, t) = \frac{1}{2\pi i} \int_{S^1} \frac{dz}{z(1 - z^{-1}t)^2(1 - z^2 t)}. \tag{4.6.13}$$

By the Residue Theorem, (4.6.13) is equal to

$$\frac{1}{2\pi i} \int_{S^1} f(z)\, dz = \sum_{a \in D^1} \mathrm{res}(f(z), a), \tag{4.6.14}$$

where D^1 is the unit disk and $f(z) = z^{-1}(1 - z^{-1}t)^{-2}(1 - z^2 t)^{-1}$. The only poles of $f(z)$ are $z = t$ and $z = \pm t^{-1/2}$. Since $|t| < 1$, the only pole in the unit disk is $z = t$. We will compute the residue. We have

$$\frac{1}{z(1 - z^{-1}t)^2(1 - z^2 t)} = \frac{1}{(z - t)^2} \frac{z}{1 - z^2 t}. \tag{4.6.15}$$

The power series of $g(z) = z/(1 - z^2 t)$ around $z = t$ gives

$$g(z) = g(t) + g'(t)(z - t) + \frac{g''(t)(z - t)^2}{2} + \cdots = \frac{t}{1 - t^3} + \frac{1 + t^3}{(1 - t^3)^2}(z - t) + \cdots. \tag{4.6.16}$$

$H(K[V]^{\mathbb{G}_m}, t)$ is the residue of

$$\frac{1}{z(1 - z^{-1}t)^2(1 - z^2 t)} = \frac{g(z)}{(z - t)^2}$$

at $z = t$ and from (4.6.16) it follows that this is equal to

$$\frac{1 + t^3}{(1 - t^3)^2}.$$

$\triangleleft$

Example 4.6.15. Let again $G = \mathrm{SL}_2$ be acting on the binary forms V_d. By Example 4.6.11, the Hilbert series $H(K[V_d]^{\mathrm{SL}_2}, t)$ of the invariant ring is the constant coefficient of

$$\frac{(1 - z^2)}{(1 - z^d t)(1 - z^{d-2} t) \cdots (1 - z^{-d} t)},$$

so

$$H(K[V_d]^{\mathrm{SL}_2}, t) = \frac{1}{2\pi i} \int_{S^1} \frac{(1 - z^2)\, dz}{z(1 - z^d t)(1 - z^{d-2} t) \cdots (1 - z^{-d} t)}. \qquad (4.6.17)$$

In this formula, we assume that $|t| < 1$.

From the Residue Theorem, it follows that (4.6.17) is equal to

$$\sum_{x \in D^1} \mathrm{res}(f_d, x),$$

where

$$f_d(t, z) = \frac{(1 - z^2)}{z(1 - z^d t)(1 - z^{d-2} t) \cdots (1 - z^{-d} t)}$$

and x runs through all the poles of f_d in the unit disk D^1. The poles of f_d in the unit disk are exactly $\zeta_{d-2j}^k t^{1/(d-2j)}$ where $0 \le j < d/2$, $0 \le k < d - 2j$ and ζ_p is a primitive p-th root of unity.

Let us consider $d = 1$. Then we have

$$f_1(t, z) = \frac{1 - z^2}{z(1 - zt)(1 - z^{-1} t)} = \frac{1 - z^2}{(1 - zt)(z - t)}.$$

Now f_1 has a pole at $z = t$. The pole at $z = t^{-1}$ lies outside the unit disk because $|t| < 1$. Let us compute the residue.

$$\mathrm{res}(f_1, t) = \lim_{z \to t} \frac{1 - z^2}{1 - zt} = 1.$$

It follows that $H(K[V_1]^{\mathrm{SL}_2}, t) = 1$. This answer was to be expected, since there are no non-trivial SL_2-invariant polynomial functions on V_1.

Let us now take $d = 2$. Then we have

$$f_2(t, z) = \frac{1 - z^2}{z(1 - z^2 t)(1 - t)(1 - z^{-2} t)}.$$

The poles of f_2 in the unit disk are $z = \sqrt{t}$ and $z = -\sqrt{t}$. We compute the residues.

$$\mathrm{res}(f_2, \sqrt{t}) = \lim_{z \to \sqrt{t}} \frac{(1 - z^2)(z - \sqrt{t})}{z(1 - z^2 t)(1 - t)z^{-2}(z + \sqrt{t})(z - \sqrt{t})} =$$

$$= \frac{(1 - t)}{\sqrt{t}(1 - t^2)(1 - t)t^{-1}(2\sqrt{t})} = \frac{1}{2(1 - t^2)}. \qquad (4.6.18)$$

In a similar way

$$\mathrm{res}(f_2, -\sqrt{t}) = \frac{1}{2(1 - t^2)},$$

so

$$H(K[V_2]^{\mathrm{SL}_2}, t) = \frac{1}{2(1 - t^2)} + \frac{1}{2(1 - t^2)} = \frac{1}{1 - t^2}.$$

Indeed, $K[V_2]^{\mathrm{SL}_2}$ is a polynomial ring in one variable of degree 2, namely the discriminant. Below is a table of the Hilbert series of $K[V_d]^{\mathrm{SL}_2}$ for $d = 1, 2, \ldots, 8$.

d	$H(K[V_d]^{\mathrm{SL}_2}, t)$
1	1
2	$\dfrac{1}{1 - t^2}$
3	$\dfrac{1}{1 - t^4}$
4	$\dfrac{1}{(1 - t^2)(1 - t^3)}$
5	$\dfrac{1 + t^{18}}{(1 - t^4)(1 - t^8)(1 - t^{12})}$
6	$\dfrac{1 + t^{15}}{(1 - t^2)(1 - t^4)(1 - t^6)(1 - t^{10})}$
7	$\dfrac{f_7(t)}{(1 - t^4)(1 - t^6)(1 - t^8)(1 - t^{10})(1 - t^{12})}$
8	$\dfrac{1 + t^8 + t^9 + t^{10} + t^{18}}{(1 - t^2)(1 - t^3)(1 - t^4)(1 - t^5)(1 - t^6)(1 - t^7)}$

with

$$f_7(t) = t^{32} - t^{26} + 2t^{24} - t^{22} + 5t^{20} + 2t^{18} + 6t^{16} + 2t^{14} + 5t^{12} - t^{10} + 2t^8 - t^6 + 1.$$

$\lhd$

Remark 4.6.16. For degree at most 16 some explicit formulas of Hilbert series for binary forms were given in Cohen and Brouwer [44]. Some computations of Hilbert series of the invariant ring for binary forms up to a high degree were done by Littelmann and Procesi [162], who used a formula in Springer [232]. If $d = 4l$, then one can prove (see Littelmann and Procesi [162]) that

$$H(K[V_d]^{\mathrm{SL}_2}, t) = \frac{a(t)}{(1 - t^2)(1 - t^3) \cdots (t^{d-1} - 1)},$$

where $a(t)$ is a polynomial in t with integer coefficients. It was conjectured by Dixmier [56] that $a(t)$ has non-negative coefficients. Moreover, it was conjectured that there exists a homogeneous system of invariants $f_2, \ldots, f_{d-1}$

where f_i is of degree i. Notice that this automatically implies that $a(t)$ has non-negative coefficients, because $K[V_d]^{\mathrm{SL}_2}$ is a free $K[f_2,\ldots,f_{d-1}]$-module.
◁

Remark 4.6.17. The *degree* of the invariant ring $K[V_d]^{\mathrm{SL}_2}$ is defined by (see Definition 1.4.7)

$$\deg(K[V_d]^{\mathrm{SL}_2}) = \lim_{t\to 1}(1-t)^{d-2}H(K[V_d]^{\mathrm{SL}_2},t).$$

If $f_1,\ldots,f_{d-2}$ is a homogeneous system of parameters with $e_i := \deg(f_i)$ for all i, then the invariant ring $K[V_d]^{\mathrm{SL}_2}$ is a *free* module over $K[f_1,\ldots,f_{d-2}]$ because $K[V_d]^{\mathrm{SL}_2}$ is Cohen-Macaulay (see Theorem 2.5.5 and Proposition 2.5.3). We have a Hironaka decomposition

$$K[V_d]^{\mathrm{SL}_2} = Rh_1 \oplus Rh_2 \oplus \cdots \oplus Rh_l,$$

where $R = K[f_1,\ldots,f_{d-2}]$, with $h_1,\ldots,h_l$ homogeneous invariants. We have $\deg(R) = (e_1 e_2 \cdots e_{d-2})^{-1}$ (see Example 1.4.8) and

$$\deg(K[V_d]^{\mathrm{SL}_2}) = \frac{l}{e_1 e_2 \cdots e_{d-2}}.$$

If we can bound the degree of the invariant ring, then we can estimate the number of secondary invariants. There is a remarkable formula for $\deg(K[V_d]^{\mathrm{SL}_2})$ proven by Hilbert [108]. If d is odd, then

$$\deg(K[V_d]^{\mathrm{SL}_2}) = -\frac{1}{4}\frac{1}{d!}\sum_{i=0}^{\frac{d-1}{2}}(-1)^i\binom{d}{i}\left(\frac{d}{2}-i\right)^{d-3},$$

and if d is even, then

$$\deg(K[V_d]^{\mathrm{SL}_2}) = -\frac{1}{2}\frac{1}{d!}\sum_{i=0}^{\frac{d}{2}-1}(-1)^i\binom{d}{i}\left(\frac{d}{2}-i\right)^{d-3}.$$

◁

The method with the Residue Theorem also works if the maximal torus of G has dimension ≥ 2. Let us start with a rational function $f \in \mathbb{C}(z_1,\ldots,z_r,t)$. We make the following assumptions.

(a) f can be written as a power series in t with coefficients in the ring of Laurent polynomials $\mathbb{C}[z_1,\ldots,z_r,z_1^{-1},\ldots,z_r^{-1}]$;
(b) f can be written as a quotient g/h where g is a Laurent polynomial in $z_1,\ldots,z_r,t$ and $h = \prod_{i=1}^{l}(1-m_i t^{d_i})$ where m_i is a Laurent monomial in $z_1,\ldots,z_r$ and d_i is a positive integer for $i=1,\ldots,r$.

If V is a representation of a torus T, then the T-Hilbert series of $K[V]$ satisfies the two conditions above.

Proposition 4.6.18. *Assume that f satisfies the two conditions above. Write f as a power series in t with coefficients in $\mathbb{C}[z_1,\ldots,z_r,z_1^{-1},\ldots,z_r^{-1}]$ and let f_0 be the coefficient of z_r^0. Then f_0 also satisfies the two conditions above.*

Proof. Clearly f_0 is a power series in t with coefficients in $\mathbb{C}[z_1,\ldots,z_{r-1},z_1^{-1},\ldots,z_{r-1}^{-1}]$.

The power series of f in t converge for $|z_1| = |z_2| = \cdots = |z_r| = 1$ and $|t| < 1$. The coefficient of f_0 is obtained by dividing f by $2\pi i z_r$ and integrating z_r over the unit circle (counterclockwise):

$$f_0 = \frac{1}{2\pi i} \int_{S^1} \frac{f(z_1,\ldots,z_r,t)}{z_r}\, dz_r.$$

By the Residue Theorem this is equal to

$$\sum_{a \in D^1} \mathrm{res}(f,a),$$

where D^1 is the unit disk and f is seen as a function in z_r. All poles have the form

$$z_r = \zeta_d^i (z_1^{b_1} z_2^{b_2} \cdots z_{r-1}^{b_{r-1}} t^c)^{1/d}$$

with $b_1,\ldots,b_{r-1},c,d \in \mathbb{Z}$, ζ_d a primitive d-th root of unity. We may assume that the greatest common divisor of $b_1,\ldots,b_{r-1},c,d$ is 1 and that $0 \le i < d$. Put

$$a_i = \zeta_d^i (z_1^{b_1} z_2^{b_2} \cdots z_{r-1}^{b_{r-1}} t^c)^{1/d}.$$

Suppose that the pole order at $z_r = a_i$ is equal to k. Then

$$\mathrm{res}(f,a_i) = \frac{1}{(k-1)!} \left(\frac{d}{dz_r}\right)^{k-1} (z_r - a_i)^k f(z_1,\ldots,z_r)\Bigg|_{z_r = a_i}.$$

From this, it is clear that $\mathrm{res}(f,a_i)$ can be written as p_i/q with

$$p_i \in \mathbb{C}[z_1,\ldots,z_{r-1},z_1^{-1},\ldots,z_{r-1}^{-1},t,t^{-1},a_i,a_i^{-1}]$$

and $q = \prod_{j=1}^m (1 - u_j)$ with u_j a Laurent monomial in $z_1,\ldots,z_{r-1},t,a_i$ for all j. Notice that

$$\frac{1}{1-u_j} = \frac{1 + u_j + \cdots + u_j^{d-1}}{1 - u_j^d}.$$

This shows that without loss of generality we may assume that u_j is a Laurent monomial in $z_1,\ldots,z_{r-1},t$. It follows that u_j and $q = \prod_{j=1}^m (1 - u_j)$ do not depend on i. We have

$$\mathrm{Res}(f,a_0) + \mathrm{Res}(f,a_1) + \cdots + \mathrm{Res}(f,a_{d-1}) = \frac{p_0 + p_1 + \cdots + p_{d-1}}{q}.$$

Now $p_0 + \cdots + p_{d-1}$ is a Laurent polynomial in $z_1, z_2, \ldots, z_{r-1}, t$. It follows that

$$f_0 = \sum_{a \in D^1} \mathrm{Res}(f, a) = \frac{g'}{h'},$$

where g' is a Laurent polynomial in $z_1, \ldots, z_{r-1}, t$ and $h' = \prod_{i=1}^{l'}(1 - m_i' t^{d_i'})$, where the m_i' are Laurent monomials in $z_1, \ldots, z_{r-1}$ and all $d_i' \in \mathbb{Z}$. We may assume that all $d_i' \geq 0$ because

$$\frac{1}{1 - m_i' t^{d_i'}} = -\frac{(m_i')^{-1} t^{-d_i'}}{1 - (m_i')^{-1} t^{-d_i'}}.$$

We know that for $|z_1| = |z_2| = \cdots = |z_{r-1}| = 1$ and $|t| < 1$ the power series for f_0 converges. In particular f_0 has no poles. This shows that $d_i' > 0$ for all i. $\qquad\square$

Remark 4.6.19. From Theorem 4.6.13, Corollary 4.6.6 and Corollary 4.6.9 follows that the Hilbert series of $K[V]^G$ can be computed using iterated applications of the Residue Theorem. $\qquad\triangleleft$

Example 4.6.20. Let $G = \mathrm{SL}_3$ and suppose that V is the adjoint representation. From Example 4.6.12 it follows that the Hilbert series of the invariant ring $H(K[V]^{\mathrm{SL}_3}, t)$ is the coefficient of $z_1^0 z_2^0$ in

$$f(z_1, z_2, t) := (1 - z_1^2 z_2^{-1})(1 - z_1^{-1} z_2^2)(1 - z_1 z_2) \big/$$
$$((1 - t)^2 (1 - z_1^2 z_2^{-1} t)(1 - z_1^{-1} z_2^2 t)(1 - z_1 z_2 t)(1 - z_1^{-2} z_2 t) \cdot$$
$$(1 - z_1 z_2^{-2} t)(1 - z_1^{-1} z_2^{-1} t)) .$$

We assume that $|z_1| = 1$ and $|t| < 1$. The poles of $f z_2^{-1}$ (as a function in z_2) inside the unit disk are $z_2 = z_1^2 t$, $z_2 = \pm\sqrt{z_1 t}$ and $z_2 = z_1^{-1} t$. We will compute the residues:

$$\mathrm{res}(z_2^{-1} f, z_1^2 t) =$$
$$\frac{z_1^{-2} t^{-1}(1 - t^{-1})(1 - z_1^3 t^2)(1 - z_1^3 t)}{(1 - t)^2 (z_1^2 t)^{-1}(1 - z_1^3 t^3)(1 - z_1^3 t^2)(1 - t^2)(1 - z_1^{-3} t^{-1})(1 - z_1^{-3})} =$$
$$= \frac{-z_1^6}{(1 - t)(1 - z_1^3 t^3)(1 - t^2)(1 - z_1^3)}, \qquad (4.6.19)$$

$$\operatorname{res}(z_2^{-1}f, \sqrt{z_1}t) = z_1^{-1/2}t^{-1/2}(1 - z_1^{3/2}t^{-1/2})(1 - t)(1 - z_1^{3/2}t^{1/2}) \Big/$$

$$\Big((1 - t)^2(1 - z_1^{3/2}t^{1/2})(1 - t^2)(1 - z_1^{3/2}t^{3/2})(1 - z_1^{-3/2}t^{3/2}).$$

$$(2z_1^{-1/2}t^{-1/2})(1 - z_1^{-3/2}t^{1/2})\Big) =$$

$$= \frac{-z_1^{3/2}t^{-1/2}}{2(1 - t)(1 - t^2)(1 - z_1^{3/2}t^{3/2})(1 - z_1^{-3/2}t^{3/2})} =$$

$$= \frac{-z_1^{3/2}t^{-1/2}(1 + z_1^{3/2}t^{3/2})(1 + z_1^{-3/2}t^{3/2})}{2(1 - t)(1 - t^2)(1 - z_1^{3}t^{3})(1 - z_1^{-3}t^{3})} =$$

$$= \frac{-z_1^{3/2}t^{-1/2} - z_1^{3}t - t - z_1^{3/2}t^{5/2}}{2(1 - t)(1 - t^2)(1 - z_1^{3}t^{3})(1 - z_1^{-3}t^{3})}. \qquad (4.6.20)$$

Similarly, we have

$$\operatorname{res}(z_2^{-1}f, -\sqrt{z_1}t) = \frac{z_1^{3/2}t^{-1/2} - z_1^{3}t - t + z_1^{3/2}t^{5/2}}{2(1 - t)(1 - t^2)(1 - z_1^{3}t^{3})(1 - z_1^{-3}t^{3})}$$

and

$$\operatorname{res}(z_2^{-1}f, \sqrt{z_1}t) + \operatorname{res}(z_2^{-1}f, -\sqrt{z_1}t) = \frac{-z_1^{3}t - t}{(1 - t)(1 - t^2)(1 - z_1^{3}t^{3})(1 - z_1^{-3}t^{3})}.$$

Furthermore we have

$$\operatorname{res}(z_2^{-1}f, z_1^{-1}t) =$$

$$\frac{z_1 t^{-1}(1 - z_1^{3}t^{-1})(1 - z_1^{-3}t^{2})(1 - t)}{(1 - t)^2(1 - z_1^{3})(1 - z_1^{-3}t^{3})(1 - t^2)(1 - z_1^{-3}t^{2})(1 - z_1^{3}t^{-1})(z_1 t^{-1})} =$$

$$= \frac{1}{(1 - t)(1 - z_1^{3})(1 - z_1^{-3}t^{3})(1 - t^2)}. \qquad (4.6.21)$$

It follows that

$$\operatorname{res}(z_2^{-1}f, z_1^{-1}t) + \operatorname{res}(z_2^{-1}f, z_1^{2}t) =$$

$$\frac{(1 - z_1^{3}t^{3}) - z_1^{6}(1 - z_1^{-3}t^{3})}{(1 - t)(1 - z_1^{3})(1 - z_1^{-3}t^{3})(1 - t^2)(1 - z_1^{3}t^{3})} =$$

$$\frac{1 - z_1^{6}}{(1 - t)(1 - z_1^{3})(1 - z_1^{-3}t^{3})(1 - t^2)(1 - z_1^{3}t^{3})} =$$

$$\frac{1 + z_1^{3}}{(1 - t)(1 - z_1^{-3}t^{3})(1 - t^2)(1 - z_1^{3}t^{3})}. \qquad (4.6.22)$$

So the sum of the four residues is

$$g := \frac{1 + z_1^{3} - z_1^{3}t - t}{(1 - t)(1 - z_1^{-3}t^{3})(1 - t^2)(1 - z_1^{3}t^{3})}.$$

So g is the coefficient of z_2^0 in f. The coefficient of z_1^0 in g is equal to the Hilbert series $H(K[V]^{\mathrm{SL}_3}, t)$. Now g has poles at $z = \zeta^j t$ where ζ is a third root of unity and $0 \leq j \leq 2$. We compute the residues

$$\mathrm{res}(z_1^{-1} g, \zeta^j t) = \frac{\zeta^{-j} t^{-1}(1 + t^3 - t^4 - t)}{(1 - t)(3\zeta^{-j} t^{-1})(1 - t^2)(1 - t^6)} = \frac{1}{3(1 - t^2)(1 - t^3)}.$$

The Hilbert series of $K[V]^{\mathrm{SL}_3}$ is the sum of the residues

$$\mathrm{res}(z_1^{-1} g, t) + \mathrm{res}(z_1^{-1} g, \zeta t) + \mathrm{res}(z_1^{-1} g, \zeta^2 t) = \frac{1}{(1 - t^2)(1 - t^3)}.$$

$\lhd$

4.7 Degree Bounds for Invariants

Suppose that G is a linearly reductive group and V is a rational finite dimensional representation of G. We have defined $\beta(K[V]^G)$ to be the smallest integer N such that all invariants of degree $\leq N$ will generate $K[V]^G$. The goal of this section is to find explicit upper bounds for $\beta(K[V]^G)$. A first step in this direction was done by Hilbert [108], who gave a constructive method for finding generators of the invariant ring using a homogeneous system of parameters of the invariant ring. However, Hilbert could not give an explicit upper bound for $\beta(K[V]^G)$. By generalizing Hilbert's constructive method and combining it with the Cohen-Macaulay property (see Theorem 2.5.5) and an estimate of the degree of the Hilbert series (see Kempf [148]), Popov found an explicit upper bound for $\beta(K[V]^G)$ (see Remark 4.7.3 if G is connected, semisimple and the base field K has characteristic 0). As shown in Derksen [54], an adaption of the method of Hilbert and Popov will give a much better upper bound. We will derive that degree bound in this section.

We already showed in Corollary 2.6.3 that whenever $f_1, \ldots, f_r \in K[V]^G$ is a homogeneous system of parameters, then

$$\beta(K[V]^G) \leq d_1 + d_2 + \cdots + d_r - r, \tag{4.7.1}$$

where $d_i := \deg(f_i)$. Recall that $f_1, \ldots, f_r \in K[V]^G$ is a homogeneous system of parameters if and only if $f_1, \ldots, f_r$ are algebraically independent and $K[V]^G$ is finite (as a module) over $K[f_1, \ldots, f_r]$ (see Definition 2.4.6).

Definition 4.7.1. *If $R = \bigoplus_{d \geq 0} R_d$ is a graded K-algebra, we define the constant $\gamma(R)$ as the smallest integer d such that there exist homogeneous $f_1, \ldots, f_l \in R$ with $\deg(f_i) \leq d$ for all i and R is finite over $K[f_1, \ldots, f_l]$.*

Remark 4.7.2. In view of Lemma 2.4.5, the property that $K[V]^G$ is finite over $K[f_1, \ldots, f_l]$ is equivalent to $\mathcal{V}(f_1, \ldots, f_l) = \mathcal{N}_V$ where $\mathcal{N}_V$ is Hilbert's nullcone (see Definition 2.4.1). So we may say that $\gamma(K[V]^G)$ is the smallest integer d such that there exist homogeneous $f_1, \ldots, f_l \in K[V]^G$ such that $\mathcal{V}(f_1, \ldots, f_l) = \mathcal{N}_V$.

$\lhd$

Remark 4.7.3. Popov [190, 191] gave an explicit upper bound for $\beta(K[V]^G)$ as follows. There exist homogeneous $g_1, \ldots, g_l$ such that $K[V]^G$ is finite over $K[g_1, \ldots, g_l]$ and $\deg(g_i) = d_i \leq \gamma(K[V]^G)$ for all i. Let $d = \operatorname{lcm}(d_1, \ldots, d_l)$ where lcm is the least common multiple, and define $g_i' = g_i^{d/d_i}$ for all i. $K[V]^G$ is finite over $S = K[g_1', g_2', \ldots, g_l']$. By the Noether Normalization Lemma, we can find linear combinations $f_1, \ldots, f_r$ of $g_1', \ldots, g_l'$ such that $f_1, \ldots, f_r$ form a homogeneous system of parameters for S and also for $K[V]^G$. From (4.7.1) it follows that

$$\beta(K[V]^G) \leq rd - r \leq nd \leq n \operatorname{lcm}(1, 2, \ldots, \gamma(K[V]^G)).$$

Hilbert found upper bounds for $\gamma(K[V]^G)$ in the case $G = \mathrm{SL}_n$. This bound was generalized by Popov to arbitrary connected semi-simple groups G. Explicit upper bounds for $\gamma(K[V]^G)$ can be found in Popov's cited papers. For a fixed semi-simple group G, $\gamma(K[V]^G)$ is bounded by a polynomial in n, the dimension of V (see Section 4.7.1).

We will give another degree bound for $\beta(K[V]^G)$ in terms of $\gamma(K[V]^G)$. In Section 4.7.1 we will study the constant $\gamma(K[V]^G)$ and give upper bounds for it.

Theorem 4.7.4. *We have*

$$\beta(K[V]^G) \leq \max\left(2, \tfrac{3}{8}r(\gamma(K[V]^G))^2\right),$$

where $r = \dim(K[V]^G)$ is the Krull dimension.

By definition there exist homogeneous $f_1, \ldots, f_l \in K[V]^G$ such that $K[V]^G$ is finite over $K[f_1, \ldots, f_l]$ and $\mathcal{V}(f_1, \ldots, f_l) = \mathcal{N}_V$. If $l = r$ where $r := \dim(K[V]^G)$, then $f_1, \ldots, f_r$ is a homogeneous system of parameters and we have a good degree bound (4.7.1). Suppose that $l > r$, i.e., the number of invariants is too large compared to $r = \dim(K[V]^G)$. In the approach of Hilbert and Popov, we would use the Noether Normalization Lemma to obtain a homogeneous system of parameters (see Remark 4.7.3). However, usually the proof of Corollary 2.4.8 would give us a homogeneous system of parameters of very high degree d because d is the least common multiple of $d_1, \ldots, d_l$ where $d_i := \deg(f_i)$. We will use a method which avoids the Noether Normalization Lemma. The idea is to construct from $f_1, \ldots, f_l$ a homogeneous system of parameters of $K[V']^G$ where V' is the direct sum of V and a number of copies of the trivial representation.

Let V be an n-dimensional rational representation of a linearly reductive group G. We identify $K[V]$ with the polynomial ring $K[x_1, \ldots, x_n]$. Let us consider the representation $V \oplus K$ where K stands for the trivial representation of G. Then $K[V \oplus K] \cong K[x_1, \ldots, x_n, y]$ and $K[V \oplus K]^G = K[V]^G[y]$.

Lemma 4.7.5. *Suppose that $f_1, \ldots, f_l \in K[V]^G$ are homogeneous invariants with*

$$\mathcal{V}(f_1, \ldots, f_l) = \mathcal{N}_V.$$

If $l > r := \dim(K[V]^G)$, then we can find homogeneous $g_1, \ldots, g_l \in K[V \oplus K]^G$ such that

(a) $g_i(x_1, \ldots, x_n, 0) = f_i(x_1, \ldots, x_n)$ *for all i;*
(b) $\mathcal{V}(g_1, \ldots, g_l) = \mathcal{N}_V \times \{0\} \subset V \oplus K$.

Proof. Consider the map $F = (f_1, \ldots, f_l) : V \to K^l$. Since $l > \dim(K[V]^G)$, we know that $f_1, \ldots, f_l$ are algebraically dependent and the map F is not dominant. We can choose $\alpha = (\alpha_1, \ldots, \alpha_l) \in K^l$ which does not lie in the image of F. We define $g_i(x_1, \ldots, x_n, y) = f_i(x_1, \ldots, x_n) - \alpha_i y^{d_i}$ with $d_i = \deg(f_i)$ for all i. Clearly $\mathcal{N} \times \{0\} \subseteq \mathcal{V}(g_1, \ldots, g_l)$. Let us prove the reverse inclusion. Suppose that $(x_1, \ldots, x_n, y) \in \mathcal{V}(g_1, \ldots, g_l)$. If $y \neq 0$, then

$$\frac{g_i(x_1, \ldots, x_n, y)}{y^{d_i}} = g_i\left(\frac{x_1}{y}, \frac{x_2}{y}, \ldots, \frac{x_n}{y}, 1\right) = f_i\left(\frac{x_1}{y}, \ldots, \frac{x_n}{y}\right) - \alpha_i.$$

This shows that $\alpha = F(x_1/y, \ldots, x_n/y)$ which is in contradiction to our assumptions. We conclude that y must be equal to 0, so $g_i(x_1, \ldots, x_n, 0) = f_i(x_1, \ldots, x_n) = 0$ for all i and this shows that $(x_1, \ldots, x_n) \in \mathcal{N}_V$. $\square$

Corollary 4.7.6. *Suppose that $f_1, \ldots, f_l \in K[V]^G$ are homogeneous invariants with*

$$\mathcal{V}(f_1, \ldots, f_l) = \mathcal{N}_V.$$

Let $V' = V \oplus K^{l-r}$ with K the trivial representation and $r = \dim(K[V]^G)$. Then there exists a homogeneous system of parameters $p_1, \ldots, p_l \in K[V']^G$ such that

$$p_i(x_1, \ldots, x_n, 0, \ldots, 0) = f_i(x_1, \ldots, x_n)$$

for all i.

Proof. We will prove this statement by induction on $l - r$. If $l - r = 0$, then $f_1, \ldots, f_l$ is already a homogeneous system of parameters. Suppose that $l > r$. Let $V'' = V \oplus K$. By Lemma 4.7.5 we have $g_1, \ldots, g_l \in K[V'']^G$ such that $g_i(x_1, \ldots, x_n, 0) = f_i(x_1, \ldots, x_n)$ for all i, and $\mathcal{V}(g_1, \ldots, g_l) = \mathcal{N}_V \times \{0\} = \mathcal{N}_{V''}$. We have $K[V'']^G \cong K[V]^G[y]$, so $\dim(K[V'']^G) = r + 1$. Since $l - (r + 1) < l - r$ we can apply the induction hypothesis to find $p_1, \ldots, p_l$ such that $p_i(x_1, \ldots, x_n, 0, \ldots, 0) = g_i(x_1, \ldots, x_n, 0) = f_i(x_1, \ldots, x_n)$ for all i and $\mathcal{V}(p_1, \ldots, p_l) = \mathcal{N}_{V'} = \mathcal{N}_{V''} \times \{0\} \subset V'' \oplus K^{l-r-1}$. Now $K[V']^G$ is a finite $K[p_1, \ldots, p_l]$ module. We have $\dim(K[V']^G) = \dim(K[V]^G[y_1, \ldots, y_{l-r}]) = r + (l - r) = l$, so $p_1, \ldots, p_l$ are algebraically independent and therefore they are a homogeneous system of parameters. $\square$

Corollary 4.7.7. *Suppose that $f_1, \ldots, f_l \in K[V]^G$ are homogeneous invariants with*

$$\mathcal{V}(f_1, \ldots, f_l) = \mathcal{N}_V.$$

Then we have

$$\beta(K[V]^G) \leq \max(d_1, d_2, \ldots, d_l, d_1 + d_2 + \cdots + d_l - l).$$

Proof. By Corollary 4.7.6 there exists a homogeneous system of parameters $p_1, \ldots, p_l \in K[V']^G$ with $d_i := \deg(p_i)$. It follows from Corollary 2.6.3 that

$$\beta(K[V']^G) \leq \max(d_1, \ldots, d_l, d_1 + d_2 + \cdots + d_l - l).$$

The inclusion $V \hookrightarrow V'$ induces a surjective ring homomorphisms $K[V'] \to K[V]$ and a surjective ring homomorphism $K[V']^G \to K[V]^G$ by 2.2.8. This shows that $\beta(K[V]^G) \leq \beta(K[V']^G)$. $\square$

Proof of Theorem 4.7.4. Let us put $\gamma = \gamma(K[V]^G)$. By Remark 4.7.2 there exist homogeneous $f_1, \ldots, f_l \in K[V]^G$ such that $\mathcal{V}(f_1, \ldots, f_l) = \mathcal{N}_V$ and $d_i \leq \gamma$ where $d_i := \deg(f_i)$. Replacing f_i by some power of itself does not change the zero set $\mathcal{V}(f_1, \ldots, f_l)$. Without loss of generality we may assume that $\gamma/2 < d_i \leq \gamma$. Suppose that $\{h_1, \ldots, h_k\} \subset \{f_1, \ldots, f_l\}$ is a subset of invariants which are all of the same degree d. By the Noether Normalization Lemma (Lemma 2.4.7) we obtain a homogeneous system of parameters $g_1, \ldots, g_j$ of $K[h_1, \ldots, h_k]$ where g_i is a linear combination of $h_1, \ldots, h_k$ for all i and $j = \dim(K[h_1, \ldots, h_k]) \leq r$, since the dimension is given by the transcendence degree. Since $K[h_1, \ldots, h_k]$ is finite as a module over $K[g_1, \ldots, g_j]$, we can replace $h_1, \ldots, h_k$ by $g_1, \ldots, g_j$. This shows that without loss of generality we may assume that for every d, at most r of the invariants $f_1, \ldots, f_l$ have degree d.

We have

$$d_1 + \cdots + d_l - l = \sum_{i=1}^{l}(d_i - 1) \leq r \sum_{i=\lceil \frac{\gamma}{2} \rceil}^{\gamma} (i - 1).$$

For even γ the right-hand side is equal to $r(\frac{3}{8}\gamma^2 - \frac{1}{4}\gamma)$ and for odd γ the right-hand side is equal to $r(\frac{3}{8}\gamma^2 - \frac{3}{8})$. In any case we have

$$d_1 + \cdots + d_l - l \leq \tfrac{3}{8}r\gamma^2.$$

We get that

$$\beta(K[V]^G) \leq \max(d_1, \ldots, d_l, d_1 + \cdots + d_l - l) \leq \max\left(\gamma, \tfrac{3}{8}r\gamma^2\right).$$

Suppose that $\gamma > \frac{3}{8}r\gamma^2$. If $r = 0$, then $\gamma = 0$ and otherwise it follows that $\gamma < \frac{8}{3}$, so $\gamma \leq 2$. It follows that we always have

$$\beta(K[V]^G) \leq \max\left(2, \tfrac{3}{8}r\gamma^2\right).$$

$\square$

4.7.1 Degree Bounds for Orbits

Popov [190, 191] gave explicit upper bounds for $\gamma(K[V]^G)$. Hiss improved Popov's bound using ideas of Knop's (see Hiss [110]). We will follow Hiss' approach. Suppose that G is a linear algebraic group and V is an n-dimensional representation of G. For linearly reductive groups G we will show the relation between $\gamma(K[V]^G)$ and the degrees of orbits.

Definition 4.7.8. *We can view V as a Zariski-open set in $\mathbb{P}^n = \mathbb{P}(V \oplus K)$. If all irreducible components of $X \subset V$ have the same dimension, then we define $\deg(X) = \deg(\overline{X})$ where $\overline{X}$ is the closure of X in $\mathbb{P}^n$.*

Remark 4.7.9. If X is a constructible set (see Hartshorne [102, Exercise II.3.18]) of dimension r, then

$$\deg(X) = \#(X \cap W_1 \cap W_2 \cap \cdots \cap W_r), \qquad (4.7.2)$$

where $W_1, W_2, \ldots, W_r \subset V$ are hyperplanes in general position. The set H of hyperplanes in V is an algebraic variety. By saying that (4.7.2) holds for $W_1, \ldots, W_r$ in general position, we mean that there exists a Zariski open, non-empty subset $U \subset H^r$ such that for all $(W_1, \ldots, W_r) \in U$, (4.7.2) is equal to $\deg(X)$.

If $W_1, \ldots, W_r$ are not in general position, then

$$\#(X \cap W_1 \cap W_2 \cap \cdots \cap W_r) \leq \deg(X)$$

whenever the left-hand side is finite (see Fulton [73, §12.3]).   ◁

Proposition 4.7.10. *Suppose that $X \subset V$ is Zariski closed and irreducible. Assume that $\psi : V \to W$ is a linear map between finite dimensional vector spaces, then*

$$\deg(X) \geq \deg(\psi(X)).$$

Proof. Let $r = \dim(X)$ and $s = \dim(\psi(X))$. We can choose $W_1, \ldots, W_s$ such that

$$\#(\psi(X) \cap W_1 \cap \cdots \cap W_s) = \deg(\psi(X)).$$

For any $y \in \psi(X) \cap W_1 \cap \cdots \cap W_s$, $\psi^{-1}(y) \cap X$ is non-empty, of dimension $r - s$. If we put $U_i := \psi^{-1}(W_i) \subset V$, then

$$X \cap U_1 \cap \cdots \cap U_s = \psi^{-1}(\psi(X) \cap W_1 \cap \cdots \cap W_s) \cap X$$

has at least $\deg(\psi(X))$ connected components.

We can take $U_{s+1}, \ldots, U_r$ such that for every connected component S of $(X \cap U_1 \cap \cdots \cap U_s)$, the intersection $S \cap U_{s+1} \cap \cdots \cap U_r$ is finite and non-empty. It follows that

$$\#(X \cap U_1 \cap \cdots \cap U_s \cap U_{s+1} \cap \cdots \cap U_r) \geq \deg(\overline{\psi(X)}).$$

On the other hand, from Remark 4.7.9 it follows that

$$\#(X \cap U_1 \cap \cdots \cap U_s \cap U_{s+1} \cap \cdots \cap U_r) \leq \deg(X).$$

$\square$

Definition 4.7.11. *We define*

$$\delta(V) = \max\{\deg(G \cdot v) \mid v \in V \setminus \mathcal{N}_V\},$$

and $\delta(V) = 0$ if $\mathcal{N}_V = V$.

It was anticipated by V. Popov that the number $\delta(V)$ may play an important role. He formulated the problem of the explicit computation of $\delta(V)$ and pointed out the connection to the formula of Kazarnovskii (see Proposition 4.7.18).

Proposition 4.7.12. *If G is linearly reductive, and V is a rational representation, then*

$$\gamma(K[V]^G) \leq \delta(V).$$

Proof. We have to show that for every $v \in V \setminus \mathcal{N}_V$ there exists a homogeneous $f \in K[V]^G$ of degree $\leq \delta(V)$ such that $f(v) \neq 0$.

So let us assume that $v \in V \setminus \mathcal{N}_V$. Define $X \subseteq V$ as the Zariski closure of the cone $G \cdot Kv$ and $Y \subset X$ as the Zariski closure of $G \cdot v$. Note that $Y \subsetneq X$ since $v \in \mathcal{N}_V$, so $0 \notin Y$. Hence $\dim(Y) = \dim(X) - 1$. Let $K[X] = K[V]/I(X)$ be the graded coordinate ring of X. Since $K[X]$ is generated in degree 1, we can find linear functions $p_1, \ldots, p_{r+1} \in V^*$ such that the images in $K[X]$ form a homogeneous system of parameters in $K[X]$. Define $W := K^{r+1}$ and let $\psi : V \twoheadrightarrow W$ be the linear projection defined by $(p_1, \ldots, p_{r+1})$. The restriction of ψ to X is a finite morphism because $p_1, \ldots, p_{r+1}$ is a homogeneous system of parameters.

Now $\psi(Y)$ defines a hypersurface in W. The vanishing ideal $I(\psi(Y)) \subset K[W]$ is generated by some polynomial g. From Proposition 4.7.10 it follows that

$$\deg(g) \leq \deg(\psi(Y)) \leq \deg(Y) \leq \delta(Y).$$

Define $f := g \circ \psi \in K[V]$. Then f is a (non-homogeneous) polynomial of degree $\leq \delta(V)$ which vanishes on Y and $f(0) \neq 0$. Define $h = \mathcal{R}(f)$ were $\mathcal{R}$ is the Reynolds operator. Notice that h vanishes on Y because $I(Y)$ is G-stable, therefore stable under $\mathcal{R}$.

Write $h = h_0 + h_1 + \cdots + h_{\delta(V)}$ with h_i homogeneous of degree i. Then h_0 is a non-zero constant because $h(0) \neq 0$. Since $h(v) = 0$, we must have $h_i(v) \neq 0$ for some $i > 0$. Since $\deg(h_i) \leq \delta(V)$, we are done. $\square$

Definition 4.7.13. *Let $\rho : G \to \mathrm{End}(V)$ be the action of G on V. We define* $\delta_{\mathrm{gen}}(V) = \deg(\rho(G))$.

Proposition 4.7.14. *Suppose that G is a linearly reductive group and V is a rational representation. We have*

$$\delta(V) \leq \delta_{\mathrm{gen}}(V).$$

Proof. Suppose that $v \in V$. We define $\psi \colon \mathrm{End}(V) \to V$, $A \mapsto Av$. Then $\overline{\psi(\rho(G))} = \overline{G \cdot v}$. From Proposition 4.7.10 it follows that

$$\deg(\overline{G \cdot v}) \leq \deg(\rho(G)) = \delta_{\mathrm{gen}}(V).$$

$\square$

Example 4.7.15. If G is finite, then we obtain $\gamma(K[V]^G) \leq |\rho(G)|$. This also follows from Proposition 3.3.2.   ◁

We have $\gamma(K[V]^G) \leq \delta(V) \leq \delta_{\text{gen}}(V)$. Using this, we will give a concrete upper bound for $\gamma(K[V]^G)$ in terms of the degrees of the polynomials defining the group G and the representation $\rho : G \to \text{GL}(V)$. We will assume that the linearly reductive group G is given as in Section 4.1.2. So G is embedded in K^l for some l which gives us a surjective homomorphism $K[z_1, \ldots, z_l] \to K[G]$. The representation $\rho : G \to \text{GL}(V)$ is given by a matrix

$$\begin{pmatrix} a_{1,1} & a_{1,2} & \cdots & a_{1,n} \\ a_{2,1} & a_{2,2} & \cdots & a_{2,n} \\ \vdots & \vdots & \ddots & \vdots \\ a_{n,1} & a_{n,2} & \cdots & a_{n,n} \end{pmatrix}$$

with $a_{i,j} \in K[z_1, \ldots, z_l]$ for all i, j.

Proposition 4.7.16. *Suppose that G is a linearly reductive group and V is a rational representation. We have an inequality*

$$\gamma(K[V]^G) \leq \delta_{\text{gen}}(V) \leq CA^m,$$

where C is the degree of $G \subset K^l$, $A = \max_{i,j}(\deg(a_{i,j}))$ and $m = \dim \rho(G)$.

Proof. We take m hyperplanes in general position in $\text{End}(V)$. Suppose these hyperplanes are given as the zero sets of the functions $h_1, \ldots, h_m$, all of degree 1. If S is the intersection of $h_1 = h_2 = \cdots = h_m = 0$ with $\rho(G)$, then by definition $\#S = \delta_{\text{gen}}(V)$. Define $u_i = \rho^*(h_i) = h_i \circ \rho$ for all i. The degree of u_i is at most A for all i. The intersection of $u_1 = u_2 = \cdots = u_m = 0$ with $G \subseteq K^l$ has at most CA^m connected components. On the other hand this intersection is equal to $\rho^{-1}(S)$, so it must have at least $\delta_{\text{gen}}(V)$ connected components.   □

Example 4.7.17. We take $G = \text{SL}(W)$ where W is a q-dimensional vector space. Let $V_d = S_d(W)$ be the d-th symmetric power. If we choose a basis of W, we get an embedding of G into the $q \times q$ matrices, which is a q^2-dimensional space. The coordinate ring of G is given by

$$K[G] = K[\{z_{i,j} \mid 1 \leq i, j \leq q\}]/(\det((z_{i,j})) - 1).$$

We want to apply Proposition 4.7.16. We have $m \leq q^2 - 1$ and $C = q$ is the degree of the determinant polynomial. The action of G on V is given by a matrix $(a_{i,j})$ where all $a_{i,j}$ have degree $\leq d$, therefore $A = d$. By Proposition 4.7.16 we have

$$\gamma(K[V_d]^{\text{SL}_q}) \leq qd^{q^2-1}.$$

Note that $r = \dim K[V_d]^{\text{SL}_q} \leq \dim V_d = \binom{q+d-1}{q-1}$. From Theorem 4.7.4 it follows that

$$\beta(K[V_d]^{\mathrm{SL}_q}) \leq \tfrac{3}{8}\binom{q+d-1}{q-1}q^2 d^{2q^2-2}.$$

Note that this upper bound depends only polynomially on d. If $q = 2$, we have binary forms of degree d and $\beta(K[V_d]^{\mathrm{SL}_2}) \leq \tfrac{3}{2}(d + 1)d^6$. This bound still seems far from sharp. For $d = 2$, for example, we have $\beta(K[V_d]^{\mathrm{SL}_2}) = 2$ and $\tfrac{3}{2}(d + 1)d^6 = 288$. For ternary forms ($q = 3$) we get $\beta(K[V_d]^{\mathrm{SL}_3}) \leq \tfrac{27}{8}(d + 2)(d + 1)d^{18}$. ◁

There also exists an exact formula for $\delta_{\mathrm{gen}}(V)$ which easily follows from a formula of Kazarnovskii which we will explain now (see Kazarnovskij [129] and Brion [27] for a generalization). We will use the notation of Section A.4. Suppose that G is a connected reductive group and put $m := \dim G$ and $r := \mathrm{rank}\, G$. Fix a Borel subgroup $B \subset G$ and $T \subset B$ a maximal torus and denote by $\alpha_1, \alpha_2, \ldots, \alpha_\ell, \ell = \tfrac{m-r}{2}$, the positive roots. Let W be the Weyl group and let $e_1, e_2, \ldots, e_r$ be the Coxeter exponents, i.e., $e_1 + 1, e_2 + 1, \ldots, e_r + 1$ are the degrees of the generating invariants of W. Let $X(T) \cong \mathbb{Z}^r$ be the group of characters of T. For a representation $\rho : G \to \mathrm{GL}(V)$ we denote by $\mathcal{C}_V \subset E := X(T) \otimes_{\mathbb{Z}} \mathbb{R}$ the convex hull of 0 and the weights of V. On E we use the volume form $d\mathcal{V}$ which is the standard volume form by an isomorphism $E \cong \mathbb{R}^r$ which identifies $X(T)$ with $\mathbb{Z}^r$. We fix a W-invariant scalar product $\langle \cdot, \cdot \rangle$ on E. We define $\widetilde{\alpha}_i = 2\langle \alpha_i, \cdot \rangle / \langle \alpha_i, \alpha_i \rangle \in E^*$ for all i.

Proposition 4.7.18. *If the representation $\rho : G \to \mathrm{GL}(V)$ has finite kernel, then we have*

$$\delta_{\mathrm{gen}}(V) = \frac{m!}{|W|(e_1! e_2! \cdots e_r!)^2 |\ker(\rho)|} \int_{\mathcal{C}_V} (\widetilde{\alpha}_1 \widetilde{\alpha}_2 \cdots \widetilde{\alpha}_\ell)^2 d\mathcal{V}.$$

Proof. This follows from the formula of Kazarnovskii (see Kazarnovskij [129] and Derksen and Kraft [55]). □

Example 4.7.19. Let V_d be the vector space of binary forms of degree d, on which SL_2 acts. With the notation of Proposition 4.7.18, we have $m = 3$, $r = 1$, $e_1 = 1$, $|W| = 2$. If d is odd, then $|\ker(\rho)| = 1$, so we get

$$\delta_{\mathrm{gen}}(V_d) = 3 \int_{-d}^{d} x^2 \, dx = 2d^3,$$

and if d is even, then $\delta_{\mathrm{gen}}(V_d) = d^3$. ◁

4.7.2 Degree Bounds for Tori

In this section we discuss an upper bound for $\beta(K[V]^G)$ in case $G = T$ is a torus. This upper bound was given by Wehlau [255] and is better than the general upper bound given in Section 4.7.1.

We will assume that $G = T$ is a torus of dimension r. Let $\rho : T \to \mathrm{GL}(V)$ be an n-dimensional faithful representation of T. We may assume that the torus acts diagonally, i.e., the representation is given by a diagonal matrix

$$\rho = \begin{pmatrix} \chi_1 & & & 0 \\ & \chi_2 & & \\ & & \ddots & \\ 0 & & & \chi_n \end{pmatrix},$$

where $\chi_1, \chi_2, \ldots, \chi_n \in X(T)$ are rational 1-dimensional characters of the torus T. The set of rational 1-dimensional characters $X(T)$ can be identified with $\mathbb{Z}^r$, and we will identify $X(T) \otimes_{\mathbb{Z}} \mathbb{R}$ with $\mathbb{R}^r$. On $\mathbb{R}^r$ we have the usual volume form dV. Notice that under all identifications $X(T)$ has covolume 1 in $X(T) \otimes_{\mathbb{Z}} \mathbb{R}$. Let $\mathcal{C}_V$ be the convex hull of $\chi_1, \ldots, \chi_n$ inside $X(T) \otimes \mathbb{R}$.

Theorem 4.7.20 (Wehlau [255]). *With the notation and assumptions above we have*

$$\beta(K[V]^T) \le \max(n - r - 1, 1)r!\,\mathrm{vol}(\mathcal{C}_V),$$

where $\mathrm{vol}(\mathcal{C}_V)$ *denotes the volume of the polytope* $\mathcal{C}_V$.

Proof. Let $x_1, \ldots, x_n$ be the coordinate functions, so $K[V] \cong K[x_1, \ldots, x_n]$. We can identify a monomial $x_1^{\alpha_1} x_2^{\alpha_2} \cdots x_n^{\alpha_n}$ with $(\alpha_1, \ldots, \alpha_n) \in \mathbb{N}^n$. A T-invariant monomial corresponds to an n-tuple $(\alpha_1, \ldots, \alpha_n)$ satisfying

$$\alpha_1 \chi_1 + \alpha_2 \chi_2 + \cdots + \alpha_n \chi_n = 0. \qquad (4.7.3)$$

Let $S \subset \mathbb{N}^n$ correspond to the set of all T-invariant monomials. Let $\mathbb{Q}_+$ be the set of non-negative rational numbers. Consider the cone $\mathbb{Q}_+ S \subset \mathbb{Q}_+^n$. If $\alpha \in \mathbb{Q}_+^n \subset \{0\}$, then we call $\mathbb{Q}_+ \alpha$ a ray. A ray $\mathbb{Q}_+ \alpha \subset \mathbb{Q}_+ S$ is called an extremal ray of $\mathbb{Q}_+ S$ if $\alpha = \beta + \gamma$, $\beta, \gamma \in \mathbb{Q}_+ S$ implies that $\beta, \gamma \in \mathbb{Q}_+ \alpha$. Now $\mathbb{Q}_+ S$ is a polyhedral cone, so it has only finitely many extremal rays, say $\ell_1, \ell_2, \ldots, \ell_s$. For each i there exists a unique monomial R_i such that $\ell_i \cap \mathbb{N}^n = \mathbb{N} R_i$. This monomial R_i is invariant because $R_i \in \mathbb{Q}_+ S \cap \mathbb{N}^n = S$. Suppose that M is a T-invariant monomial. Since the dimension of $\mathbb{Q}_+ S$ is $n - r$, there exist indices $j_1, \ldots, j_{n-r}$ such that M lies in the convex hull of the rays $\ell_{j_1}, \ell_{j_2}, \ldots, \ell_{j_{n-r}}$. We can write

$$M = \alpha_1 R_{j_1} + \alpha_2 R_{j_2} + \cdots + \alpha_{n-r} R_{j_{n-r}}$$

with $\alpha_1, \ldots, \alpha_{n-r} \in \mathbb{Q}_+$. If we write $\alpha_i = a_i + \gamma_i$ where a_i is a non-negative integer and $0 \le \gamma_i < 1$ for all i, then we get

$$M = R_{j_1}^{a_1} R_{j_2}^{a_2} \cdots R_{j_{n-r}}^{a_{n-r}} N,$$

where the degree of N satisfies

$$\deg(N) = \gamma_1 \deg(R_{j_1}) + \gamma_2 \deg(R_{j_2}) + \cdots + \gamma_{n-r} \deg(R_{j_{n-r}}) \le$$

$$\le (n - r) \max\{\deg(R_i) \mid i = 1, 2, \ldots, s\}.$$

By Ewald and Wessels [65, Theorem 2], N is decomposable into smaller invariant monomials if

$$\gamma_1 + \cdots + \gamma_{n-r} > n - r - 1 \geq 1.$$

It follows that

$$\beta(K[V]^T) \leq \max\{n - r - 1, 1\} \max\{\deg(R_i) \mid i = 1, 2 \ldots, s\}.$$

We would like to bound $\deg(R_i)$. After a permutation of the variables we can assume that $R_i = (\mu_1, \ldots, \mu_t, 0, 0, \ldots, 0)$ with $\mu_1, \ldots, \mu_t \in \mathbb{N} \setminus \{0\}$. We claim that the characters $\chi_1, \chi_2, \ldots, \chi_t$ span a $(t - 1)$ dimensional vector space. If they spanned a space of dimension $< t - 1$, then there would be a solution $T = (\tau_1, \ldots, \tau_t, 0, \ldots, 0) \in \mathbb{Q}^n$ to (4.7.3) independent of R_i and $R_i \pm \epsilon T \in \mathbb{Q}_+ S$ for small ϵ. This contradicts the extremality of the ray ℓ_i. After another permutation of $x_{t+1}, \ldots, x_n$ we may assume that $\chi_1, \chi_2, \ldots, \chi_{r+1}$ span an r-dimensional vector space. The equations

$$\alpha_1 \chi_1 + \alpha_2 \chi_2 + \cdots + \alpha_{r+1} \chi_{r+1} = \alpha_{r+2} = \cdots = \alpha_n = 0 \qquad (4.7.4)$$

have a one-dimensional solution space. By Cramer's rule, we can find a non-zero solution $A = (\alpha_1, \ldots, \alpha_{r+1}, 0, \ldots, 0)$ to (4.7.4):

$$\alpha_i = (-1)^i \det(\chi_1, \ldots, \chi_{i-1}, \chi_{i+1}, \ldots, \chi_{r+1}) =$$

$$= \pm r! \operatorname{vol}(\mathcal{C}(0, \chi_1, \ldots, \chi_{i-1}, \chi_{i+1}, \ldots, \chi_{r+1})).$$

Here $\mathcal{C}$ denotes the convex hull and vol denotes the volume. Since A is a rational (even an integral) multiple of R_i, we obtain

$$\deg(R_i) \leq |\alpha_1| + |\alpha_2| + \cdots + |\alpha_{r+1}| = r! \sum_{i=1}^{r+1} \operatorname{vol}(\mathcal{C}(0, \chi_1, \ldots, \widehat{\chi_i}, \ldots, \chi_{r+1})) =$$

$$= r! \operatorname{vol}(\mathcal{C}(\chi_1, \chi_2, \ldots, \chi_{r+1})) \leq r! \operatorname{vol}(\mathcal{C}_V)$$

so we have

$$\beta(K[V]^T) \leq \max\{n - r - 1, 1\} r! \operatorname{vol}(\mathcal{C}_V).$$

$$\square$$

4.8 Properties of Invariant Rings

Let K be a field of characteristic 0, and let V be a representation of a linearly reductive group G. Let $\pi : V \to V /\!/ G$ be the quotient map. For finite groups, the ring of invariants is a polynomial ring if and only if G is a generalized reflection group. For infinite linearly reductive groups there is not such a simple answer to the question when the invariant ring is a polynomial ring. In this section we will briefly discuss various properties of invariant rings, interactions between these properties, and some of the classifications which can be found in the literature.

We consider the following properties:

(FA) (*Free Algebra*): $K[V]^G$ is a polynomial ring.
(ED) (*EquiDimensional*): All fibers of π have the same dimension.
(FO) (*Finitely many Orbits*): Each fiber of π has finitely many orbits.
(FM) (*Free Module*): The ring $K[V]$ is a free $K[V]^G$-module.

The following implications hold:

(FO)$\Rightarrow$(ED): Let d be the dimension of a general fiber. A general orbit has dimension $\leq d$ and by semicontinuity of dimensions (see Hartshorne [102, Exercise II.3.22]) of stabilizers, every orbit has dimension $\leq d$. If we assume that every fiber has finitely many orbits, then every fiber has dimension $\leq d$. By semicontinuity of the dimension of the fiber, every fiber must have dimension d.

(FA) and (ED) $\Leftrightarrow$ (FM): See Popov and Vinberg [194, II.8.1] and the citations there.

Without loss of generality we may assume that G acts faithfully on V. Various classifications have been made. We mention a few.

(i) Assume G is connected and simple, and V is irreducible.
   a) All (G, V) satisfying (FA) were classified by Kac et al. [125].
   b) All (G, V) satisfying (ED) were classified by Popov [188].
   c) All (G, V) satisfying (FO) were classified by Kac [122].
   d) Moreover, all (G, V) satisfying
        (ST) Every point in V has a nontrivial stabilizer
      were classified by Popov [187,186].
   It turns out that for irreducible representations of connected simple groups G, the lists for all classifications coincide. In particular (FA), (ED), (FO), (FM) and (ST) are all equivalent for these examples.
(ii) Assume G is connected and simple, and V is arbitrary.
   a) All (G, V) satisfying (FA) were classified by Adamovich and Golovina [3] and by Schwarz [212] independently.
   b) All (G, V) satisfying (ED) were classified by Adamovich [2].
(iii) Assume G is connected and *semisimple*, and V is irreducible.
   a) All (G, V) satisfying (FA) were classified by Littelmann [161].
   b) All (G, V) satisfying (DE) also can be found by Littelmann [161].
   c) All (G, V) satisfying (FO) were classified by Kac [123] (see Dadoc and Kac [51] for the corrections in the tables).

Popov [188] made the following so-called *Russian Conjecture*:

Conjecture 4.8.1. *If G is connected and semisimple and V is a representation, then (ED) implies (FA).*

Despite all classifications, no counterexample has been found yet. For a more detailed survey on the Russian Conjecture we refer the reader to Wehlau [256].

A more complete discussion of these classifications can be found in Popov and Vinberg [194, II.8.].

5 Applications of Invariant Theory

In this chapter we give a survey of some applications of invariant theory. The selection of topics is very incomplete, and so are certainly the references given for each topic. For example, we omit applications to projective geometry, which are very well explained in Sturmfels [239, Chapter 3]. We try to present a wide range of applications from different fields, and exemplify the use of invariant theory in each case.

5.1 Cohomology of Finite Groups

The cohomology of finite groups has been a very active field of research in recent years, and it has used invariant theory as a tool for computing cohomology rings (see Adem and Milgram [8]). The most important way invariant theory comes in is the following. Suppose that G is a finite group and V is a KG-module, and we want to know the cohomology $H^*(G, V)$. Furthermore, suppose that we have an exact sequence of groups

$$1 \longrightarrow N \longrightarrow G \longrightarrow H \longrightarrow 1$$

such that the characteristic of K does not divide the order of H. Then there is a natural action of H on $H^*(N, V)$ (see Evens [64, p. 35]), and the restriction $\mathrm{res}_{G,N}$ provides an isomorphism

$$H^*(G, V) \xrightarrow{\sim} H^*(N, V)^H. \tag{5.1.1}$$

This is well known and follows from the fact that $\mathrm{cores}_{N,G} \circ \mathrm{res}_{G,N}$ is multiplication by $|H|$ (see Evens [64, Proposition 4.2.2]) and that $\mathrm{res}_{G,N} \circ \mathrm{cores}_{N,G} = \sum_{\sigma \in H} \sigma$ by the Mackey formula (see Evens [64, Theorem 4.2.6]). A particularly interesting case is $V = K$, the trivial module. The cup product gives $H^*(G, K)$ the structure of a (not quite commutative) K-algebra which is respected by the H-action, so the right-hand side of the isomorphism (5.1.1) is an invariant ring. The isomorphism does not hold if we drop the assumption that $\mathrm{char}(K) \nmid |H|$, but then there is still a strong relation between $H^*(G, K)$ and $H^*(N, K)^H$ (see Adem and Milgram [7]).

There are two difficulties in applying the standard techniques of invariant theory to obtain $H^*(N, K)^H$. The first is that $H^*(N, K)$ is not commutative but instead $\alpha\beta = (-1)^{de}\beta\alpha$ for α and β homogeneous of degree d and e, respectively. This difficulty vanishes in characteristic 2, or if we consider only the even part of the cohomology. The second difficulty is that $H^*(N, K)$ is in general not a standard graded algebra , i.e., not generated in degree 1. Nevertheless, techniques from the above sections can be used in many examples to facilitate the computation of cohomology rings.

Example 5.1.1 (cf. Adem and Milgram [8]). We want to compute the mod 2 cohomology $H^*(A_4, \mathbb{F}_2)$ of the alternating group on four symbols. Let $N = \langle(1\ 2)(3\ 4), (1\ 3)(2\ 4)\rangle$ be the Sylow 2-subgroup. By Evens [64, p. 33] we have $H^*(N, \mathbb{F}_2) = \mathbb{F}_2[\alpha, \beta]$ with α, β algebraically independent of degree 1. A generator of G/N acts on $H^*(N, \mathbb{F}_2)$ by $\alpha \mapsto \beta$ and $\beta \mapsto \alpha + \beta$. In this situation we can use the standard algorithms from Sections 3.3 and 3.5 and obtain

$$H^*(A_4, \mathbb{F}_2) = \mathbb{F}_2[A, B, C]$$

with

$$A = \alpha^2\beta + \alpha\beta^2,$$
$$B = \alpha^3 + \alpha\beta^2 + \beta^3,$$
$$C = \alpha^2 + \alpha\beta + \beta^2,$$

and, using the methods of Section 3.6, the relation

$$C^3 + A^2 + AB + B^2 = 0.$$

$\lhd$

5.2 Galois Group Computation

If $f \in K[X]$ is a separable polynomial over a field K, then the Galois group $\mathrm{Gal}(N/K)$ of the splitting field N of f over K acts on the zeros $\alpha_1, \ldots, \alpha_n \in N$ of f. This yields a faithful permutation representation $\mathrm{Gal}(N/K) \to S_n$, whose image is denoted by $\mathrm{Gal}(f)$, the Galois group of f. Of course $\mathrm{Gal}(f)$ is only determined up to conjugacy in S_n. In this section we look at methods for computing the Galois group. We will discuss the basic ideas of Stauduhar's algorithm [237], which is the one that is most widely used. Our presentation is strongly influenced by the article of Geißler and Klüners [82].

Suppose we know already that $\mathrm{Gal}(f) \leq G$ for a subgroup $G \leq S_n$ (this is always true for $G = S_n$). Given a smaller group $H \leq G$ (often a maximal subgroup of G), we would like to check whether $\mathrm{Gal}(f) \leq H$. The basic idea is to use a polynomial $F \in K[x_1, \ldots, x_n]^H$ such that $\sigma \cdot F \neq F$ for all $\sigma \in G \setminus H$. Such a polynomial is called a **G-relative H-invariant**. In other words, we are looking for a polynomial $F \in K[x_1, \ldots, x_n]$ with $\mathrm{Stab}_G(F) = H$. G-relative H-invariants always exist, one (standard) example being provided by

$$F := \sum_{\sigma \in H} \left(\sigma \cdot \prod_{i=1}^{n-1} x_i^i \right).$$

Since it is important for the efficiency of Stauduhar's algorithm to obtain "simple" G-relative H-invariants, Geißler and Klüners [82] propose to compare the Hilbert series of $K[x_1, \ldots, x_n]^G$ and $K[x_1, \ldots, x_n]^H$, which can be computed by Molien's formula (see Theorem 3.2.2 and Section 3.10.4). If H is maximal in G, the minimal degree of a G-relative H-invariant can be read off from this comparison, and then the actual construction of F can be done by forming H-orbit sums of monomials. In some special situations it is also possible to obtain G-relative H-invariants whose evaluation requires a fairly small number of arithmetic operations (see Eichenlaub [58] or Geißler and Klüners [82]). Geißler's master thesis [81, Section 5.2] has some examples where a good choice of invariants dramatically reduces the number of arithmetic operations. The following proposition explains why G-relative H-invariants are useful for Galois group computation.

Proposition 5.2.1. *Suppose that in the above situation F is a G-relative H-invariant and*

$$F(\alpha_{\sigma(1)}, \ldots, \alpha_{\sigma(n)}) \neq F(\alpha_1, \ldots, \alpha_n) \quad \text{for all} \quad \sigma \in G \setminus H, \qquad (5.2.1)$$

where $\alpha_1, \ldots, \alpha_n$ are the roots of f, and $\mathrm{Gal}(f) \leq G$. Then

$$\mathrm{Gal}(f) \leq H \quad \Longleftrightarrow \quad F(\alpha_1, \ldots, \alpha_n) \in K.$$

Proof. Write $\gamma := F(\alpha_1, \ldots, \alpha_n)$. If $\mathrm{Gal}(f) \leq H$, then for every $\sigma \in \mathrm{Gal}(f)$ we have $\sigma \cdot F = F$ and therefore $\sigma \cdot \gamma = \gamma$ for every $\sigma \in \mathrm{Gal}(N/K)$, since the Galois action on the α_i is the same as the permutation action on the x_i. Thus $\gamma \in N^{\mathrm{Gal}(N/K)} = K$. On the other hand, if $\mathrm{Gal}(f) \not\leq H$, there exists $\sigma \in \mathrm{Gal}(N/K)$ whose action on the α_i lies in $G \setminus H$. Hence by (5.2.1) we have $\sigma \cdot \gamma \neq \gamma$, so $\gamma \notin K$. $\qquad \square$

We will explain later how the condition $F(\alpha_1, \ldots, \alpha_n) \in K$ can be tested. First we discuss how the hypothesis (5.2.1) can be achieved. Fortunately, one does not have to change F for this, but can use Tschirnhausen transformations.

Lemma 5.2.2 (Girstmair [84]). *Suppose that K is an infinite field. Then in the situation of Proposition 5.2.1 there exist $c_0, \ldots, c_{n-1} \in K$ such that for $\beta_i := \sum_{j=0}^{n-1} c_j \alpha_i^j$ we have*

$$F(\beta_{\sigma(1)}, \ldots, \beta_{\sigma(n)}) \neq F(\beta_1, \ldots, \beta_n) \quad \text{for all} \quad \sigma \in G \setminus H$$

and $\beta_i \neq \beta_j$ for $i \neq j$. In fact, if $S \subset K$ is any set with $|S| > \deg(F) \cdot \binom{[G:H]}{2} + \binom{n}{2}$, the c_i can be chosen from S.

Proof. Let $C_0, \ldots, C_{n-1}$ be indeterminates over N. Then the $B_i :=$ $\sum_{j=0}^{n-1} C_j \alpha_i^j$ are algebraically independent over N (Vandermonde determinant). Set

$$D(C_0, \ldots, C_{n-1}) :=$$
$$\prod_{1 \leq i < j \leq n} (B_i - B_j) \prod_{1 \leq i < j \leq m} \left(F(B_{\sigma_i(1)}, \ldots, B_{\sigma_i(n)}) - F(B_{\sigma_j(1)}, \ldots, B_{\sigma_j(n)}) \right),$$

where $\sigma_1, \ldots, \sigma_m$ is a set of left coset representatives of H in G. Since $\sigma \cdot F \neq \tau \cdot F$ for $\sigma H \neq \tau H$, it follows that $D(C_0, \ldots, C_{n-1}) \neq 0$. Since $|S| > \deg(D)$, there exists $c_0 \in S$ with $D(c_0, C_1, \ldots, C_{n-1}) \neq 0$. Continuing this way, we find $c_0, \ldots, c_{n-1} \in K$ such that $D(c_0, \ldots, c_{n-1}) \neq 0$. But this means that $F(\beta_{\sigma(1)}, \ldots, \beta_{\sigma(n)}) \neq F(\beta_1, \ldots, \beta_n)$ for $\sigma \in G \setminus H$, and $\beta_i \neq \beta_j$ for $i \neq j$. $\square$

If $\hat{f} := \prod_{i=1}^{n} (X - \beta_i) \in K[X]$ with the β_i as in Lemma 5.2.2, then of course $\mathrm{Gal}(\hat{f}) = \mathrm{Gal}(f)$. The rough algorithm for computing $\mathrm{Gal}(f)$ is now clear: First (pre-)compute the subgroup lattice of S_n, and for each pair $H \leq G$ of subgroups with H maximal in G find a G-relative H-invariant. Then start with S_n and use Proposition 5.2.1 and if necessary Tschirnhausen transformations to walk down the subgroup lattice until a group $G \leq S_n$ is reached such that $\mathrm{Gal}(f) \leq G$ but $\mathrm{Gal}(f) \not\leq H$ for every maximal subgroup $H \leq G$. Then $\mathrm{Gal}(f) = G$. Apart from a number of implementation issues, there is still one fundamental problem with this idea: We do not usually know (the exact values of) the roots α_i of f. We will discuss two basic approaches to handle this problem, both of which involve the resolvent, defined as follows: For $F \in K[x_1, \ldots, x_n]$ a G-relative H-invariant polynomial, define the **"resolvent form"** as

$$R_{G,H,F}(X) := \prod_{\sigma \in G/H} (X - \sigma \cdot F) \in K[x_1, \ldots, x_n]^G[X],$$

where the σ run through a set of left coset representatives. The **resolvent** is then defined by

$$\bar{R}_{G,H,F}(X) := \prod_{\sigma \in G/H} \left(X - F(\alpha_{\sigma(1)}, \ldots, \alpha_{\sigma(n)}) \right) \in K[X],$$

i.e., by substituting $x_i = \alpha_i$ in $R_{G,H,F}(X)$. Proposition 5.2.1 now translates to:

Theorem 5.2.3. *Let $f \in K[x]$ be a separable polynomial of degree n and $G \leq S_n$ a subgroup with $\mathrm{Gal}(f) \leq G$. Moreover, let $H < G$ be a proper subgroup and $F \in K[x_1, \ldots, x_n]$ a G-relative H-invariant. Suppose that the resolvent $\bar{R}_{G,H,F}(X) \in K[X]$ has non-zero discriminant. Then the following statements are equivalent:*

(a) There exists $\sigma \in G$ such that $\mathrm{Gal}(f) \leq \sigma H \sigma^{-1}$.

(b) $\bar{R}_{G,H,F}(X)$ has a zero in K.

In this case, after a suitable renumbering of the zeros of f we may assume that $\mathrm{Gal}(f) \leq H$.

Example 5.2.4. Suppose that $\mathrm{char}(K) \neq 2$. Then

$$F := \prod_{1 \leq i < j \leq n} (x_i - x_j)$$

is an S_n-relative A_n-invariant. The resolvent is

$$\bar{R}_{S_n,A_n,F}(X) = X^2 - D(f),$$

where $D(f)$ is the discriminant. In this case, Theorem 5.2.3 yields the well-known criterion that $\mathrm{Gal}(f) \leq A_n$ if and only if $D(f)$ is a square in K.
◁

We now present two approaches how to handle the fact that the zeros α_i of f are not known.

5.2.1 Approximating Zeros

The most obvious and most practical approach is to approximate the α_i and work with approximations instead of exact values. Of course it depends on the ground field K whether and in what sense we can talk about "approximation" at all. Fields like $\mathbb{Q}$ or rational function fields $k(t)$ lend themselves for approximating zeros. Moreover, we need to be able to detect whether for a G-relative H-invariant $F \in K[x_1, \ldots, x_n]$ we have $F(\alpha_1, \ldots, \alpha_n) \in K$ by evaluating F at the approximations of the α_i. The most important case is $K = \mathbb{Q}$. For this case a p-adic approximation of the α_i, using a suitable prime p, turns out to be practical (see Geißler and Klüners [82]). Of course numeric approximations also work. Moreover, the α_i can be assumed to be integral over $\mathbb{Z}$ (simply by multiplying them with a suitable integer). If the G-relative H-invariant polynomials F are chosen to lie in $\mathbb{Z}[x_1, \ldots, x_n]$, then it only has to be tested whether $F(\alpha_1, \ldots, \alpha_n) \in \mathbb{Z}$. This makes it possible for Geißler and Klüners to derive criteria for the accuracy of the approximation of the α_i. If the accuracy meets these criteria, then it is possible to decide whether $F(\alpha_1, \ldots, \alpha_n)$ lies in $\mathbb{Z}$. It can also be decided from the approximation whether $\bar{R}_{G,H,F}(X)$ is separable.

These remarks should give the reader a general ideal how algorithms for computing $\mathrm{Gal}(f)$ by approximating zeros can be set up. There are many details and implementation issues that we omit. For these, we refer the reader to Geißler and Klüners [82].

5.2.2 The Symbolic Approach

A different approach which avoids approximations was proposed by Colin [45]. The idea uses the fact that the resolvent form $R_{G,H,F}(X)$ has coefficients in $K[x_1,\ldots,x_n]^G$. Therefore these coefficients can be expressed as a polynomial in generating invariants for $K[x_1,\ldots,x_n]^G$. This can be applied most readily in the case $G = S_n$, i.e., if we are dealing with absolute resolvents. Then

$$R_{S_n,H,F}(X) = X^n + \sum_{i=1}^{n} H_i(s_1,\ldots,s_n)X^{n-i}$$

with s_i the i-th elementary symmetric polynomial and $H_i \in K[x_1,\ldots,x_n]$. The H_i can be found by Algorithm 3.10.2. If

$$f = X^n + \sum_{i=1}^{n}(-1)^i a_i X^{n-i},$$

then $a_i = s_i(\alpha_1,\ldots,\alpha_n)$, and therefore the resolvent can be expressed as

$$\bar{R}_{S_n,H,F}(X) = X^n + \sum_{i=1}^{n} H_i(a_1,\ldots,a_n)X^{n-i}.$$

Thus it is not necessary to know the zeros α_i in order to find the resolvent. It is also easy to apply Tschirnhausen transformations to arrange that $\bar{R}_{S_n,H,F}(X)$ is separable (using Lemma 5.2.2). In fact, if $\beta_i = g(\alpha_i)$ with $g \in K[X]$ a polynomial whose coefficients were chosen randomly, then

$$\prod_{i=1}^{n}(X - g(x_i)) = X^n + \sum_{i=1}^{n} h_i(s_1,\ldots,s_n)X^{n-i},$$

where the $h_i \in K[x_1,\ldots,x_n]$ can again be computed by Algorithm 3.10.2. Then for $\hat{f} := \prod_{i=1}^{n}(X - \beta_i)$ we have

$$\hat{f} = X^n + \sum_{i=1}^{n} h_i(a_1,\ldots,a_n)X^{n-i},$$

where the a_i are again the coefficients of the original polynomial f.

Things become much more delicate if we are dealing with the relative case, i.e., $G \lneqq S_n$. Again the task is to find the (relative) resolvent $\bar{R}_{G,H,F}(X)$ without knowing the α_i. Suppose that we have descended along a chain of subgroups $S_n = G_0 \gneqq G_1 \gneqq \cdots \gneqq G_k = G$ and proved, using Theorem 5.2.3, that $\mathrm{Gal}(f) \leq G$. If F_i is a G_{i-1}-relative G_i-invariant, this amounts to having found the values $b_i := F_i(\alpha_1,\ldots,\alpha_n) \in K$ (after a suitable renumbering of the unknown zeros). By Galois theory we know that $K(x_1,\ldots,x_n)^{G_i} = K(x_1,\ldots,x_n)^{G_{i-1}}(F_i)$. Thus $K(x_1,\ldots,x_n)^G = K(s_1,\ldots,s_n,F_1,\ldots,F_k)$. In order to find $\bar{R}_{G,H,F}(X)$, we want to write the resolvent form $R_{G,H,F}(X)$ in terms of the s_i and F_i. Each coefficient of $R_{G,H,F}(X)$ lies in $K(x_1,\ldots,x_n)^G = K(s_1,\ldots,s_n,F_1,\ldots,F_k)$, hence we can write

$$R_{G,H,F}(X) = X^n + \sum_{i=1}^{n} H_i(s_1,\ldots,s_n,F_1,\ldots,F_k)X^{n-i} \qquad (5.2.2)$$

with H_i rational functions in $n+k$ arguments. In this case Algorithm 3.10.2 or any other standard algorithm using linear algebra does not help to find the H_i. Instead an algorithm of Sweedler [242], which involves Gröbner bases, can be used. Substituting the α_i into (5.2.2) yields

$$\bar{R}_{G,H,F}(X) = X^n + \sum_{i=1}^{n} H_i(a_1,\ldots,a_n,b_1,\ldots,b_k)X^{n-i},$$

but only if no zero-division occurs on the right hand side. Such a zero-division can be avoided by using Tschirnhausen transformations again, in the spirit of Lemma 5.2.2. Then of course applying a Tschirnhausen transformation changes the values of the a_i and b_i. All the questions arising here can be solved, but at this point we prefer to stop and refer the reader to Colin [45] for further details.

Vectorial polynomials. Suppose that we have a Galois field extension N/K with $\mathbb{F}_q \subseteq K$. Let $\mathcal{M} \subset N$ be a finite set such that $N = K(\mathcal{M})$ and $\mathrm{Gal}(N/K)$ maps $\mathcal{M}$ to itself. Then $\mathrm{Gal}(N/K)$ also acts on the $\mathbb{F}_q$-span V of $\mathcal{M}$, and $N = K(V)$, so the action is faithful. The polynomial $f := \prod_{\alpha \in V}(X - \alpha) \in K[X]$ has the form

$$f = X^{q^n} + a_1 X^{q^{n-1}} + \cdots + a_{n-1}X^q + a_n X$$

(see Wilkerson [260]). Polynomials of this form are called **vectorial** (see Abhyankar [1]). Obviously the set of zeros of *any* vectorial polynomial is a vector space over $\mathbb{F}_q$. In the same way as $\mathrm{Gal}(f)$ is the image of $\mathrm{Gal}(N/K)$ under the permutation representation on the roots of f, we define $\mathrm{Gal}_{\mathrm{vect}}(f)$ to be the image of $\mathrm{Gal}(N/K)$ under the representation $\mathrm{Gal}(N/K) \to \mathrm{GL}(V)$. So $\mathrm{Gal}_{\mathrm{vect}}(f) \leq \mathrm{GL}(V)$ is a finite linear group, and we can try to formulate an analogue of Stauduhar's method for the determination of $\mathrm{Gal}_{\mathrm{vect}}(f)$. This method will use invariants in $\mathbb{F}_q[V]^G$ for subgroups $G \leq \mathrm{GL}(V)$ to test whether $\mathrm{Gal}_{\mathrm{vect}}(f) \leq G$.

5.3 Noether's Problem and Generic Polynomials

In inverse Galois theory (see Malle and Matzat [165]) one is interested in obtaining polynomials which have a given group as Galois group. It is even more desirable to have a polynomial which parametrizes all polynomials with a given group, or at least all Galois field extensions having this group. A typical example is the polynomial $X^2 - t$, which parametrizes all C_2-extensions over a field of characteristic not 2. Such polynomials are called generic. More precisely, we define:

Definition 5.3.1. *Let K be a field and G a finite group. A separable polynomial $g(t_1, \ldots, t_m; X) \in K(t_1, \ldots, t_m)[X]$ with coefficients in the rational function field $K(t_1, \ldots, t_m)$ is called* **generic** *for G over K if the following two properties hold:*

(a) The Galois group of g (as a polynomial in X) is G;
(b) if L is an infinite field containing K and N/L is a Galois field extension with group G, then there exist $\lambda_1, \ldots, \lambda_m \in L$ such that N is the splitting field of $g(\lambda_1, \ldots, \lambda_m; X)$ over L.

A connection between generic polynomials and invariant theory was discovered by Emmy Noether [184], who proved that if the invariant field $K(x_1, \ldots, x_n)^G$ of a permutation group $G \le S_n$ is purely transcendental as a field extension of K, then a generic polynomial for G exists and has n parameters. A generic polynomial constructed in this way even parametrizes (almost) all polynomials with Galois group G. Because of this result, the question whether the invariant field of a finite linear group is purely transcendental is known as **Noether's problem**. Apart from a complete answer in the case of abelian groups given by Lenstra [160] and a few examples and counter-examples (see Swan [241], Saltman [206]), not much is known about Noether's problem. In particular, we have no algorithm for deciding whether the invariant field of a given finite linear group is purely transcendental or not. For a survey on Noether's problem we refer the reader to Saltman [207] or Kemper [131].

The following theorem gives a somewhat more general construction method for generic polynomials, which is not restricted to permutation representations.

Theorem 5.3.2 (see Kemper and Mattig [145]). *Let G be a finite group and V an m-dimensional, faithful linear representation of G over a field K. If the invariant field $K(V)^G$ is purely transcendental over K, then there exists a generic polynomial with m parameters for G over K.*

The generic polynomial whose existence is guaranteed by Theorem 5.3.2 can be constructed as follows: Assume that $K(V)^G = K(\varphi_1, \ldots, \varphi_m)$ (which implies that the φ_i are algebraically independent). Choose a finite, G-stable subset $\mathcal{Y} \subset K(V)$ on which G acts faithfully, and set

$$f(X) := \prod_{y \in \mathcal{Y}} (X - y) \in K(V)[X].$$

The coefficients of f are G-invariant, so f can be written as $f(X) = g(\varphi_1, \ldots, \varphi_m; X)$ with $g \in K(t_1, \ldots, t_m)[X]$. Then $g(t_1, \ldots, t_m; X)$ is the desired generic polynomial for G over K. We say that $\varphi_1, \ldots, \varphi_m$ form a **minimal basis** of the invariant field. Thus the knowledge of a minimal basis leads directly to the construction of a generic polynomial. The advantage of working with linear representations instead of permutation representations is that they usually lead to simpler generic polynomials. Moreover, considering a representation of small degree often makes it easier to give a positive answer to Noether's problem and to find a minimal basis.

Example 5.3.3 ([145]). We would like to construct a generic polynomial for the cyclic group $G = C_4$ over $K = \mathbb{Q}$. The smallest faithful representation is given by sending a generator of G to $\left(\begin{smallmatrix} 0 & 1 \\ -1 & 0 \end{smallmatrix}\right)$. It is easy to compute generating invariants using the methods of Chapter 3. We obtain

$$f_1 = x_1^2 + x_2^2, \quad f_2 = x_1^2 x_2^2, \quad \text{and} \quad f_3 = x_1 x_2(x_1^2 - x_2^2)$$

as generating invariants, subject to the relation

$$f_3^2 = f_1^2 f_2 - 4 f_2^2. \tag{5.3.1}$$

The invariant ring is not a polynomial ring, but dividing (5.3.1) by f_2^2 tells us that $\varphi_1 := f_1$ and $\varphi_2 := f_3/f_2$ form a minimal basis:

$$K(V)^G = K(f_1, f_2, f_3) = K(\varphi_1, \varphi_2).$$

We choose $\mathcal{Y} = \{\pm x_1, \pm x_2\}$ and obtain

$$f(X) = \prod_{y \in \mathcal{Y}} (X - y) = X^4 - f_1 X^2 + f_2 = X^4 - \varphi_1 X^2 + \frac{\varphi_1^2}{\varphi_2^2 + 4}.$$

This yields the generic polynomial $g(t_1, t_2; X) = X^4 - t_1 X^2 + t_1^2/(t_2^2 + 4)$. Alternatively, taking $\mathcal{Y} = \{\pm \varphi_1/x_1, \pm \varphi_1/x_2\}$ and replacing φ_1 and φ_2 by $-\varphi_1/2$ and $2\varphi_2$, respectively, yields the nicer generic polynomial

$$g(t_1, t_2; X) = X^4 + 2t_1(t_2^2 + 1)X^2 + t_1^2(t_2^2 + 1).$$

This is much simpler than the "classical" generic polynomial given by Seidelmann [213]. ◁

Similarly, one gets nice generic polynomials over $K = \mathbb{Q}$ for groups like C_3, the Klein 4-group, the dihedral group D_4 of order 8, and many more, in particular for groups which become reflection groups after adding some scalar matrices (see Kemper and Mattig [145]). The approach given by Theorem 5.3.2 turns out to be particular successful in positive characteristic. For example, the fact that the invariant ring $\mathbb{F}_q[V]^{\mathrm{GL}(V)}$ is generated by the Dickson invariants (see Wilkerson [260]) leads to the generic polynomial

$$g(t_1, \ldots, t_m; X) = X^{q^m - 1} + t_1 X^{q^{m-1} - 1} + \cdots + t_{m-1} X^{q-1} + t_m.$$

for $G = \mathrm{GL}_m(\mathbb{F}_q)$ over $K = \mathbb{F}_q$.

Example 5.3.4 (Artin-Schreier polynomials). Let $G = C_p$ be the cyclic group of prime order p and $K = \mathbb{F}_p$. A faithful representation of G over K is given by sending a generator of G to $\left(\begin{smallmatrix} 1 & 1 \\ 0 & 1 \end{smallmatrix} \right)$. The invariant ring is a polynomial ring generated by

$$f_1 = x_1 \quad \text{and} \quad f_2 = x_2^p - x_1^{p-1} x_2,$$

which follows by Theorem 3.7.5. Hence f_1 and f_2 also form a minimal basis of $K(V)^G$. A more convenient minimal basis is given by $\varphi_1 = x_1$ and $\varphi_2 = f_2/x_1^p$. We choose $\mathcal{Y} = \{(ax_1 + x_2)/x_1 \mid a \in \mathbb{F}_p\}$. Then

$$f(X) = \prod_{a \in \mathbb{F}_p} \left(X - \frac{x_2}{x_1} - a \right) = \left(X - \frac{x_2}{x_1} \right)^p - \left(X - \frac{x_2}{x_1} \right) =$$

$$= X^p - X - \frac{x_2^p - x_1^{p-1} x_2}{x_1^p} = X^p - X - \varphi_2.$$

Thus we get $g(t, X) = X^p - X - t$ as a generic polynomial for C_p over $\mathbb{F}_p$. This is the well-known Artin-Schreier polynomial (see Lang [158, Chapter VIII, Theorem 6.4]).   ◁

Many more examples of generic polynomials in positive characteristic (for example for classical groups) can be obtained from the fact that for every finite, irreducible reflection group the invariant field is purely transcendental (Kemper and Malle [144]).

5.4 Systems of Algebraic Equations with Symmetries

We already mentioned in Section 1.2 that Gröbner bases can be used to solve systems of algebraic equations. In practice, such systems very often have symmetries, i.e., the solution set or the given equations are invariant under the action of a linear group (which may be finite or infinite). The disadvantage of Gröbner bases is that they destroy the symmetries of the system, instead of using them. For example, a lexicographic Gröbner basis of the ideal $(x + y - 1, xy)$ (invariant under $x \leftrightarrow y$) is $\{x + y - 1, y^2 - y\}$. In this section we will explain how invariants (in combination with Gröbner bases) can be used to rectify this shortcoming of Gröbner bases. The goal of these techniques is to facilitate (or make possible at all) practical computations. The ideas of this exposition are drawn from Sturmfels [239, Section 2.6].

Let $G \leq \mathrm{GL}(V)$ be a finite group acting on a finite dimensional vector space V over an algebraically closed field K. From the finiteness of G it follows that the categorical quotient $\pi_G \colon V \to V /\!\!/ G$ is in fact a geometric quotient (see Section 2.3.1). This implies that

$$\pi_G^{-1}(\pi_G(X)) = G \cdot X \tag{5.4.1}$$

for any affine variety $X \subseteq V$. We are interested in the situation in which X is G-stable, so we have $X = \pi_G^{-1}(\pi_G(X))$. The idea of using π_G is that it shrinks a G-orbit of points in V into a single point in $V /\!\!/ G$. For this reason the image $\pi_G(X)$ is also called the **orbit variety** of X. Suppose X is given by an ideal $I = (f_1, \ldots, f_m)$. How can we use (5.4.1) for the computation of X? We first have to calculate the orbit variety $\pi_G(X)$, which is equal to $\pi_G(G \cdot X)$ and hence closed by Corollary 2.3.4. Therefore we can use the method of Section 1.2.1 to compute $\pi_G(X)$. Let $g_1, \ldots, g_r \in K[V]^G$ be generators for $K[V]^G$ as a K-algebra. Then π_G is given by $\pi_G(v) = (g_1(v), \ldots, g_r(v))$. Thus we obtain the following algorithm.

Algorithm 5.4.1 (Orbit variety). Given polynomials $f_1, \ldots, f_m \in K[V] = K[x_1, \ldots, x_n]$ defining a variety $X \subseteq V$ and generating invariants $g_1, \ldots, g_r$ for the invariant ring $K[V]^G$ of a finite group $G \leq \mathrm{GL}(V)$, perform the following steps to compute an ideal $J \subseteq K[y_1, \ldots, y_r]$ such that $\mathcal{V}(J) = \pi_G(X)$.

(1) Form the ideal

$$\bar{J} := (f_1, \ldots, f_m, g_1 - y_1, \ldots, g_r - y_r) \subseteq K[x_1, \ldots, x_n, y_1, \ldots, y_r].$$

(2) Use Algorithm 1.2.1 to compute J as the elimination ideal

$$J := \bar{J} \cap K[y_1, \ldots, y_r].$$

If $J = (h_1, \ldots, h_k)$ is the result of Algorithm 5.4.1, then

$$G \cdot X = \pi_G^{-1}(\pi_G(X)) = \mathcal{V}_V\left(h_1(g_1, \ldots, g_r), \ldots, h_k(g_1, \ldots, g_r)\right).$$

Therefore we have a way to compute the G-orbit of a variety. If I is a zero-dimensional G-invariant ideal, we are interested in explicitly calculating the points of $X = \pi_G^{-1}(\pi_G(X))$. Thus for $(\eta_1, \ldots, \eta_r) \in \pi_G(X)$ we want to compute the preimage

$$\pi_G^{-1}(\eta_1, \ldots, \eta_r) = \{v \in V \mid g_i(v) = \eta_i \ \forall i\}.$$

This can be done by calculating a lexicographic Gröbner basis of the ideal

$$(g_1 - \eta_1, \ldots, g_r - \eta_r) \subseteq K[x_1, \ldots, x_n].$$

Thus we have split the process of finding X into two parts, both of which involve Gröbner basis calculations: the finding of $\pi_G(X)$ by Algorithm 5.4.1 and the computation of $\pi_G^{-1}(\eta_1, \ldots, \eta_r)$ for a point $(\eta_1, \ldots, \eta_r) \in \pi_G(X)$. Apart from this, we need generators for the invariant ring, which only have to be calculated once for the group G. In many practical examples this divide and conquer approach leads to much better running times, or makes computations possible which would otherwise be out of reach (see Worfolk [262]).

5.5 Graph Theory

Let K be a field and g a graph with vertices $\{1, \ldots, n\}$ and edges which are weighted by values of K. We say that g is a K-**weighted graph** with n vertices. Thus g is given by the function m_g associating to each subset $\{i, j\} \subseteq \{1, \ldots, n\}$ of size two the weight of the edge between the vertices i and j. The set of all K-weighted graphs with n vertices may be identified with the vector space $V = K^{\binom{n}{2}}$. If g' is another K-weighted graph with n vertices, we say that g and g' are isomorphic if there exists a permutation $\sigma \in S_n$ such that $m_g(\{\sigma(i), \sigma(j)\}) = m_{g'}(\{i, j\})$ for all $i \neq j$. This means that m_g and $m_{g'}$ lie in the same orbit under the action of $G = S_n$ on the two-sets $\{i, j\}$. Assume for the moment that K is algebraically closed. Then the G-orbits in V are in bijective correspondence to points in the categorical quotient $V /\!/ G$, since G is finite (see Section 2.3.1). Thus we have a bijection

$$\{\text{isomorphism classes of } K\text{-weighted graphs with } n \text{ vertices}\} \leftrightarrow V /\!/ G,$$

which endows the set of isomorphism classes of graphs with the structure of an algebraic variety. If we drop the assumption that K be algebraically closed, then we still have an injective map from the set of isomorphism classes of K-weighted graphs with n vertices into $V /\!/ G$.

Let us put this injection into more explicit terms. Consider the polynomial ring $R := K[x_{\{i,j\}} \mid 1 \leq i, j \leq n, i \neq j]$ in $\binom{n}{2}$ variables. For a polynomial $f \in R$ and a K-weighted graph g with n vertices, define $f(g)$ by sending $x_{\{i,j\}}$ to $m_g(\{i, j\})$. The symmetric group $G = S_n$ acts on R by $\sigma \cdot x_{\{i,j\}} = x_{\{\sigma(i), \sigma(j)\}}$. Suppose we have invariants $f_1, \ldots, f_m \in R^{S_n}$ which generate the invariant ring or a separating subalgebra. Then we have the following simple test whether two graphs are isomorphic.

Lemma 5.5.1. *With the above notation, two K-weighted graphs g and g' with n vertices are isomorphic if and only if*

$$f_i(g) = f_i(g') \quad for \quad i = 1, \ldots, m. \tag{5.5.1}$$

This approach requires the pre-computation of R^{S_n} for given values of n. The action of S_n on the indeterminates of R is the permutation action on subsets of size two of $\{1, \ldots, n\}$. Unfortunately, the computation of these invariant rings turns out to be harder than one might expect. The computation is easy for $n \leq 3$, and was done for $n = 4$ by Aslaksen et al. [11] (see Example 3.5.4(b)). In [11] the authors also considered the corresponding permutation representation of S_5, which is 10-dimensional. This representation has long been resistant against all existing algorithms for the computation of generating invariants. Using his library PerMuVAR [245], Thiéry [247] studied the case $n = 5$. He was able to find (optimal) primary invariants of degrees 1, 2, 2, 3, 3, 4, 4, 5, 5, 6, and to check that the invariants of degree up to 17 are generated by invariants of degree at most 9. He used algorithms which are based on the concept of SAGBI-Gröbner bases (see Thiéry [248]). It was later confirmed by the second author of this book that the invariants for $n = 5$ are indeed generated in degrees at most 9. The computations for this confirmation are very demanding and were done in MAGMA. Thiéry conducted some experiments for $n = 6$ and thinks that the degree bound should be 11.

Let us remark that graphs with discretely weighted edges inject into the space V of K-weighted graphs, provided that K is large enough to permit an (arbitrary) injection of the weights into K. Thus Lemma 5.5.1 can also be used to test discretely weighted graphs for isomorphism. Moreover, if the vertices are also weighted, this translates into an action of S_n on $K^{\binom{n}{2}+n}$, so in this situation invariants can be used as well. This is also possible for directed graphs. In that case we will get an action of S_n by permutations combined with changes of signs.

Ulam's conjecture. We return to applications of invariant theory in graph theory itself. Let g be a **multigraph**, i.e., an undirected graph with vertices $\{1, \ldots, n\}$ and (possibly) multiple edges. We might say that g is an $\mathbb{N}_0$-weighted graph, and therefore the set of these graphs is a subset of the $\mathbb{Q}$-weighted graphs. So again Lemma 5.5.1 applies and gives a test whether two multigraphs are isomorphic.

A multigraph is called **simple** if there is at most one edge between two vertices. An interesting conjecture about simple graphs was formulated by Ulam [249]. Let $\mathcal{G}_n$ be the set of isomorphism classes of simple graphs with n vertices. For a simple graph g with vertices $\{1, \ldots, n\}$, consider the function $S_g \colon \mathcal{G}_{n-1} \to \mathbb{N}_0$ which assigns to each isomorphism class C of graphs with $n-1$ vertices the number of indices $i \in \{1, \ldots, n\}$ such that the graph obtained by deleting the vertex i from g lies in the class C. Clearly if g and g' are isomorphic, then $S_g = S_{g'}$. With this notation Ulam's conjecture reads as follows.

Conjecture 5.5.2. *Let g and g' be two simple graphs with vertices $\{1, \ldots, n\}$ with $n \geq 3$. Then $S_g = S_{g'}$ implies that g and g' are isomorphic.*

A stronger version of this conjecture can be translated into algebraic terms and leads to the following conjecture about the invariant rings R^{S_n}.

Conjecture 5.5.3 (Pouzet [195]). *Let R^{S_n} be the invariant ring of the symmetric group acting on $R = \mathbb{Q}[x_{\{i,j\}} \mid 1 \leq i < j \leq n]$ by $\sigma \cdot x_{\{i,j\}} = x_{\{\sigma(i),\sigma(j)\}}$. Then for $n \geq 3$, R^{S_n} is generated as a $\mathbb{Q}$-algebra by sums over S_n-orbits of monomials in the $x_{\{i,j\}}$ with $1 \leq i < j < n$ (i.e., by orbit-sums over monomials not involving the indeterminates $x_{\{i,n\}}$).*

This conjecture would imply Conjecture 5.5.2, even without the hypothesis that the graphs be simple. Unfortunately, however, Conjecture 5.5.3 was recently disproved by Thiéry [247], who used Hilbert series and counting arguments to show that the conjecture is false for $11 \leq n \leq 18$ (and probably for larger values of n as well). See also Thiéry [246] for some experimental results, and Pouzet and Thiéry [196] for more background and some related questions.

5.6 Combinatorics

Invariant theory has also been applied to combinatorics. Quite a few interesting examples can be found in Stanley's survey article [236] or in Stanley [235]. In some of these, combinatorial quantities are in some way encoded into a power series, which is then recognized as the Hilbert series of some invariant ring (e.g. by comparing with Molien's formula). Any knowledge about the invariant rings in question then leads to results about combinatorics.

Example 5.6.1 (Stanley [236]). Suppose we want to evaluate the sum

$$S(k) := \sum_{j=1}^{k} |1 - \zeta^j|^{-2},$$

where $\zeta = e^{\frac{2\pi i}{k}} \in \mathbb{C}$. In order to write $S(k)$ as a limit, define

$$F_k(t) := \sum_{j=0}^{k} \frac{1}{(1 - \zeta^j t)(1 - \zeta^{-j} t)}. \tag{5.6.1}$$

Then

$$S(k) = \lim_{t \to 1} \left(F_k(t) - \frac{1}{(1-t)^2} \right). \tag{5.6.2}$$

By comparing Equation (5.6.1) with Molien's formula (Theorem 3.2.2), we see that $\frac{1}{k} F_k(t)$ is the Hilbert series of the invariant ring $\mathbb{C}[x,y]^G$ of the cyclic group

$$G = \langle \begin{pmatrix} \zeta & 0 \\ 0 & \zeta^{-1} \end{pmatrix} \rangle.$$

But the invariants of G are easily determined: They are $\mathbb{C}$-linear combinations of invariant monomials, and a monomial $x^i y^j$ is invariant if and only if $i \equiv j$ mod k. Thus $\mathbb{C}[x,y]^G$ is generated by x^k, y^k, and xy with the sole relation $(xy)^k = x^k y^k$. This yields

$$H\left(\mathbb{C}[x,y]^G, t \right) = \frac{1 - t^{2k}}{(1 - t^2)(1 - t^k)^2}.$$

Equating this to $\frac{1}{k} F_k(t)$ and using Equation (5.6.2), we obtain

$$S(k) = \lim_{t \to 1} \left(k \frac{1 - t^{2k}}{(1 - t^2)(1 - t^k)^2} - \frac{1}{(1-t)^2} \right) = \frac{k^2 - 1}{12},$$

where the last step is performed by l'Hôpital's rule, or by forming the limit with MAPLE (Char et al. [42]). ◁

Example 5.6.2. This example is due to Solomon [230], and our account here is largely drawn from Stanley [236]. Given non-negative integers $d_1, \ldots, d_r \in \mathbb{N}_0$, we are interested in the number $P_n(\mathbf{d})$ of ways to write $\mathbf{d} := (d_1, \ldots, d_r)$ as a sum of n vectors from $\mathbb{N}_0^r$, where the order of the summation is disregarded. One says that $P_n(\mathbf{d})$ is the number of **multipartite partitions** of $\mathbf{d}$. Introduce the polynomial ring $R := K[x_{i,j} \mid 1 \le i \le r, 1 \le j \le n]$ in $r \cdot n$ variables over a field K of characteristic 0. We give R an $\mathbb{N}_0^r$-grading by defining the homogeneous component $R_{\mathbf{d}}$ of multi-degree $\mathbf{d} = (d_1, \ldots, d_r)$ as the K-span of the monomials $\prod_{i=1}^{r} \prod_{j=1}^{n} x_{i,j}^{e_{i,j}}$ with $\sum_{j=1}^{n} e_{i,j} = d_i$ for all i. Obviously the number of monomials of degree $\mathbf{d}$ is equal to the number of ways how $\mathbf{d}$ can be written as a sum of n vectors in $\mathbb{N}_0^r$, but now the order of the summation matters. More precisely, two summations having the same summands but in a different order correspond to two monomials lying in the same orbit under the action of the symmetric group S_n by $\sigma \cdot x_{i,j} := x_{i,\sigma(j)}$. Therefore $P_n(\mathbf{d})$ is equal to the number of S_n-orbits of monomials of multi-degree $\mathbf{d}$. But this is the same as the dimension of $R_{\mathbf{d}}^{S_n}$, the homogeneous part of the invariant ring of multi-degree $\mathbf{d}$ (see the remark at the beginning of Section 3.10.4). Thus $P_n(\mathbf{d}) = \dim_K(R_{\mathbf{d}}^{S_n})$, and assembling this into a formal power series yields

$$\sum_{\mathbf{d} \in \mathbb{N}_0^r} P_n(\mathbf{d}) \cdot t_1^{d_1} \cdots t_r^{d_r} = \sum_{\mathbf{d} \in \mathbb{N}_0^r} \dim_K(R_{\mathbf{d}}^{S_n}) \cdot t_1^{d_1} \cdots t_r^{d_r} =: H(R^{S_n}; t_1, \ldots, t_r),$$

$$(5.6.3)$$

where the right-hand side is the multi-graded Hilbert series of R^{S_n}. Thus our interest lies with R^{S_n}, the ring of vector invariants of the symmetric group. Quite a bit is known about this invariant ring. In particular, it is a result of Hermann Weyl that R^{S_n} is generated by the polarized elementary symmetric polynomials (see Smith [225, Theorem 3.4.1]). However, this is not enough information to derive a nice expression for the Hilbert series $H(R^{S_n}; t_1, \ldots, t_r)$. Something can nevertheless be said about its structure. In fact, the elementary symmetric polynomials $s_{i,j}$ ($1 \le i \le r$, $1 \le j \le n$), as defined in Equation (3.10.2), form a system of primary invariants (see Theorem 3.10.1). Moreover, $s_{i,j}$ is multi-homogeneous of degree $(0, \ldots, 0, j, 0, \ldots, 0)$, where the j appears in the i-th position. Therefore the subalgebra $A := K[s_{i,j} \mid 1 \le i \le r, 1 \le j \le n]$ has the multi-graded Hilbert series $\prod_{i=1}^{r} \prod_{j=1}^{n} (1 - t_i^j)^{-1}$. Since R^{S_n} is Cohen-Macaulay (Theorem 3.4.1), it is a free module over A. By the homogeneous Nakayama Lemma (Lemma 3.5.1), free generators can be chosen as a subset of any set of homogeneous generators. But there exist multi-homogeneous generators for R^{S_n}, and therefore

$$H(R^{S_n}; t_1, \ldots, t_r) = \frac{\mathbf{t}^{\mathbf{e}_1} + \cdots + \mathbf{t}^{\mathbf{e}_m}}{\prod_{i=1}^{r} \prod_{j=1}^{n} (1 - t_i^j)},$$

where the $\mathbf{e}_i$ are the multi-degrees of multi-homogeneous free generators. Putting this together with Equation (5.6.3), we see that

$$\sum_{d_1,\ldots,d_r \in \mathbb{N}_0} P_n(\mathbf{d}) \cdot t_1^{d_1} \cdots t_r^{d_r} = \frac{f(t_1,\ldots,t_r)}{\prod_{i=1}^{r} \prod_{j=1}^{n} (1 - t_i^j)},$$

where f is a polynomial in $t_1,\ldots,t_r$ with non-negative integers as coefficients.
◁

A more recent connection between combinatorics and invariant theory was discovered by Elashvili and Jibladze [61]. The authors consider the invariants of the cyclic group $G = C_n$ acting by the regular representation. Let $a(n,m)$ be the dimension of the subspace of invariants in $K[V_{\mathrm{reg}}]^G$ of degree m, so

$$H(K[V_{\mathrm{reg}}]^G, t) = \sum_{m=0}^{\infty} a(n,m) t^m.$$

The Hilbert series $H(K[V_{\mathrm{reg}}]^G, t)$ is independent of the choice of the field K (see the remark at the beginning of Section 3.10.4). Thus K can be chosen as $\mathbb{C}$. Then the regular representation is isomorphic to the diagonal representation where a generator of G acts by $\mathrm{diag}(\zeta^0,\ldots,\zeta^{n-1})$ with $\zeta = e^{\frac{2\pi i}{n}}$. In this representation, the invariants are precisely the $\mathbb{C}$-linear combination of invariant monomials, and a monomial $x_1^{e_1} \cdots x_n^{e_n}$ is invariant if and only if $\sum_{i=0}^{n-1} i e_i \equiv 0 \mod n$. Thus $a(n,m)$ counts the number of solutions of

$$\sum_{i=0}^{n-1} i e_i \equiv 0 \mod n, \quad \sum_{i=0}^{n-1} e_i = m. \tag{5.6.4}$$

But this is the same as the number of partitions of multiples of n into m summands which do not exceed $n - 1$, thereby providing a combinatorial interpretation of the numbers $a(n,m)$. In Elashvili et al. [62], another combinatorial interpretation in terms of so-called "necklaces" was given to the $a(n,m)$. On the other hand, $H(K[V_{\mathrm{reg}}]^G, t)$ can be evaluated by Molien's formula. The result is

$$H(K[V_{\mathrm{reg}}]^G, t) = \frac{1}{n} \sum_{d|n} \varphi(d)(1 - t^d)^{-\frac{n}{d}},$$

where $\varphi(d)$ is the Euler totient function (see [61]). From this, the authors of [61] proceeded to derive a more explicit formula for $a(n,m)$, from which the interesting reciprocity law

$$a(n,m) = a(m,n)$$

can be read off.

5.7 Coding Theory

An interesting application of invariant theory to coding theory can be found in the very nice survey article by Sloane [223]. Let $C \subseteq \mathbb{F}_q^n$ be a code of length n defined over the finite field $\mathbb{F}_q$. If a_i is the number of codewords in C of weight i (i.e., codewords having exactly i non-zero coordinates), then the **weight enumerator** of C is defined as

$$W_C(x,y) := \sum_{i=0}^{n} a_i x^{n-i} y^i \in \mathbb{C}[x,y].$$

Suppose C is linear (i.e., a linear subspace of $\mathbb{F}_q^n$), and $C^{\perp}$ is the dual code, which by definition consists of all vectors in $\mathbb{F}_q^n$ whose standard scalar product with all codewords from C is zero. Then a celebrated theorem of MacWilliams [164] says that

$$q^k W_{C^{\perp}}(x,y) = W_C(x + (q-1)y, x - y), \tag{5.7.1}$$

where k is the dimension of C. An area of great interest in coding theory is the study of self-dual codes, i.e., linear codes C with $C^{\perp} = C$. This condition implies that $k = \dim(C) = n/2$. Since $W_C(x,y)$ is homogeneous of degree n, we obtain

$$W_C(\sqrt{q}\cdot x, \sqrt{q}\cdot y) = q^k W_C(x,y) = q^k W_{C^{\perp}}(x,y).$$

Thus for a self-dual code C, Equation (5.7.1) becomes

$$W_C\left(\frac{x + (q-1)y}{\sqrt{q}}, \frac{x - y}{\sqrt{q}}\right) = W_C(x,y).$$

This can be expressed by saying that W_C is invariant under the group G generated by the linear transformation

$$\sigma = \frac{1}{\sqrt{q}}\begin{pmatrix} 1 & 1 \\ q-1 & -1 \end{pmatrix}$$

occurring above. The invariant ring of G is a polynomial ring generated by $f = x + (\sqrt{q}-1)y$ and $g = y(x-y)$. Therefore $W_C(x,y)$ must be a polynomial in f and g. But since the degree n of W_C is even, it must in fact be a polynomial in f^2 and g, or, somewhat simpler, in $f^2 + 2(1 - \sqrt{q})g = x^2 + (q-1)y^2$ and $g = y(x-y)$. Thus invariant theory gives strong restrictions on possible weight enumerators of self-dual codes.

Assume that we have the additional restriction that the code is binary ($q = 2$), and all occurring weights are divisible by 4. This means that $W_C(x,y)$ is also invariant under the transformation given by $\tau = \begin{pmatrix} 1 & 0 \\ 0 & i \end{pmatrix}$. Let $\hat{G}$ be the group generated by the above transformation σ (for $q = 2$) and by τ. It is found that $\hat{G}$ is a complex reflection group of order 192 (G_9 in the classification of Shephard and Todd [221]). The invariant ring $\mathbb{C}[x,y]^{\hat{G}}$ is a polynomial ring generated by the invariants

$$\theta = x^8 + 14x^4y^4 + y^8,$$
$$\varphi = x^4 y^4 (x^4 - y^4)^4.$$

of degrees 8 and 24. (The generators can either be taken from Sloane [223] or computed with MAGMA, which takes about half a second.) Thus every weight enumerator of a self-dual binary code whose weights are all divisible by 4 must be a polynomial in θ and φ. This result goes back to Gleason [85]. Moreover, the coefficients of W_C must be non-negative integers, and the coefficient of x^n must be 1. All this imposes strong restrictions on the possible weight enumerators. These restrictions can be used to determine the weight enumerator of specific codes using only very little further knowledge (see Sloane [223, Section I.F]), or to rule out the existence of certain codes. For example, it can be proved that the minimum distance of a self-dual binary code of length n with all weights divisible by 4 cannot exceed $4[n/24] + 4$, and only a finite number of codes exist where this upper bound is attained (see Sloane [223, Section IV.B]).

A huge number of other, more recent examples where invariant theory is used for coding theory can be found in Rains and Sloane [197] and Bannai et al. [12]. The applications are all in the same spirit as the one explained above, but the invariant rings that emerge are considerably more complicated.

5.8 Equivariant Dynamical Systems

In the theory of dynamical systems one studies differential equations of the type

$$\frac{d}{dt}\mathbf{x} = \mathbf{f}(\mathbf{x}), \tag{5.8.1}$$

where $\mathbf{f}\colon \mathbb{R}^n \to \mathbb{R}^n$ is a function ("vector field"), and solutions $\mathbf{x}$ are functions $\mathbb{R} \to \mathbb{R}^n$ of t. Often the vector field $\mathbf{f}$ also depends on a tuple $\lambda \in \mathbb{R}^l$, so we have the equation $\frac{d}{dt}\mathbf{x} = \mathbf{f}(\mathbf{x}, \lambda)$. It is quite common that a dynamical system has symmetries, such as rotational symmetries when the flow of some liquid through a (motionless) cylinder is considered. Such symmetries translate into the condition that the vector field $\mathbf{f}$ is equivariant under the action of some group $G \leq \mathrm{GL}_n(\mathbb{R})$, i.e.,

$$\mathbf{f} \circ \sigma = \sigma \circ \mathbf{f} \quad \text{for all } \sigma \in G.$$

In this context G may or may not be finite. In bifurcation theory one studies the question of how the behavior of the system changes under small perturbations of the parameter λ. If $\mathbf{f}$ is equivariant with respect to some group, we speak of equivariant bifurcation theory. Of particular interest is the study of stationary states, i.e., solutions with constant $\mathbf{x}$, which are given by

$$\mathbf{f}(\mathbf{x}, \lambda) = 0. \tag{5.8.2}$$

Again the solutions $\mathbf{x}$ depend on λ, and (equivariant) bifurcation theory studies those values of λ where branches of solutions meet. For a detailed introduction into equivariant bifurcation theory we refer the reader to the books by Sattinger [208], Golubitsky and Schaeffer [91], and Golubitsky et al. [92]. Information on equivariant dynamical systems and many more references can be found in Gatermann [75], Gatermann and Guyard [77].

If $\mathbf{f}$ is G-equivariant (for every λ), then the solution set of Equation (5.8.2) is G-stable, hence the methods of Section 5.4 apply if $\mathbf{f}$ consists of (or can be approximated by) polynomials, and G is finite. On the other hand, if G is equivariant and infinite, and $G \subseteq O_n(\mathbb{R})$, suppose that $\pi_1, \ldots, \pi_k \colon \mathbb{R}^n \to \mathbb{R}^n$ are equivariants generating the module of equivariants over the invariant ring. Then the scalar products $\pi_i \cdot \mathbf{f}$ are G-invariant functions $\mathbb{R}^n \to \mathbb{R}$. It is easy to see that

$$\mathcal{V}_{\mathbb{R}^n}(\mathbf{f}) = \mathcal{V}_{\mathbb{R}^n}(\pi_1 \cdot \mathbf{f}, \ldots, \pi_k \cdot \mathbf{f})$$

(Jarić et al. [120]). Hence in this situation the methods of Section 5.4 can also be used to try to facilitate the solution of Equation (5.8.2). This approach was carried out successfully by Worfolk [262], for example.

A closely related technique used in the study of equivariant dynamical systems is called orbit space reduction, and is based on the following observation. Let $g_1, \ldots, g_r \in K[x_1, \ldots, x_n]^G$ be generators for the invariant ring. We view the g_i as functions $\mathbb{R}^n \to \mathbb{R}$. For $\sigma \in G$ we have

$$\operatorname{grad} g_i = \operatorname{grad}(g_i \circ \sigma) = (\operatorname{grad} g_i \circ \sigma) \cdot \sigma$$

by the invariance and by the chain rule. Since $\mathbf{f}$ is G-equivariant, if follows that the product $(\operatorname{grad} g_i) \cdot \mathbf{f}$ (with the gradient written as a row and $\mathbf{f}$ as a column) is G-invariant as a function $\mathbb{R}^n \to \mathbb{R}$. Let us assume for simplicity that $\mathbf{f}$ consists of polynomials. Then $(\operatorname{grad} g_i) \cdot \mathbf{f} = h_i(g_1, \ldots, g_r)$ with $h_i \in \mathbb{R}[y_1, \ldots, y_r]$ a polynomial in new variables y_j. Now if $\mathbf{x} \colon \mathbb{R} \to \mathbb{R}^n$ is a solution of Equation (5.8.1), then by the chain rule

$$\frac{d}{dt}(g_i \circ \mathbf{x}) = (\operatorname{grad} g_i \circ \mathbf{x}) \cdot \frac{d}{dt}\mathbf{x} = (\operatorname{grad} g_i \circ \mathbf{x}) \cdot \mathbf{f}(\mathbf{x}) = h_i(g_1 \circ \mathbf{x}, \ldots, g_r \circ \mathbf{x}).$$

Summarizing the $g_i \circ \mathbf{x}$ into a function $\mathbf{g} \colon \mathbb{R} \to \mathbb{R}^r$ and the h_i into a vector field $\mathbf{h} \colon \mathbb{R}^r \to \mathbb{R}^r$, we can write this as

$$\frac{d}{dt}\mathbf{g} = \mathbf{h}(\mathbf{g}). \tag{5.8.3}$$

The process of obtaining (5.8.3) from (5.8.1) is called orbit space reduction since this method contracts orbits into points. For example, a periodic solution which moves along a G-orbit is contracted into a stationary solution. For any solution $\mathbf{x}$ of (5.8.1), the corresponding $\mathbf{g}$ satisfies (5.8.3), and it also satisfies the algebraic relations between the g_i (see Section 3.6). Thus we have a differential equation with algebraic side conditions, or, equivalently, a differential equation on the quotient variety $\mathbb{R}^n /\!/ G$. The relation between solutions of (5.8.3) and (5.8.1) is quite intricate, since a point $\mathbf{g}(t)$ has many preimages $\mathbf{x}(t)$ in $\mathbb{R}^n$, and it is not clear how they can be put together for different t to obtain a solution of (5.8.1). But the advantage of (5.8.3) is that its investigation is much easier.

Example 5.8.1. Consider the dynamical system given by

$$\frac{d}{dt}\begin{pmatrix} x \\ y \end{pmatrix} = \begin{pmatrix} y \\ -x \end{pmatrix} + \left(1 - \lambda(x^2 + y^2)\right)\begin{pmatrix} x \\ y \end{pmatrix}.$$

The vector field is equivariant under $G = \mathrm{SO}_2(\mathbb{R})$. It is easily seen that the invariant ring $\mathbb{R}[x, y]^G$ is generated by the single invariant $g = x^2 + y^2$. We perform orbit space reduction and obtain

$$\frac{d}{dt}g = 2x\left(y + \left(1 - \lambda(x^2 + y^2)\right)x\right) +$$

$$2y\left(-x + \left(1 - \lambda(x^2 + y^2)\right)y\right) = 2g(1 - \lambda g).$$

This equation is much more tractable that the original one, and has the general solution

$$g(t) = \frac{c}{\lambda c + (1 - \lambda c)e^{-2t}}$$

with $c = g(0) \in \mathbb{R}$ non-negative. (For $\lambda < 0$ and $c > 0$, $g(t)$ is only defined for $0 \leq t < \ln\left(1 - 1/(\lambda c)\right)/2$.) However, this solution does not capture the angular movement of (x, y), for example the circular movement in the case of the stationary solution $g(t) = 1/\lambda$ (for $\lambda > 0$).   ◁

Orbit space reduction has been studied and used by many authors. We content ourselves in giving the references Gatermann and Lauterbach [78], Lari-Lavassani et al. [159] here, where the symbolic computation of invariants is used. It is clear that the methods discussed here find applications in physics, chemistry, and engineering. Three examples are given by the references Jarić et al. [120], Campbell and Holmes [39], and Collins and Thompson [47].

5.9 Material Science

For fuel or water tanks in aircrafts and rockets one frequently uses textile reinforced composites to achieve high durability at a low weight. It is important to find failure conditions for such materials, i.e., functions in terms of the stresses (usually given by a symmetric tensor with components $\sigma_x, \sigma_y, \sigma_z, \tau_{yz}, \tau_{zx}, \tau_{yx}$) which describe under what stress the material will break. Since at present there is no valid theory to compute such failure conditions from the geometry and basic ingredients of the material, experiments are necessary for their determination. More precisely, one writes (approximate) failure conditions as polynomials (or other simple functions) in the stresses and then performs a few experiments to determine the coefficients of these polynomials. In this process it is important to incorporate any additional information on the material into the initial polynomials. In particular, failure conditions are invariant under any orthogonal transformation of coordinates which respects the symmetries of the material. Therefore the failure conditions should be chosen as invariants under the symmetry group of the material. This drastically reduces the number of coefficients to be determined by experiments, and thus the number of experiments that are necessary. An invariant theoretic approach has been taken by materials researchers for quite a while (see, for example, Hashin [103], Helisch [104]).

The symmetry group is given as a subgroup of $O_3(\mathbb{R})$. Since we are considering polynomials in the stresses, we have to determine the action of a matrix $A \in O_3(\mathbb{R})$ on the tensor of stresses. This is given by

$$A \cdot \begin{pmatrix} \sigma_x & \tau_{yx} & \tau_{zx} \\ \tau_{yx} & \sigma_y & \tau_{yz} \\ \tau_{zx} & \tau_{yz} & \sigma_z \end{pmatrix} = A^{-1} \begin{pmatrix} \sigma_x & \tau_{yx} & \tau_{zx} \\ \tau_{yx} & \sigma_y & \tau_{yz} \\ \tau_{zx} & \tau_{yz} & \sigma_z \end{pmatrix} A.$$

Observe that $-1 \in O_3(\mathbb{R})$ lies in the kernel of the action. Let us consider a few examples of material symmetries.

Example 5.9.1. Let us consider the case of a "unidirectional reinforced composite", where the fibers all lie parallel to each other, say along the x-coordinate axis. The symmetry group is $O_2(\mathbb{R}) \times \{\pm 1\}$, with $O_2(\mathbb{R})$ acting on the y-z-plane and $\{\pm 1\}$ acting as a reflection at this plane. Hence

$$G = \left\{ \begin{pmatrix} z_1 & 0 & 0 \\ 0 & z_2 z_3 & -z_4 \\ 0 & z_2 z_4 & z_3 \end{pmatrix} \mid z_1^2 = z_2^2 = z_3^2 + z_4^2 = 1 \right\}. \tag{5.9.1}$$

An element $g \in G$ acts on the σ's and τ's by

$$g \cdot \begin{pmatrix} \sigma_x \\ \sigma_y \\ \sigma_z \\ \tau_{yx} \\ \tau_{zx} \\ \tau_{yz} \end{pmatrix} = \begin{pmatrix} 1 & 0 & 0 & 0 & 0 & 0 \\ 0 & z_3^2 & z_4^2 & 0 & 0 & 2z_3 z_4 \\ 0 & z_4^2 & z_3^2 & 0 & 0 & -2z_3 z_4 \\ 0 & 0 & 0 & z_1 z_2 z_3 & z_1 z_2 z_4 & 0 \\ 0 & 0 & 0 & -z_1 z_4 & z_1 z_3 & 0 \\ 0 & -z_2 z_3 z_4 & z_2 z_3 z_4 & 0 & 0 & z_2 - 2z_2 z_4^2 \end{pmatrix} \begin{pmatrix} \sigma_x \\ \sigma_y \\ \sigma_z \\ \tau_{yx} \\ \tau_{zx} \\ \tau_{yz} \end{pmatrix}. \tag{5.9.2}$$

G is an infinite, reductive group, so we can use the methods of Section 4.1 to compute generators of the invariant ring. The ideal $I(G)$ defining the group and the matrix A defining the action are given by Equations (5.9.1) and (5.9.2), respectively. We form the ideal in step (2) of Algorithm 4.1.9, compute the elimination ideal $\mathfrak{b}$ as in step (3), and set the second set of variables equal to zero. The resulting ideal is

$$I = (\underbrace{\sigma_x}_{=:f_{11}}, \underbrace{\sigma_y + \sigma_z}_{=:f_{12}}, \underbrace{\tau_{yx}^2 + \tau_{zx}^2}_{=:f_{21}}, \underbrace{\sigma_z^2 + \tau_{yz}^2}_{=:f_{22}}, \underbrace{\sigma_z \tau_{zx}^2 + \tau_{yx} \tau_{zx} \tau_{yz}}_{=:f_3}), \tag{5.9.3}$$

which is equal to the ideal generated by all invariants of positive degree. It is easy to check that the generators f_{11}, f_{12}, and f_{21} are already invariants. We have to replace the generators f_{22} and f_3 by invariants. In Algorithm 4.1.9, this is done by applying the Reynolds operator. Here we choose a different approach. First we substitute the f_{22} by

$$\widetilde{f_{22}} := f_{22} - \sigma_z f_{12} = \tau_{yz}^2 - \sigma_y \sigma_z,$$

which is seen to be an invariant. It remains to find an irreducible invariant of degree 3 in I. To this end, we write down a basis of the space of elements of I of degree 3, modulo the space generated by products of invariants of degree less than 3. This basis has 37 elements. We apply a general element of G, given by the matrix in (5.9.2), to a linear combination with unknown coefficients of these 37 polynomials. From this we subtract the original linear combination, and then form the normal form of the result with respect to $I(G) = (z_1^2 - 1, z_2^2 - 1, z_3^2 + z_4^2 - 1)$. A necessary and sufficient condition for the linear combination to be G-invariant is that this normal form is zero. Setting all coefficients of the normal form equal to zero yields a system of 148 linear equations in 37 unknowns. The solution space is one-dimensional, as expected, and we obtain the invariant

$$\widetilde{f_3} = 2f_3 - \tau_{zx}^2 f_{12} - \sigma_z f_{21} = 2\tau_{yx}\tau_{zx}\tau_{yz} - \sigma_y \tau_{zx}^2 + \sigma_z \tau_{yx}^2.$$

Thus we have found that the invariant ring is generated by f_{11}, f_{12}, f_{21}, $\widetilde{f_{22}}$, and $\widetilde{f_3}$. This confirms a result of Hashin [103]. The computation in this example was done in MAGMA and took about 5 seconds.   ◁

In the following example we consider material structures which have a finite symmetry group.

Example 5.9.2. (a) A very common case is a woven textile reinforcement with warp and weft inclined at $90°$, but made of different materials. Then the symmetry group is generated by reflections at the coordinate planes. Since $-1 \in O_3(\mathbb{R})$ acts trivially, the symmetry group acts as a Klein 4-group. The invariant ring is a hypersurface generated by invariants of degree at most 3 (see Boehler [22]). It can easily be calculated with the methods from Chapter 3.

(b) Almost as common is the case of a woven textile with identical fibers in warp and weft, inclined at $90°$. Here the symmetry group is generated by a rotation by $90°$ about an axis orthogonal to the textile layer and reflections at the coordinate planes. The resulting group is a dihedral group of order 8. The invariant ring is a complete intersection generated by invariants of maximal degree 4 (see Smith et al. [224]).

(c) A more exotic structure is a textile layer with a $45°$-symmetry. Here the symmetry group is a dihedral group of order 16. The invariant ring was calculated by Meckbach and Kemper [166] using the methods of Chapter 3. It is a Gorenstein ring but considerably more complicated than the ones in the previous examples. There are 10 generators of maximal degree 8.   ◁

5.10 Computer Vision

A very rich field of applications of invariant theory is computer vision. Philosophically, this comes from the fact that the camera image of an object (two or three dimensional) depends very much on the angle from which it is viewed, as well as many other factors. Therefore the immediate parameters that can be measured from the image are almost never inherent object characteristics. It is therefore an obvious approach to use invariant theory in order to extract inherent characteristics from the image parameters, i.e., characteristics that remain unchanged if the camera angle or other factors are changed. It is not surprising that there is a multitude of ways in which invariant theory is used in computer vision. We found a rich source of information in the book by Mundy and Zisserman [170], which has a nice introductory chapter written by the editors. The relevant literature on this field is vast, as can be seen from the references given in [170]. Let us also direct the reader to the references Reiss [199], Kanatani [126], and Florack [70]. In keeping with the general style of this chapter, we will only give a very limited number of applications to computer vision in this section.

5.10.1 View Invariants of 3D Objects

A very important and difficult issue in computer vision is the recognition of three-dimensional objects or scenes from their two-dimensional images given by camera views. Quite a number of sophisticated techniques are applied to interpret 2D projections, such as reconstruction from multiple or stereo views, shape from motion, shape from shading, or shape from texture. Most of these techniques use image features as an input, which are produced by preprocessing the camera data with some standard tools, such as edge detectors or detectors of points with maximal brightness or highest curvature. Once an edge is detected in an image, it can be classified as being a straight line, a conic, etc., and for each such category of curves the defining parameters can be determined. No classification is necessary for points, where the defining parameters are just the two coordinates of the measured 2D projection of the point. This second step is also referred to as the extraction of geometric primitives, where points, straight lines, conics etc. are regarded as geometric primitives.

The problem is that the position and therefore also the defining parameters of the detected geometric primitives change when the same 3D object or scene is rotated or viewed from a different angle or distance. This is where invariant theory comes in as a helpful tool. The idea is to form expressions from the feature parameters which remain invariant under changes of viewing angle or distance. The question how many such invariants exist and what discriminatory power they have cannot be answered in complete generality. One has to restrict to certain configurations of geometric primitives which either have been detected or are a priori known to be present in the image.

Thus one can talk, for instance, of invariants of a configuration of two straight lines and two conics. This configuration would be detected when viewing a cylindrical object (with one conic partially occluded). Rothwell et al. [205] report on an experimental recognition system which uses such invariants. The system receives a 2D view of several objects, typically metallic plates and other workpieces, and compares these objects to models from a database.

Let us consider a simple example of single view invariants. Suppose that we have detected n points in a 3D scene, and suppose we know for some reason that these points are coplanar. What we have measured are the $2n$ coordinates of the projections of the points onto the camera plane, which is in general different from the plane containing the points. Changes in camera position and orientation are adequately represented by the action of the group $G = \mathbb{R}^3 \rtimes (\mathbb{R}^* \times \mathrm{SO}_3(\mathbb{R}))$ generated by all translations, rotations, and scalars. If we exclude the possibility that the n points lie on a plane perpendicular to the camera plane, then it is easy to see that the 2D projection of the G-orbit of the original points coincides with the orbit of the projected points under the two-dimensional affine linear group $H = \mathrm{AGL}_2(\mathbb{R}) = \mathbb{R}^2 \rtimes \mathrm{GL}_2(\mathbb{R})$. If we also assume that no three points are collinear, then Theorem 4.4.4 provides the rational invariants

$$f_{i,j} = \frac{\det \begin{pmatrix} x_i & x_j \\ y_i & y_j \end{pmatrix}}{\det \begin{pmatrix} x_1 & x_2 \\ y_1 & y_2 \end{pmatrix}}, \quad (1 \le i \le 2,\ 3 \le j \le n-1),$$

where (x_i, y_i) is the vector connecting the camera images of the i-th and the n-th point. It would be interesting to obtain invariants which also ignore the action of the symmetric group S_n by permuting the n points. This is our next topic, but we will restrict ourselves to the 2D case, i.e., all points lie on the camera plane.

5.10.2 Invariants of n Points on a Plane

In a way this is a continuation of the discussion in Section 5.5. Suppose we have measured the mutual distances of n points on a plane which coincides with the camera plane (one could also think of the angular distances of n stars). In order to recognize the object we have measured, we wish to compare the measured distances with the distances of configurations of points stored in a database (such as a database of constellations). But we do not know how the order in which the measured points appear in our measurement corresponds to the order of points chosen for an item in the database. The proper translation of this problem into graph theory is that we are given an $\mathbb{R}$-weighted graph g (the weights being the measured distances), and we want to find a graph in the database that is isomorphic to g. Thus Lemma 5.5.1 can be used to obtain a relatively easy recognition procedure for objects made of n

points. Unfortunately the computational problems mentioned in 5.5 severely limit the scope of this approach. However, even having some invariants will provide necessary conditions for the congruence of two objects made up of n points.

Let (x_i, y_i) be the measured coordinates of the points $P_i \in \mathbb{R}^2$. Then the squared distances $d_{i,j} = (x_i - x_j)^2 + (y_i - y_j)^2$ are invariants under the affine orthogonal group $\mathrm{AO}_2(\mathbb{R}) = \mathbb{R}^2 \rtimes \mathrm{O}_2(\mathbb{R})$ generated by all translations and orthogonal transformations. In fact, the $d_{i,j}$ generate the invariant ring $\mathbb{R}[x_1, \ldots, x_n, y_1, \ldots, y_n]^{\mathrm{AO}_2(\mathbb{R})}$, and they separate all orbits (since we are only working with real coordinates). This is the reason why it is enough to compare the mutual distances of two objects for deciding whether the objects are "the same". But the invariants $d_{i,j}$ are not independent. Indeed they satisfy relations that come from the fact that the Gram matrix of any number of vectors in $\mathbb{R}^2$ has rank at most 2. Fix the point P_n, for example, and consider the connecting vectors $v_i := (x_i - x_n, y_i - y_n)$ for $1 \leq i \leq n - 1$. Then the (3×3)-minors of the Gram matrix $(\langle v_i, v_j \rangle)_{i,j=1,\ldots,n-1}$ (with scalar products as entries) are zero. But $\langle v_i, v_j \rangle = (d_{i,n} + d_{j,n} - d_{i,j})/2$. For $n = 4$, for example, we get the relation

$$\det (d_{i,4} + d_{j,4} - d_{i,j})_{i,j=1,\ldots,3} = 0.$$

Thus the $d_{i,j}$ lie on a variety $X \subseteq V = \mathbb{R}^{\binom{n}{2}}$, on which S_n acts. Computing graph invariants as in Section 5.5 and then evaluating them at the $d_{i,j}$ amounts to computing $\mathbb{R}[V]^{S_n}$ and then applying the canonical epimorphism $\mathbb{R}[V]^{S_n} \to \mathbb{R}[X]^{S_n}$. But it might be more efficient to compute generating invariants for $K[X]^{S_n}$ directly.

5.10.3 Moment Invariants

Let us look at a third way of applying invariant theory to computer vision. Our presentation is motivated by the article of Taubin and Cooper [243]. Suppose we have a 2D grey scale image given by a function $f \colon \mathbb{R}^2 \to \mathbb{R}$ with compact support. One could think of an image obtained by scanning a post stamp, for example. The goal would then be to identify this post stamp (by comparing with a database representing the currently valid post stamps). An approach that does not require any edge detection or extraction of geometric primitives is provided by moments, i.e., integrals of the form

$$\langle f, g \rangle = \int_{\mathbb{R}^2} g(x, y) f(x, y) dx \, dy$$

where $g \in \mathbb{R}[x, y]$ is a polynomial. It suffices to do this for monomials. For non-negative integers i and j, write

$$a_{i,j} = \int_{\mathbb{R}^2} x^i y^j f(x, y) dx \, dy.$$

The idea is to compute these $a_{i,j}$ for a limited range of i's and j's until enough data is collected to efficiently discriminate between different post stamps, for example. Again, the problem is that the values $a_{i,j}$ change drastically when the image is rotated or shifted by a vector in $\mathbb{R}^2$. It is easy to see that any affine linear map applied to the grey-scale image results in a linear transformation of the $a_{i,j}$. Thus the idea to form invariant polynomials from the $a_{i,j}$ is very natural. Such invariants are called **moment invariants**. As a first step we want to form invariants under translations by vectors from $\mathbb{R}^2$. The idea is to shift the image in such a way that its center of mass is sent to the coordinate origin, and then to compute the moments. Explicitly, set $\bar{x} := a_{1,0}/a_{0,0}$ and $\bar{y} := a_{0,1}/a_{0,0}$. Then the **normalized moments** may be defined by

$$\bar{a}_{i,j} := a_{0,0}^{i+j-1} \cdot \int_{\mathbb{R}^2} (x - \bar{x})^i (y - \bar{y})^j f(x,y)\, dx\, dy.$$

It is clear that $\bar{a}_{i,j}$ is translation-invariant. It is also clear from the construction that the normalized moments generate a separating algebra (in the sense of Definition 2.3.2) for the action of the translation group. We inserted the factor $a_{0,0}^{i+j-1}$ to make sure that $\bar{a}_{i,j}$ is a polynomial in the non-normalized moments $a_{\nu,\mu}$. For example, we obtain

$$\begin{aligned}
\bar{a}_{1,0} &= \bar{a}_{0,1} = 0, \\
\bar{a}_{2,0} &= a_{0,0}a_{2,0} - a_{1,0}^2, \\
\bar{a}_{1,1} &= a_{0,0}a_{1,1} - a_{1,0}a_{0,1}, \\
\bar{a}_{0,2} &= a_{0,0}a_{0,2} - a_{0,1}^2,
\end{aligned} \qquad (5.10.1)$$

and so on. After normalizing the moments we are left with linear actions. A matrix $\sigma = \left(\begin{smallmatrix} \alpha & \beta \\ \gamma & \delta \end{smallmatrix} \right)$ acts on the $\bar{a}_{i,j}$ in the same way as on the $a_{i,j}$. The action is closely related to the action on binary forms (see Example 2.1.2). For the $\bar{a}_{i,j}$ with $i + j = 2$ we obtain

$$\sigma : \begin{pmatrix} \bar{a}_{2,0} \\ \bar{a}_{1,1} \\ \bar{a}_{0,2} \end{pmatrix} \mapsto \det(\sigma)^2 \begin{pmatrix} \alpha^2 & 2\alpha\beta & \beta^2 \\ \alpha\gamma & \alpha\delta + \beta\gamma & \beta\delta \\ \gamma^2 & 2\gamma\delta & \delta^2 \end{pmatrix} \begin{pmatrix} \bar{a}_{2,0} \\ \bar{a}_{1,1} \\ \bar{a}_{0,2} \end{pmatrix}.$$

We can now use Algorithm 4.1.9 to compute generating invariants in $\mathbb{R}[\bar{a}_{2,0}, \bar{a}_{1,1}, \bar{a}_{0,2}]$ under the special orthogonal group $SO_2(\mathbb{R})$ (regarding the $\bar{a}_{i,j}$ as indeterminates). We find two generators:

$$g_1 = \bar{a}_{2,0} + \bar{a}_{0,2} \quad \text{and} \quad g_2 = \bar{a}_{2,0}\bar{a}_{0,2} - \bar{a}_{1,1}^2.$$

Using (5.10.1) to express the g_i in terms of the non-normalized moments yields

$$\begin{aligned}
g_1 &= a_{0,0}(a_{2,0} + a_{0,2}) - a_{1,0}^2 - a_{0,1}^2, \\
g_2/a_{0,0} &= a_{0,0}\left(a_{2,0}a_{0,2} - a_{1,1}^2\right) + 2a_{1,1}a_{1,0}a_{0,1} - a_{1,0}^2 a_{0,2} - a_{0,1}^2 a_{2,0}.
\end{aligned}$$

It may be surprising that g_2 turns out to be divisible by $a_{0,0}$. This shows that although the normalized moments are separating invariants for the translation action, they are not generating invariants. In addition to the above invariants, we have $a_{0,0}$ as an invariant under translations and rotations. It is clear that if we compute more moments $a_{i,j}$ we can also expect a larger number of moment invariants that can be formed from them.

Of course the Euclidean group ($\mathbb{R}^2 \rtimes \mathrm{SO}_2(\mathbb{R})$) that we used above is not the only group relevant for the identification of 2D images. Depending on the situation, other groups such as the scale Euclidean group $\mathbb{R}^2 \rtimes (\mathbb{R}^* \times \mathrm{SO}_2(\mathbb{R}))$ generated by translations, rotations, and scalar matrices may be appropriate. In that case, we will get fewer moment invariants.

A Linear Algebraic Groups

A.1 Linear Algebraic Groups

Let us fix an algebraically closed field K. We work in the category of affine varieties over K. The purpose of this section is to give a brief exposition on the basic facts of algebraic groups.

Definition A.1.1. *A **linear algebraic group** is an affine algebraic variety G together with a unit element $e \in G$ and morphisms $m : G \times G \to G$ and $i : G \to G$ satisfying the group axioms*

(a) $m(\sigma, e) = m(e, \sigma) = \sigma$ for all $\sigma \in G$;
(b) $m(\sigma, i(\sigma)) = m(i(\sigma), \sigma) = e$ for all $\sigma \in G$;
(c) $m(\sigma, m(\tau, \gamma)) = m(m(\sigma, \tau), \gamma)$ (associativity) for all $\sigma, \tau, \gamma \in G$.

Often we will just write $\sigma\tau$ and σ^{-1} instead of $m(\sigma, \tau)$ and $i(\sigma)$, respectively.

Example A.1.2. The group GL_n, the set of invertible $n \times n$ matrices over K, is a linear algebraic group. Any Zariski closed subgroup $G \subset \mathrm{GL}_n$ is a linear algebraic group. In fact, every linear algebraic group is isomorphic to a Zariski closed subgroup of GL_n (see Borel [23, Proposition I.1.10]). This justifies calling it a *linear* algebraic group. ◁

The axioms for a linear algebraic group can also be stated in terms of the coordinate ring. The maps $m : G \times G \to G$ and $i : G \to G$ correspond to ring homomorphisms $m^* : K[G] \to K[G] \otimes K[G]$ and $i^* : K[G] \to K[G]$. The unit element $e \in G$ corresponds to a ring homomorphism $\epsilon : K[G] \to K$ defined by $f \mapsto f(e)$. The axioms for a linear algebraic group translate to

(a) $(\mathrm{id} \otimes \epsilon) \circ m^* = (\epsilon \otimes \mathrm{id}) \circ m^* = \mathrm{id}$;
(b) $(i^* \otimes \mathrm{id}) \circ m^* = (\mathrm{id} \otimes i^*) \circ m^* = \epsilon$ (using the inclusion $K \subset K[G]$ we can view $\epsilon : K[G] \to K$ as a map $K[G] \to K[G]$);
(c) $(\mathrm{id} \otimes m^*) \circ m^* = (m^* \otimes \mathrm{id}) \circ m^*$.

Example A.1.3. The additive group $\mathbb{G}_a$ is defined as the additive group of K. We define $m(x, y) = x + y$, $i(x) = -x$ for all $x, y \in K$ and $e = 0 \in K$. The coordinate ring $K[\mathbb{G}_a]$ is equal to the polynomial ring $K[t]$. The homomorphism $m^* : K[t] \to K[t] \otimes K[t]$ is defined by $m^*(t) = t \otimes 1 + 1 \otimes t$,

$i^* : K[t] \to K[t]$ is defined by $t \mapsto -t$ and $\epsilon : K[t] \to K$ is defined by $f \mapsto f(0)$.   ◁

Example A.1.4. The group GL_n is an algebraic group. The coordinate ring $K[\mathrm{GL}_n]$ is isomorphic to

$$K[\{z_{i,j} \mid 1 \le i, j \le n\}, T]/(f),$$

where $f = T \det(Z) - 1$ with Z the matrix $(z_{i,j})_{i,j=1}^n$. Now $m^* : K[\mathrm{GL}_n] \to K[\mathrm{GL}_n] \otimes K[\mathrm{GL}_n]$ is defined by $z_{i,j} \mapsto \sum_{k=1}^n z_{i,k} \otimes z_{k,j}$ for all i, j. Notice that $m^*(\det(Z)) = \det(Z) \otimes \det(Z)$.   ◁

Example A.1.5. Let $T = (K^*)^r$. The coordinate ring $K[T]$ is isomorphic to

$$K[z_1, \ldots, z_r, z_1^{-1}, \ldots, z_r^{-1}].$$

The ring homomorphisms $m^* : K[T] \to K[T] \otimes K[T]$, $i^* : K[T] \to K[T]$ and $\epsilon : K[T] \to K$ are given by $m^*(z_i) = z_i \otimes z_i$, $i^*(z_i) = z_i^{-1}$ and $\epsilon(z_i) = 1$ for all i.   ◁

Definition A.1.6. *Suppose that X is an affine variety. A regular action of G on X is a morphism $\mu : G \times X \to X$ satisfying the axioms for an action*

(a) $\mu(e, x) = x$ *for all* $x \in X$;
(b) $\mu(\sigma, \mu(\tau, x)) = \mu(m(\sigma, \tau), x)$ *for all* $\sigma, \tau \in G$ *and all* $x \in X$.

We will often just write $\sigma \cdot x$ instead of $\mu(\sigma, x)$. Instead of saying that G acts regularly on X, we often just say that X is a G-variety.

Definition A.1.7. *Suppose that V is a (possibly infinite dimensional) vector space over K. A linear action $\mu : G \times V \to V$ of G on a vector space is called* **rational** *if there is a map $\mu^\star : V \to V \otimes K[G]$ satisfying $\mu(\sigma, v) = \sum_{i=1}^l v_i f_i(\sigma)$ whenever $\mu^\star(v) = \sum_{i=1}^l v_i \otimes f_i$*

Lemma A.1.8. *Suppose that a linear action of a linear algebraic group G on a vector space V is rational. For every $v \in V$ there exists a finite dimensional G-stable subspace $W \subseteq V$ containing v. The restriction of the G-action to W is also rational.*

Proof. Write $\mu^\star(v) = \sum_{i=1}^l v_i \otimes f_i$ with $f_1, \ldots, f_l$ linearly independent over K. Let $W = K(G \cdot v)$, the linear span of the orbit $G \cdot v$. It is not hard to see that W is a subspace of the space spanned by $v_1, \ldots, v_l$. This shows that $\dim W < \infty$.

We can restrict $\mu^\star$ to W to obtain a map $W \to W \otimes K[G]$. Therefore the G-action on W is rational.   □

The content of Lemma A.1.8 can be expressed by saying that a rational action is "locally finite" . If G acts regularly on an affine variety X, then G also acts on the coordinate ring $K[X]$. For $\sigma \in G$ and $f \in K[X]$ we define $\sigma \cdot f \in K[X]$ by

$$(\sigma \cdot f)(x) = f(\sigma^{-1} \cdot x)$$

for all $x \in X$.

A map $X \to Y$ between two sets on which G acts is called G-equivariant if $\varphi(\sigma \cdot x) = \sigma \cdot \varphi(x)$ for all $x \in X$ and all $\sigma \in G$.

Lemma A.1.9. *Suppose that X is a G-variety. Then there exists a finite dimensional rational representation V and a G-equivariant closed embedding $X \hookrightarrow V$.*

Proof. Choose generators $f_1, \ldots, f_r$ of $K[X]$ and let W be a finite dimensional G-stable subspace of $K[X]$ containing $f_1, \ldots, f_r$. The inclusion $W \to K[X]$ extends to a surjective G-equivariant ring homomorphism $S(W) \twoheadrightarrow K[X]$, where $S(W)$ is the symmetric algebra. This ring homomorphism corresponds to a G-equivariant closed embedding $X \hookrightarrow W^*$ where W^* is the dual space of W. $\qquad\square$

The following two propositions can be found in Borel [23, Proposition I.1.2].

Proposition A.1.10. *A linear algebraic group G is smooth.*

Proposition A.1.11. *Let G° be the connected component of $e \in G$. Then G° is a normal subgroup of G of finite index.*

A.2 The Lie Algebra of a Linear Algebraic Group

We will first study the dual vector space $K[G]^*$ of $K[G]$. As we will see, $K[G]^*$ contains the Lie algebra $\mathfrak{g}$ as a finite dimensional subspace. For every $\sigma \in G$ we can define $\epsilon_\sigma : K[G] \to K$ by $f \mapsto f(\sigma)$. In this way we can view G as a subset of $K[G]^*$. By Definition A.2.1 below, the group structure of G equips $K[G]^*$ with an associative algebra structure with identity element $\epsilon = \epsilon_e$. This even allows us to see $K[G]^*$ as an enveloping algebra of $\mathfrak{g}$ (containing the universal enveloping algebra of $\mathfrak{g}$). Our approach is similar to Borel [23, §I.3].

Definition A.2.1. *Suppose that $\delta, \gamma \in K[G]^*$. Then we define the convolution $\gamma * \delta \in K[G]^*$ as the composition of $\gamma \otimes \delta : K[G] \otimes K[G] \to K \otimes K = K$ and $m^* : K[G] \to K[G] \otimes K[G]$. In other words, suppose $f \in K[G]$ and $m^*(f) = \sum_i g_i \otimes h_i$, then*

$$(\gamma * \delta)(f) = \sum_i \gamma(g_i)\delta(h_i).$$

Proposition A.2.2. *The space $K[G]^*$ is an associative algebra with the multiplication $*$ and unit element $\epsilon = \epsilon_e$.*

Proof. From the axiom

$$m(m(\sigma, \tau), \mu) = m(\sigma, m(\tau, \mu)),$$

it follows that

$$(m^* \otimes \mathrm{id}) \circ m^* = (\mathrm{id} \otimes m^*) \circ m^*.$$

This multiplication is associative, because

$$(\delta * \gamma) * \varphi = (((\delta \otimes \gamma) \circ m^*) \otimes \varphi) \circ m^* = (\delta \otimes \gamma \otimes \varphi) \circ (m^* \otimes \mathrm{id}) \circ m^* =$$
$$= (\delta \otimes \gamma \otimes \varphi) \circ (\mathrm{id} \otimes m^*) \circ m^* = (\delta \otimes ((\gamma \otimes \varphi) \circ m^*)) \circ m^* = \delta * (\gamma * \varphi).$$
$$(\mathrm{A.2.1})$$

Notice that from $m(e, \sigma) = \sigma$ it follows that $(\epsilon \otimes \mathrm{id}) \circ m^* = \mathrm{id}$. We get

$$\delta * \epsilon = (\delta \otimes \epsilon) \circ m^* = \delta \circ (\mathrm{id} \otimes \epsilon) \circ m^* = \delta \circ \mathrm{id} = \delta,$$

and similarly $\epsilon * \delta = \delta$. $\qquad\qquad\qquad\square$

Example A.2.3. If $f \in K[G]$ and $\sigma \in G$, we define $L_\sigma(f)$ and $R_\sigma(f)$ by $L_\sigma(f)(\tau) = f(\sigma\tau)$ and $R_\sigma(f)(\tau) = f(\tau\sigma)$. Notice that $(\mathrm{id} \otimes \epsilon_\sigma) \circ m^* = R_\sigma$ and $(\epsilon_\sigma \otimes \mathrm{id}) \circ m^* = L_\sigma$. From this it follows that

$$\delta * \epsilon_\sigma = \delta \circ (\mathrm{id} \otimes \epsilon_\sigma) \circ m^* = \delta \circ R_\sigma$$

and

$$\epsilon_\sigma * \delta = \delta \circ (\epsilon_\sigma \otimes \mathrm{id}) \circ m^* = \delta \circ L_\sigma.$$

$\qquad\qquad\qquad\qquad\qquad\qquad\qquad\qquad\qquad\qquad\qquad\qquad\qquad\qquad\triangleleft$

Definition A.2.4. *We define the* **Lie algebra** $\mathfrak{g}$ *of G as the set of all $\delta \in$ $K[G]^*$ satisfying*

$$\delta(fg) = \delta(f)g(e) + f(e)\delta(g). \qquad\qquad (\mathrm{A.2.2})$$

for all $f, g \in K[G]$.

An element $\delta \in K[G]$ satisfying (A.2.2) is called a point derivation of $K[G]$ at e. Let $\mathfrak{m}_e$ be the maximal ideal of $K[G]$ vanishing at $e \in G$. The Zariski tangent space $T_e(G)$ of G at e is defined as $(\mathfrak{m}_e/\mathfrak{m}_e^2)^*$ which is the dual space of $\mathfrak{m}_e/\mathfrak{m}_e^2$. If $\delta \in (\mathfrak{m}_e/\mathfrak{m}_e^2)^*$, we can define $\rho(\delta) \in K[G]^*$ by

$$\rho(\delta)f = \delta(f - f(e) + \mathfrak{m}_e^2).$$

It is not hard to show that $\rho(\delta)$ is a point derivation at e and that $\delta \mapsto \rho(\delta)$ is an isomorphism between $T_e(G)$ and $\mathfrak{g}$. Because G is smooth (Proposition A.1.10) we have $\dim(\mathfrak{g}) = \dim(T_e(G)) = \dim(G)$.

Proposition A.2.5. *The space $\mathfrak{g}$ is a Lie algebra with the Lie bracket*

$$[\delta, \gamma] := \delta * \gamma - \gamma * \delta.$$

Proof. The only thing we have to show is that $[\delta, \gamma] \in \mathfrak{g}$ if $\delta, \gamma \in \mathfrak{g}$. But this is checked by a straightforward calculation. $\qquad\square$

Definition A.2.6. *For $\delta \in K[G]^*$, we define $\bar{\delta} \in K[G]^*$ to be the composition $\delta \circ i^* : K[G] \to K[G] \to K$.*

Lemma A.2.7. *The map $\delta \mapsto \bar{\delta}$ is an involution, and a skew algebra homomorphism $K[G]^* \to K[G]^*$, i.e., $\overline{\delta * \gamma} = \bar{\gamma} * \bar{\delta}$ for all $\delta, \gamma \in K[G]^*$ and $\bar{\epsilon} = \epsilon$.*

Proof. The map $\delta \to \bar{\delta}$ is an involution because i and i^* are involutions.

Let $S : G \times G \to G \times G$ be defined by $S(\sigma, \tau) \to (\tau, \sigma)$ and let $S^* : K[G] \otimes K[G] \to K[G] \otimes K[G]$ be the dual ring homomorphism. From $i(m(\sigma, \tau)) = m(i(\tau), i(\sigma))$ it follows that

$$(i^* \otimes i^*) \circ m^* = S^* \circ m^* \circ i^*.$$

We can deduce from this that

$$\bar{\delta} * \bar{\gamma} = (\delta \otimes \gamma) \circ (i^* \otimes i^*) \circ m^* = (\delta \otimes \gamma) \circ S^* \circ m^* \circ i^* =$$
$$= (\gamma \otimes \delta) \circ m^* \circ i^* = (\gamma * \delta) \circ i^* = \overline{\gamma * \delta}. \quad (A.2.3)$$

From $e^{-1} = e$ it follows that $\epsilon \circ i^* = \epsilon$ so $\bar{\epsilon} = \epsilon$. $\qquad\square$

Remark A.2.8. An action $\mu : G \times V \to V$ of G on a vector space is a rational representation if there is a map $\mu^* : V \to V \otimes K[G]$ with $\mu(\sigma, v) = ((\mathrm{id} \otimes \epsilon_\sigma) \circ \mu^*)v$. $\qquad\triangleleft$

Definition A.2.9. *For $\delta \in K[G]^*$ and $v \in V$ we define $\delta \cdot v$ by $((\mathrm{id} \otimes \delta) \circ \mu^*)v$. In other words, if $\mu^*(v) = \sum_i v_i \otimes h_i$, then*

$$\delta \cdot v = \sum_i v_i \delta(h_i).$$

In particular, we have $\sigma \cdot v = \epsilon_\sigma \cdot v$ for all $\sigma \in G$ and $v \in V$.

Proposition A.2.10. *The map $(\delta, v) \mapsto \delta \cdot v$ gives V the structure of a $K[G]^*$-module.*

Proof. From the axiom

$$\mu(\sigma, \mu(\tau, v)) = \mu(m(\sigma, \tau), v)$$

for all $\sigma, \tau \in G$ and $v \in V$, it follows that

$$(\mathrm{id} \otimes m^*) \circ \mu^* = (\mu^* \otimes \mathrm{id}) \circ \mu^*.$$

We have

$$\gamma \cdot (\delta \cdot v) = ((\mathrm{id} \otimes \gamma) \circ \mu^* \circ (\mathrm{id} \otimes \delta) \circ \mu^*)v =$$

$$((\mathrm{id} \otimes \gamma) \circ (\mathrm{id} \otimes \mathrm{id} \otimes \delta) \circ (\mu^* \otimes \mathrm{id}) \circ \mu^*)v = ((\mathrm{id} \otimes \gamma \otimes \delta) \circ (\mathrm{id} \otimes m^*) \circ \mu^*)v$$

$$= ((\mathrm{id} \otimes ((\gamma \otimes \delta) \circ m^*)) \circ \mu^*)v = (\gamma * \delta) \cdot v. \quad (A.2.4)$$

From $\mu(e, x) = x$ it follows that

$$(\mathrm{id} \otimes \epsilon) \circ \mu^* = \mathrm{id},$$

so

$$\epsilon \cdot v = ((\mathrm{id} \otimes \epsilon) \circ \mu^*)v = \mathrm{id}\, v = v.$$

$\square$

Example A.2.11. The group G acts on $K[G]$ as follows: Let $\widetilde{\mu} : G \times G \to G$ defined by $(\tau, \sigma) \mapsto (\sigma^{-1}\tau)$ and let $\mu^* : K[G] \to K[G] \otimes K[G]$ be the dual homomorphism. The maps m and m^* are as usual, the multiplication map $G \times G \to G$ and its dual homomorphism respectively. We have

$$(\delta \cdot f)(e) = ((\epsilon \otimes \delta) \circ \mu^*)f = ((\epsilon \otimes (\delta \circ i^*)) \circ m^*)f = (\epsilon * \overline{\delta})f = \overline{\delta}f.$$

$\triangleleft$

Example A.2.12. Let X be an affine G-variety. Suppose that $\mu^* : K[X] \to K[X] \otimes K[G]$ is the dual homomorphism of $\widetilde{\mu} : X \times G \to X$ defined by $(x, \sigma) \mapsto \sigma^{-1} \cdot x$. For $f \in K[X]$ we already defined

$$(\sigma \cdot f)(x) = f(\sigma^{-1} \cdot x) = f(\widetilde{\mu}(x, \sigma)) = \mu^*(f)(x, \sigma) = ((\mathrm{id} \otimes \epsilon_\sigma) \circ \mu^*)(f)(x)$$

for all $f \in K[X], x \in X, \sigma \in G$. This shows that $K[X]$ is a rational representation of G. Now also $K[G]^*$ acts on $K[X]$ and in particular $\mathfrak{g}$ acts on $K[X]$.
$\triangleleft$

Lemma A.2.13. *The Lie algebra $\mathfrak{g}$ acts on $K[X]$ by derivations.*

Proof. Let $f, u \in K[X]$, and write $\mu^*(f) = \sum_i g_i \otimes h_i$ and $\mu^*(u) = \sum_j v_j \otimes w_j$. Then we have

$$\delta \cdot fu = \sum_{i,j} g_i v_j \delta(h_i w_j) = \sum_{i,j} \big(g_i v_j \delta(h_i) w_j(e) + g_i v_j h_i(e) \delta(w_j)\big) =$$

$$= (\delta \cdot f)(\epsilon \cdot u) + (\epsilon \cdot f)(\delta \cdot u) = (\delta \cdot f)u + f(\delta \cdot u). \quad (A.2.5)$$

$\square$

Definition A.2.14. *For $\sigma \in G$ and $\delta \in K[G]^*$, we define*

$$\mathrm{Ad}(\sigma)\delta := \epsilon_\sigma * \delta * \epsilon_{\sigma^{-1}}.$$

Notice that $\epsilon_{\sigma^{-1}}$ is the inverse of ϵ_σ in the algebra $K[G]^*$. This shows that $\mathrm{Ad}(\sigma)$ is an algebra automorphisms of $K[G]^*$ for every $\sigma \in G$.

Lemma A.2.15. *The map* Ad *defines a rational action of G on* $\mathfrak{g}$.

Proof. If $f \in K[G]$, $\sigma \in G$ and $\delta \in K[G]^*$, then from Example A.2.3 it follows that

$$(\mathrm{Ad}(\sigma)\delta)f = (\epsilon_\sigma * \delta * \epsilon_{\sigma^{-1}})f = L_\sigma R_{\sigma^{-1}} f.$$

Now $\mathfrak{g}$ can be characterized as the set of all $\delta \in K[G]^*$ such that $\delta(\mathfrak{m}_e^2) = \{0\}$ and $\delta(1) = 1$ where $\mathfrak{m}_e$ is the maximal ideal in $K[G]$ corresponding to $e \in G$. Since both $\mathfrak{m}_e^2$ and 1 are stable under $L_\sigma R_{\sigma^{-1}}$, it follows that $\mathrm{Ad}(\sigma)\mathfrak{g} \subseteq \mathfrak{g}$. We can let G act on $K[G]$ by

$$(\sigma \cdot f)(\tau) = f(\sigma^{-1}\tau\sigma) = (L_{\sigma^{-1}} R_\sigma f)(\tau).$$

This action of G on $K[G]^*$ via Ad is dual to this action. In particular the action of G via Ad on $\mathfrak{g}$ is dual to the action of G on $\mathfrak{m}_e/\mathfrak{m}_e^2$ and therefore defines a rational representation. $\qquad\square$

A.3 Reductive and Semi-simple Groups

We now introduce some structure theory of linear algebraic groups. Let G be a linear algebraic group.

Definition A.3.1. *A* **Borel subgroup** B *of G is a maximal connected solvable subgroup.*

We also often will need a maximal torus T of G. Since tori are connected and solvable, clearly T is contained in some Borel subgroup B and it is a maximal torus of B as well.

Theorem A.3.2. *All Borel subgroups of G are conjugate.*

Proof. See Borel [23, IV.11.1]. $\qquad\square$

Definition A.3.3. *An endomorphism $A \in \mathrm{GL}(V)$ is called* **unipotent** *if $A - I$ is nilpotent, i.e., $(A - I)^N = 0$ for some positive integer N where I is the identity of $\mathrm{GL}(V)$. An algebraic group U is called* **unipotent** *if for every finite dimensional representation $\rho : U \to \mathrm{GL}(V)$ and every $\sigma \in U$ we have that $\rho(\sigma)$ is a unipotent endomorphism.*

Proposition A.3.4. *Suppose that B is a solvable connected linear algebraic group. Let T be a maximal torus of B. Then there exists a unique connected unipotent normal subgroup B_u such that B is a semidirect product of T and B_u.*

Proof. See Borel [23, III.10.6]. $\qquad\square$

Definition A.3.5. *Suppose that G is a linear algebraic group and let $\mathcal{B}$ be the set of all Borel subgroups of G. The* **radical** *$R(G)$ of G is defined as the connected component of the intersection of all Borel subgroups, i.e.,*

$$R(G) = \left(\bigcap_{B \in \mathcal{B}} B \right)^{\circ}.$$

The unipotent part $R(G)_u$ of $R(G)$ is called the **unipotent radical** *of G.*

Definition A.3.6. *A linear algebraic group G is called* **reductive** *if the unipotent radical $R(G)_u$ is trivial, or equivalently, if $R(G)$ is a torus. If $R(G)$ is trivial, then G is called* **semi-simple**.

A.4 Roots

Suppose that G is a connected reductive group, $T \subseteq G$ is a maximal torus and $B \subseteq G$ is a Borel subgroup containing T. Let $m = \dim G$ and $r = \dim T$. Let $X(T) \cong \mathbb{Z}^r$ be the set of 1-dimensional characters, i.e., the set of algebraic group homomorphisms $T \to \mathbb{G}_m$, where $\mathbb{G}_m$ is the multiplicative group. Usually, we use additive notation for the group $X(T)$. We let $X^{\vee}(T) \cong \mathbb{Z}^r$ be the set of algebraic group homomorphisms $\mathbb{G}_m \to T$. We have a pairing $\langle \cdot, \cdot \rangle : X(T) \times X^{\vee}(T) \to \mathbb{Z}$ which is defined as follows. If $\chi : T \to \mathbb{G}_m$ and $\lambda : \mathbb{G}_m \to T$ are algebraic group homomorphisms, then $\chi(\lambda(\sigma)) = \sigma^a$ for some $a \in \mathbb{Z}$ and all $\sigma \in \mathbb{G}_m$. We define $\langle \chi, \lambda \rangle = a$.

The Weyl group W is the group $N_G(T)/Z_G(T)$ where $N_G(T)$ is the normalizer of T in G and $Z_G(T)$ is the centralizer of T in G. The Weyl groups acts in a natural way on $X(T)$ and $X^{\vee}(T)$.

Consider the adjoint representation of G on $\mathfrak{g}$. For an element $\alpha \in X(T)$ we define

$$\mathfrak{g}_\alpha = \{ v \in \mathfrak{g} \mid \mathrm{Ad}(\sigma)v = \alpha(\sigma)v \text{ for all } \sigma \in T \}.$$

A non-zero element $\alpha \in X(T)$ is called a **root** if $\mathfrak{g}_\alpha \neq 0$. We write Φ for the set of all roots. Since the action of the torus is diagonalizable (see Borel [23, Proposition III.8.4]), we have a decomposition

$$\mathfrak{g} = \mathfrak{g}^T \oplus \bigoplus_{\alpha \in \Phi} \mathfrak{g}_\alpha$$

where $\mathfrak{g}^T = \mathfrak{g}_0$ is the T-invariant subspace. For each root α, the space $\mathfrak{g}_\alpha$ is 1-dimensional (this follows from Humphreys [117][8.4.]).

Let α be a root. The kernel T_α of α is a torus of dimension $r - 1$. There is a unique non-trivial element $s_\alpha \in W$ which fixes T_α. The group W is generated by all reflections $s_\alpha, \alpha \in \Phi$.

We define $\alpha^{\vee} \in X^{\vee}(T)$ by

$$s_\alpha(\beta) = \beta - \langle \beta, \alpha^{\vee} \rangle \alpha.$$

Let $\mathfrak{t} \subset \mathfrak{g}$ be the Lie algebra of the torus T and let $\mathfrak{u} \subset \mathfrak{g}$ be the Lie algebra of the unipotent complement U of T in B. We have $\mathfrak{t} = \mathfrak{g}^T$, and $\mathfrak{u}$ is a direct sum of weight spaces $\mathfrak{g}_\alpha$. Let us define the set of positive roots Φ_+ by

$$\mathfrak{u} = \bigoplus_{\alpha \in \Phi_+} \mathfrak{g}_\alpha.$$

We define $\Phi_- := -\Phi_+$. It is known that Φ is the disjoint union of Φ_+ and Φ_-.

We can uniquely choose a subset $\Delta = \{\alpha_1, \ldots, \alpha_l\}$ of Φ_+ with the following properties:

(a) $\alpha_1, \ldots, \alpha_l$ is an $\mathbb{R}$-basis of $E = X(T) \otimes_{\mathbb{Z}} \mathbb{R}$;
(b) $\langle \alpha_i, \alpha_j^\vee \rangle \le 0$ for all $i \ne j$.

Elements of Δ are called **simple roots**. Every $\alpha \in \Phi_+$ can uniquely be written as a sum

$$m_1 \alpha_1 + m_2 \alpha_2 + \cdots + m_l \alpha_l$$

for some non-negative integers $m_1, \ldots, m_l$.

A.5 Representation Theory

Let G be a connected reductive linear algebraic group over an algebraically closed field K. Let $\Delta = \{\alpha_1, \ldots, \alpha_l\}$ be the set of simple roots. A weight $\lambda \in X(T)$ is called **dominant** if $\langle \lambda, \alpha_i^\vee \rangle \ge 0$ for all simple roots α_i. There exist unique so-called **fundamental weights** $\lambda_1, \ldots, \lambda_l \in X(T) \otimes_{\mathbb{Z}} \mathbb{Q}$ such that $\langle \lambda_i, \alpha_j^\vee \rangle = \delta_{i,j}$ where $\delta_{i,j}$ is the Kronecker delta symbol.

Theorem A.5.1. *Choose a Borel subgroup $B \subset G$, a maximal torus $T \subset G$ and a unipotent subgroup $U \subset B$ such that $B = T \ltimes U$. If V is an irreducible representation of G, then $V^U = Kv$ for some non-zero vector $v \in V$. We have $\sigma \cdot v = \lambda(\sigma)v$ for all $\sigma \in T$ for some dominant weight $\lambda \in X(T)$ called the **highest weight** of V. For each dominant weight $\lambda \in X(T)$ there exists a unique irreducible representation V_λ with highest weight λ.*

Proof. See Humphreys [118, Chapter IX]. $\square$

A vector $v_\lambda \in V_\lambda$ with $V_\lambda^U = Kv_\lambda$ and $\sigma \cdot v_\lambda = \lambda(\sigma)v_\lambda$ for all $\sigma \in T$ is called a **highest weight vector** of V_λ.

Definition A.5.2. *Suppose $\mu : G \to \mathrm{GL}(V)$ is a rational representation of G. Then the character $\chi^V : T \to K$ is defined by*

$$\chi^V(\sigma) = \mathrm{Tr}(\mu(\sigma))$$

for all $\sigma \in T$.

In the sequel, we would rather use multiplicative notation in $X(T)$. To each fundamental weight λ_i $(1 \leq i \leq l)$ in the additive notation we associate an indeterminate z_i (i.e., a character in the multiplicative notation). If $\lambda = \sum_{i=1}^{l} a_i \lambda_i$ is a weight, we define $z^\lambda := z_1^{\lambda_1} z_2^{\lambda_2} \cdots z_l^{\lambda_l}$. The action of the torus T is diagonalizable, so after a convenient choice of basis, the action is given by a matrix

$$\mu = \begin{pmatrix} m_1(z) & 0 & \cdots & 0 \\ 0 & m_2(z) & \cdots & 0 \\ \vdots & & \ddots & \vdots \\ 0 & 0 & \cdots & m_n(z) \end{pmatrix}.$$

where $m_1, \ldots, m_n$ are monomials in $z_1, \ldots, z_l$. The character χ^V is then given by the formal sum

$$m_1(z) + m_2(z) + \cdots + m_n(z).$$

Let us write $\rho = \lambda_1 + \cdots + \lambda_l$.

Theorem A.5.3 (Weyl's Theorem, see Humphreys [117, 24.3]).

$$\chi^{V_\lambda} \sum_{w \in W} \mathrm{sgn}(w) z^{w(\rho)} = \sum_{w \in W} \mathrm{sgn}(w) z^{w(\rho + \lambda)}.$$

References

In the square bracket at the end of each reference we give the pages where the reference is cited.

[1] Shreeram S. Abhyankar, *Galois embeddings for linear groups*, Trans. Amer. Math. Soc. **352** (2000), 3881–3912 [215].

[2] O. M. Adamovich, *Equidimensional representations of simple algebraic groups*, in: *Geom. Metod. Zadach. Algebry Anal.*, vol. 2, pp. 120–125, 1980, English transl.: Transl., II. Ser., Ann. Math. Soc. **128** (1986), 25–29 [206].

[3] O. M. Adamovich, E. .O. Golovina, *Simple linear Lie groups having a free algebra of invariants*, in: *Vopr. Teor. Grupp. Gomologicheskoj Algebry*, vol. 2, pp. 3–41, 1979, English transl.: Sel. Math. Sov. **3 (2)** (1984), 183–220 [206].

[4] F. Adams, *Lectures on Lie Groups*, W.A. Benjamin, New York, Amsterdam 1969 [181].

[5] J. F. Adams, C. W. Wilkerson, *Finite H-spaces and algebras over the Steenrod algebra*, Ann. of Math. (1980), 95–143 [132].

[6] William W. Adams, Phillippe Loustaunau, *An Introduction to Gröbner Bases*, Graduate Studies in Mathematics **3**, Americal Mathematical Society, Providence, RI 1994 [7].

[7] Alejandro Adem, R. James Milgram, *Invariants and cohomology of groups*, Bol. Soc. Mat. Mex. **37** (1992), 1–25 [209].

[8] Alejandro Adem, R. James Milgram, *Cohomology of Finite Groups*, Springer-Verlag, Berlin, Heidelberg, New York 1994 [209, 210].

[9] Gert Almkvist, Robert M. Fossum, *Decompositions of exterior and symmetric powers of indecomposable $\mathbb{Z}/p\mathbb{Z}$-modules in characteristic p and relations to invariants*, in: *Sém. d'Algèbre P. Dubreil*, Lecture Notes in Math. **641**, pp. 1–111, Springer-Verlag, Berlin, Heidelberg, New York 1976–1977 [79, 129].

[10] Maria Emilia Alonso, Teo Mora, Mario Raimondo, *Local decomposition algorithms*, in: Shojiro Sakata, ed., *Applied Algebra, Algebraic Algorithms and Error-Correcting Codes (AAECC-8)*, Lect. Notes Comput. Sci. **508**, pp. 208–221, Springer-Verlag, Berlin, Heidelberg, New York 1991 [27].

[11] Helmer Aslaksen, Shih-Piug Chan, Tor Gulliksen, *Invariants of S_4 and the shape of sets of vectors*, Appl. Algebra Engrg. Comm. Comput. **7** (1996), 53–57 [92, 220].

[12] Eiichi Bannai, Steven T. Dougherty, Masaaki Harada, Manabu Oura, *Type II codes, even unimodular lattices, and invariant rings*, IEEE Trans. Inf. Theory **45** (1999), 1194–1205 [226].

[13] Dave Bayer, Mike Stillman, *Computation of Hilbert functions*, J. Symbolic Computation **14** (1992), 31–50 [26].

[14] Jozsef Beck, Vera T. Sós, *Discrepancy theory*, in: R. L. Graham, Martin Grötschel, Laszlo Lovász, eds., *Handbook of Combinatorics*, vol. 2, pp. 1405–1488, North-Holland, 1995 [161].

[15] Thomas Becker, Volker Weispfenning, *Gröbner Bases*, Springer-Verlag, Berlin, Heidelberg, New York 1993 [7, 9, 12–17, 27–31].

[16] David J. Benson, *Representations and Cohomology I*, Cambridge Studies in Advanced Mathematics **30**, Cambridge Univ. Press, Cambridge 1991 [130].

[17] David J. Benson, *Representations and Cohomology II*, Cambridge Studies in Advanced Mathematics **31**, Cambridge Univ. Press, Cambridge 1991 [88].

[18] David J. Benson, *Polynomial Invariants of Finite Groups*, Lond. Math. Soc. Lecture Note Ser. **190**, Cambridge Univ. Press, Cambridge 1993 [2, 63, 73, 99, 105, 107, 137].

[19] David J. Benson, *Noether's degree bound*, private communication, 2000 [108–110].

[20] Michel Van den Bergh, *Modules of covariants*, in: *Proceedings of the International Congress of Mathematics, ICM '94*, vol. 1, pp. 352–362, Birkhäuser, Basel 1995 [87].

[21] Anna Maria Bigatti, Massimo Caboara, Lorenzo Robbiano, *Computation of Hilbert-Poincaré series*, Applicable Algebra in Engineering, Communication and Computing **2** (1993), 21–33 [26].

[22] J. P. Boehler, ed., *Applications of Tensor Functions in Solid Mechanics*, CISM Courses and Lectures **292**, Springer-Verlag, Wien, New York 1987 [230].

[23] Armand Borel, *Linear Algebraic Groups*, vol. 126 of *Graduate Texts in Mathematics*, Springer Verlag, New York 1991 [39, 50, 237, 239, 243, 244].

[24] Wieb Bosma, John J. Cannon, Catherine Playoust, *The Magma algebra system I: The user language*, J. Symbolic Comput. **24** (1997), 235–265 [7, 73, 83].

[25] Dorra Bourguiba, Said Zarati, *Depth and the Steenrod algebra*, Invent. math. **128** (1997), 589–602 [132].

[26] Jean-Francois Boutot, *Singularités rationnelles et quotients par les groupes réductifs.*, Invent. Math. **88** (1987), 65–68 [64, 70].

[27] Michel Brion, *Groupe de Picard et nombres caractéristiques des variétés sphériques*, Duke Math. J. **58** (1989), 397–424 [203].

[28] Theodor Bröcker, Tammo tom Dieck, *Representations of Compact Lie Groups*, vol. 98 of *Graduate Texts in Mathematics*, Springer-Verlag, New York–Berlin 1985 [187].

[29] Abraham Broer, *A new method for calculating Hilbert series*, J. of Algebra **168** (1994), 43–70 [188].

[30] Abraham Broer, *Remarks on invariant theory of finite groups*, preprint, Université de Montréal, Montréal, 1997 [115, 116].

[31] Winfried Bruns, Jürgen Herzog, *Cohen-Macaulay Rings*, Cambridge University Press, Cambridge 1993 [63, 64, 99].

[32] Bruno Buchberger, *An algorithm for finding a basis for the residue class ring of a zero-dimensional polynomial ideal* (German), Dissertation, Institute for Mathematics, University of Innsbruck 1965 [12].

[33] H. E. A. Campbell, I. P. Hughes, *Vector invariants of $U_2(\mathbb{F}_p)$: A proof of a conjecture of Richman*, Adv. in Math. **126** (1997), 1–20 [111, 112, 120].

[34] H. E. A. Campbell, I. P. Hughes, R. D. Pollack, *Rings of invariants and p-Sylow subgroups*, Canad. Math. Bull. **34(1)** (1991), 42–47 [87].

[35] H. E. A. Campbell, I. P. Hughes, R. J. Shank, *Preliminary notes on rigid reflection groups*, available at `http://www.ukc.ac.uk/ims/maths/people/R.J.Shank/`, 1996 [106, 107].

[36] H. E. A. Campbell, A. V. Geramita, I. P. Hughes, G. G. Smith, D. L. Wehlau, *Hilbert functions of graded algebras*, The Curves Seminar at Queen's, Volume XI, in: Queen's Papers in Pure and Applied Math. **105** (1997), 60–74 [83].

[37] H. E. A. Campbell, A. V. Geramita, I. P. Hughes, R. J. Shank, D. L. Wehlau, *Non-Cohen-Macaulay vector invariants and a Noether bound for a Gorenstein ring of invariants*, Canad. Math. Bull. **42** (1999), 155–161 [89].

[38] H. E. A. Campbell, I. P. Hughes, G. Kemper, R. J. Shank, D. L. Wehlau, *Depth of modular invariant rings*, Transformation Groups **5** (2000), 21–34 [100, 127].

[39] Sue Ann Campbell, Philip Holmes, *Heteroclinic cycles and modulated travelling waves in a system with D_4 symmetry*, Physica D **59** (1992), 52–78 [228].

[40] A. Capani, G. Niesi, L. Robbiano, *CoCoA: A system for doing computations in commutative algebra*, available via anonymous ftp from `cocoa.dima.unige.it`, 2000 [7].

[41] David Carlisle, Peter H. Kropholler, *Modular invariants of finite symplectic groups*, preprint, Queen Mary, University of London, 1992 [107, 137].

[42] B. Char, K. Geddes, G. Gonnet, M. Monagan, S. Watt, *Maple Reference Manual*, Waterloo Maple Publishing, Waterloo, Ontario 1990 [73, 222].

[43] Claude Chevalley, *Invariants of finite groups generated by reflections*, Amer. J. Math. **77** (1955), 778–782 [104].

[44] A. M. Cohen, A. E. Brouwer, *The Poincaré series of the polynomials invariant under SU_2 in its irreducible representation of degree ≤ 17*, Math. Centrum Amsterdam Afd. Zuivere Wiskunde **ZW 134/79** (1979), 1–20 [191].

[45] Antoine Colin, *Formal computation of Galois groups with relative resolvents*, in: Gérard Cohen, Marc Giusti, Teo Mora, eds., *Applied Algebra, Algebraic Algorithms and Error-Correcting Codes (AAECC-11)*, Lecture Notes in Computer Science **948**, pp. 169–182, Springer-Verlag, Berlin 1995 [214, 215].

[46] Stéphane Collart, Michael Kalkbrenner, Daniel Mall, *Converting bases with the Gröbner walk*, J. Symbolic Comput. **24** (1997), 465–469 [13].

[47] Michael A. Collins, Keiran C. Thompson, *Group theory and the global functional shapes for molecular potential energy surfaces*, in: Danail Bonchev, Dennis H. Rouvray, eds., *Chemical Group Theory: Techniques and Applications*, pp. 191–234, Gordon and Breach Publishers, Reading 1995 [228].

[48] David Cox, John Little, Donal O'Shea, *Ideals, Varieties, and Algorithms*, Springer-Verlag, New York, Berlin, Heidelberg 1992 [7, 16, 17].

[49] David Cox, John Little, Donal O'Shea, *Using Algebraic Geometry*, Springer-Verlag, New York 1998 [7, 21].

[50] Charles W. Curtis, Irving Reiner, *Methods of Representation Theory I*, J. Wiley & Sons, New York 1981 [77, 78, 130].

[51] Jiri Dadoc, Victor G. Kac, *Polar representations*, J. Algebra **92 (2)** (1985), 504–524 [206].

[52] Daniel Daigle, Gene Freudenburg, *A counterexample to Hilbert's fourteenth problem in dimension five*, J. Algebra **128** (1999), 528–525 [44].

[53] Harm Derksen, *Computation of invariants for reductive groups*, Adv. in Math. **141** (1999), 366–384 [139].

[54] Harm Derksen, *Polynomial bounds for rings of invariants*, Proc. Amer. Math. Soc. **129** (2001), 955–963 [196].

[55] Harm Derksen, Hanspeter Kraft, *Constructive Invariant Theory*, in: *Algèbre non commutative, groupes quantiques et invariants (Reims, 1995)*, vol. 36 of *Sémin. Congr.*, pp. 221–244, Soc. Math. France, Paris 1997 [40, 203].

[56] Jacques Dixmier, *Quelques résultats et conjectures concerant les séries de Poincaré des invariants des formes binaires*, in: *Sém. d'Algèbre P. Dubreil and M. P. Malliavin*, vol. 1146 of *Lecture Notes in Mathematics*, pp. 127–160, Springer-Verlag, Berlin 1985 [191].

[57] Mátyás Domokos, Pál Hegedűs, *Noether's bound for polynomial invariants of finite groups*, Arch. Math. **74** (2000), 161–167 [111, 112].

[58] Yves Eichenlaub, *Problèmes effectifs de la théorie de Galois en degrés 8 à 11*, Dissertation, Université Bordeaux 1, 1996 [211].

[59] David Eisenbud, *Commutative Algebra with a View Toward Algebraic Geometry*, Springer-Verlag, New York 1995 [7, 15, 17, 21, 22, 24, 35, 61, 63, 68, 80, 99].

[60] David Eisenbud, Craig Huneke, Wolmer V. Vasconcelos, *Direct methods for primary decomposition*, Invent. Math. **110** (1992), 207–235 [27].

[61] A. Elashvili, M. Jibladze, *Hermite reciprocity for the regular representations of cyclic groups*, Indag. Math., New Ser. **9** (1998), 233–238 [224].

[62] A. Elashvili, M. Jibladze, D. Pataraia, *Combinatorics of necklaces and "Hermite reciprocity"*, J. Algebr. Comb. **10** (1999), 173–188 [224].

[63] Geir Ellingsrud, Tor Skjelbred, *Profondeur d'anneaux d'invariants en caractéristique p*, Compos. Math. **41** (1980), 233–244 [87, 100, 121].

[64] Leonard Evens, *The Cohomology of Groups*, Oxford University Press, Oxford 1991 [209, 210].

[65] Guenter Ewald, Uwe Wessels, *On the ampleness of invertible sheaves in complete projective toric varieties*, Result. Math. **19** (1991), 275–278 [204].

[66] J. C. Faugère, P. Gianni, D. Lazard, T. Mora, *Efficient computation of zero-dimensional Gröbner bases by change of ordering*, J. Symbolic Comput. **16** (1993), 329–344 [13, 30].

[67] Peter Fleischmann, *A new degree bound for vector invariants of symmetric groups*, Trans. Amer. Math. Soc. **350** (1998), 1703–1712 [115].

[68] Peter Fleischmann, *The Noether bound in invariant theory of finite groups*, Adv. in Math. **156** (2000), 23–32 [108, 110, 111].

[69] Peter Fleischmann, Wolfgang Lempken, *On degree bounds for invariant rings of finite groups over finite fields*, in: *Finite Fields: Theory, Applications, and Algorithms (Waterloo, ON, 1997)*, pp. 33–41, Amer. Math. Soc., Providence, RI 1999 [108].

[70] Luc Florack, *Image Structure*, Computational Imaging and Vision **10**, Kluwer, Dordrecht, Boston, London 1997 [231].

[71] John Fogarty, *On Noether's bound for polynomial invariants of a finite group*, Electron. Res. Announc. Amer. Math. Soc. **7** (2001), 5–7 [108–110].

[72] Benno Fuchssteiner, Waldemar Wiwianka, Klaus Gottheil, Andreas Kemper, Oliver Kluge, Karsten Morisse, Holger Naundorf, Gudrun Oevel, Thorsten Schulze, *MuPAD. Multi Processing Algebra Data Tool. Tutorial. Benutzerhandbuch. MuPAD Version 1.1*, Birkhäuser, Basel 1993 [74].

[73] William Fulton, *Intersection Theory*, Springer-Verlag, Berlin, Heidelberg, New York 1984 [104, 200].

[74] William Fulton, Joe Harris, *Representation Theory: A first Course*, vol. 129 of *Graduate Texts in Mathematics*, Springer-Verlag, New York–Berlin–Heidelberg 1991 [47].

[75] Karin Gatermann, *Computer Algebra Methods for Equivariant Dynamical Systems*, Lecture Notes in Mathematics **1728**, Springer-Verlag, Berlin, Heidelberg 2000 [227].

[76] Karin Gatermann, F. Guyard, *The Symmetry package in Maple*, Available by www at http://www.zib.de/gatermann/symmetry.html, 1996 [73].

[77] Karin Gatermann, F. Guyard, *Gröbner bases, invariant theory and equivariant dynamics*, J. Symbolic Comput. **28** (1999), 275–302 [227].

[78] Karin Gatermann, Reiner Lauterbach, *Automatic classification of normal forms*, Nonlinear Analysis, Theory, Methods, and Applications **34** (1998), 157–190 [228].

[79] Joachim von zur Gathen, Jürgen Gerhard, *Modern Computer Algebra*, Cambridge University Press, Cambridge 1999 [13].

[80] K. O. Geddes, S. R. Czapor, G. Labahn, *Algorithms for Computer Algebra*, Kluver, Boston, Dordrecht, London 1992 [29, 30].

[81] Katharina Geißler, *Zur Berechnung von Galoisgruppen*, Diplomarbeit, Technische Univerität Berlin, 1997 [211].

[82] Katharina Geißler, Jürgen Klüners, *Galois group computation for rational polynomials*, J. Symb. Comput. **30** (2000), 653–674 [210, 211, 213].

[83] Patrizia Gianni, Barry Trager, Gail Zacharias, *Gröbner bases and primary decomposition of polynomial ideals*, J. Symbolic Comput. **6** (1988), 149–267 [27].

[84] Kurt Girstmair, *On the computation of resolvents and Galois groups*, Manuscr. Math. **43** (1983), 289–307 [211].

[85] Andrew M. Gleason, *Weight polynomials of self-dual codes and the MacWilliams identities*, in: *Actes Congr. internat. Math., 3, 1970*, pp. 211–215, Gauthier-Villars, Paris 1971 [226].

[86] Manfred Göbel, *Computing bases for rings of permutation-invariant polynomials*, J. Symbolic Comput. **19** (1995), 285–291 [124–126, 128, 129].

[87] Manfred Göbel, *The invariant package of MAS*, in: Hubert Comon, ed., *Rewriting Techniques and Applications, 8th International Conference*, Lecture Notes in Computer Science **1232**, pp. 327–330, Springer-Verlag, Berlin, Heidelberg, New York 1997 [73, 127].

[88] Manfred Göbel, *On the number of special permutation-invariant orbits and terms*, Appl. Algebra Engrg. Comm. Comput. **8** (1997), 505–509 [127].

[89] Manfred Göbel, *A constructive description of SAGBI bases for polynomial invariants of permutation groups*, J. Symbolic Comput. **26** (1998), 261–272 [128].

[90] David M. Goldschmidt, *Lectures on Character Theory*, Publish or Perish, Inc., Berkeley 1980 [79].

[91] Martin Golubitsky, David G. Schaeffer, *Singularities and Groups in Bifurcation Theory I*, Applied Mathematical Sciences **51**, Springer-Verlag, New York 1985 [227].

[92] Martin Golubitsky, Ian Stewart, David G. Schaeffer, *Singularities and Groups in Bifurcation Theory II*, Applied Mathematical Sciences **69**, Springer, New York 1988 [227].

[93] Roe Goodman, Nolan R. Wallach, *Representations and Invariants of the Classical Groups*, Encyclopedia of Mathematics and its Applications **68**, Cambridge University Press, Cambridge 1998 [2].

[94] Paul Gordan, *Beweis, dass jede Covariante und Invariante einer binären Form eine ganze Funktion mit numerischen Coefficienten einer endlichen Anzahl solcher Formen ist*, J. Reine Angew. Math. **69** (1868), 323–354 [41, 49].

[95] Paul Gordan, *Neuer Beweis des Hilbertschen Satzes über homogene Funktionen*, Nachrichten König. Ges. der Wiss. zu Gött. (1899), 240–242 [1].

[96] Nikolai L. Gordeev, *Finite linear groups whose algebras of invariants are complete intersections*, Izv. Akad. Nauk SSSR Ser. Mat. **50** (1986), 343–392, English translation in: Math. USSR, Izv. **28** (1987), 335–379 [97].

[97] Daniel R. Grayson, Michael E. Stillman, *Macaulay 2, a software system for research in algebraic geometry*, available at http://www.math.uiuc.edu/Macaulay2, 1996 [7].

[98] Gert-Martin Greuel, Gerhard Pfister, *Gröbner bases and algebraic geometry*, in: Bruno Buchberger, Franz Winkler, eds., *Gröbner Bases and Applications*, Cambridge University Press, Cambridge 1998 [34].

[99] Gert-Martin Greuel, Gerhard Pfister, Hannes Schönemann, *Singular version 1.2 user manual*, Reports On Computer Algebra **21**, Centre for Computer Algebra, University of Kaiserslautern, 1998, available at http://www.mathematik.uni-kl.de/~zca/Singular [7, 73].

[100] Frank Grosshans, *Observable groups and Hilbert's fourteenth problem*, Am. J. Math. **95 (1)** (1973), 229–253 [157].

[101] William J. Haboush, *Reductive groups are geometrically reductive*, Ann. of Math. **102** (1975), 67–83 [50].

[102] Robin Hartshorne, *Algebraic Geometry*, Springer-Verlag, New York, Heidelberg, Berlin 1977 [25, 54, 200, 206].

[103] Z. Hashin, *Failure criteria for unidirectional fiber composites*, J. Appl. Mech. **47** (1980), 329–334 [229, 230].

[104] W. Helisch, *Invariantensysteme und Tensorgeneratoren bei Materialtensoren zweiter und vierter Stufe*, Dissertation, RWTH Aachen, Aachen 1993 [229].

[105] Grete Hermann, *Die Frage der endlich vielen Schritte in der Theorie der Polynomideale*, Math. Ann. **95** (1926), 736–788 [117, 118].

[106] Agnes Eileen Heydtmann, finvar.lib: *A Singular-library to compute invariant rings and more*, available at http://www.mathematik.uni-kl.de/ftp/pub/Math/Singular/bin/Singular-1.1-share.tar.gz, 1997 [73].

[107] David Hilbert, *Über die Theorie der algebraischen Formen*, Math. Ann. **36** (1890), 473–534 [1, 21, 49].

[108] David Hilbert, *Über die vollen Invariantensysteme*, Math. Ann. **42** (1893), 313–370 [1, 49, 60, 192, 196].

[109] David Hilbert, *Mathematische Probleme*, Archiv für Math. und Physik **1** (1901), 44–63, Gesammelte Abhandlungen Band III (1970), Springer Verlag, Berlin–Heidelberg–New York, 290–329. [40].

[110] Karin Hiss, *Constructive invariant theory for reductive algebraic groups*, preprint, 1993 [199].

[111] Melvin Hochster, John A. Eagon, *Cohen-Macaulay rings, invariant theory, and the generic perfection of determinantal loci*, Amer. J. of Math. **93** (1971), 1020–1058 [86].

[112] Melvin Hochster, Craig Huneke, *Tight cosure, invariant theory, and the Briançon-Skoda teorem*, J. Amer. Math. Soc. **3** (1990), 31–116 [64, 65].

[113] Melvin Hochster, Joel L. Roberts, *Rings of invariants of reductive groups acting on regular rings are Cohen-Macaulay*, Adv. in Math. **13** (1974), 115–175 [64, 87].

[114] W. V. D. Hodge, D. Pedoe, *Methods of Algebraic Geometry*, Cambridge University Press, Cambridge 1947 [164].

[115] Ian Hughes, Gregor Kemper, *Symmetric powers of modular representations, Hilbert series and degree bounds*, Comm. in Algebra **28** (2000), 2059–2088 [80, 117].

[116] Ian Hughes, Gregor Kemper, *Symmetric powers of modular representations for groups with a Sylow subgroup of prime order*, J. of Algebra **241** (2001), 759–788 [80, 100].

[117] James E. Humphreys, *Introduction to Lie Algebras and Representation Theory*, Springer-Verlag, Berlin, Heidelberg, New York 1980 [169, 171, 187, 244, 246].

[118] James E. Humphreys, *Linear Algebraic Groups*, second edn., Springer-Verlag, Berlin, Heidelberg, New York 1981 [39, 50, 56, 245].

[119] Christoph Jansen, Klaus Lux, Richard Parker, Robert Wilson, *An Atlas of Brauer Characters*, Clarendon Press, Oxford 1995 [85].

[120] M. V. Jarić, L. Michel, R. T. Sharp, *Zeros of covariant vector fields for the point groups: Invariant formulation*, J. Physique **45** (1984), 1–27 [227, 228].

[121] Theo de Jong, *An algorithm for computing the integral closure*, J. Symbolic Comput. **26** (1998), 273–277 [33, 35].

[122] Victor G. Kac, *On the question of describing the orbit space of linear algebraic groups*, Usp. Mat. Nauk **30 (6)** (1975), 173–174, (Russian) [206].

[123] Victor G. Kac, *Some remarks on nilpotent orbits*, J. Algebra **64** (1980), 190–213 [206].

[124] Victor G. Kac, Kei-Ichi Watanabe, *Finite linear groups whose ring of invariants is a complete intersection*, Bull. Amer. Math. Soc. **6** (1982), 221–223 [88].

[125] Victor G. Kac, Vladimir L. Popov, Ernest B. Vinberg, *Sur les groupes linéaires algébriques dont l'algèbre des invariants est libre*, C. R. Acad. Sci., Paris, Ser. A **283** (1976), 875–878 [206].

[126] Kenichi Kanatani, *Group-Theoretical Methods in Image Understanding*, Springer-Verlag, Berlin, Heidelberg, New York 1990 [231].

[127] William M. Kantor, *Subgroups of classical groups generated by long root elements*, Trans. Amer. Math. Soc. **248** (1979), 347–379 [105].

[128] D. Kapur, K. Madlener, *A completion procedure for computing a canonical basis of a k-subalgebra*, in: E. Kaltofen, S. Watt, eds., *Proceedings of Computers and Mathematics* **89**, pp. 1–11, MIT, Cambridge, Mass. 1989 [123].

[129] B. Kazarnovskij, *Newton polyhedra and the Bezout formula for matrix-valued functions of finite-dimensional representations*, Functional Analysis and its Applications **21(4)** (1987), 73–74 [203].

[130] Gregor Kemper, *The* Invar *package for calculating rings of invariants*, Preprint **93-34**, IWR, Heidelberg, 1993 [73].

[131] Gregor Kemper, *Das Noethersche Problem und generische Polynome*, Dissertation, Universität Heidelberg, 1994, also available as: Preprint **94-49**, IWR, Heidelberg, 1994 [216].

[132] Gregor Kemper, *Calculating invariant rings of finite groups over arbitrary fields*, J. Symbolic Comput. **21** (1996), 351–366 [103].

[133] Gregor Kemper, *Lower degree bounds for modular invariants and a question of I. Hughes*, Transformation Groups **3** (1998), 135–144 [83, 115].

[134] Gregor Kemper, *Computational invariant theory*, The Curves Seminar at Queen's, Volume XII, in: Queen's Papers in Pure and Applied Math. **114** (1998), 5–26 [63].

[135] Gregor Kemper, *An algorithm to calculate optimal homogeneous systems of parameters*, J. Symbolic Comput. **27** (1999), 171–184 [81–83].

[136] Gregor Kemper, *Die Cohen-Macaulay-Eigenschaft in der modularen Invariantentheorie*, Habilitationsschrift, Universität Heidelberg, 1999 [137].

[137] Gregor Kemper, *Hilbert series and degree bounds in invariant theory*, in: B. Heinrich Matzat, Gert-Martin Greuel, Gerhard Hiss, eds., *Algorithmic Algebra and Number Theory*, pp. 249–263, Springer-Verlag, Heidelberg 1999 [85, 128, 130].

[138] Gregor Kemper, *On the Cohen-Macaulay property of modular invariant rings*, J. of Algebra **215** (1999), 330–351 [88, 107, 137].

[139] Gregor Kemper, *The calculation of radical ideals in positive characteristic*, Preprint **2000-58**, IWR, Heidelberg, 2000, submitted [27, 30–32].

[140] Gregor Kemper, *A characterization of linearly reductive groups by their invariants*, Transformation Groups **5** (2000), 85–92 [64].

[141] Gregor Kemper, *The depth of invariant rings and cohomology*, with an appendix by Kay Magaard, J. of Algebra **245** (2001), 463–531 [88, 100, 101, 107].

[142] Gregor Kemper, *Loci in quotients by finite groups, pointwise stabilizers and the Buchsbaum property*, J. reine angew. Math. (2001), to appear [137].

[143] Gregor Kemper, Gunter Malle, *The finite irreducible linear groups with polynomial ring of invariants*, Transformation Groups **2** (1997), 57–89 [105–107].

[144] Gregor Kemper, Gunter Malle, *Invariant fields of finite irreducible reflection groups*, Math. Ann. **315** (1999), 569–586 [108, 218].

[145] Gregor Kemper, Elena Mattig, *Generic polynomials with few parameters*, J. Symbolic Comput. **30** (2000), 843–857 [216, 217].

[146] Gregor Kemper, Allan Steel, *Some algorithms in invariant theory of finite groups*, in: P. Dräxler, G.O. Michler, C. M. Ringel, eds., *Computational Methods for Representations of Groups and Algebras, Euroconference in Essen, April 1-5 1997*, Progress in Mathematics **173**, pp. 267–285, Birkhäuser, Basel 1999 [73, 76, 94, 96].

[147] Gregor Kemper, Elmar Körding, Gunter Malle, B. Heinrich Matzat, Denis Vogel, Gabor Wiese, *A database of invariant rings*, Exp. Math. **10** (2001), 537–542 [74].

[148] George Kempf, *The Hochster-Roberts theorem of invariant theory*, Michigan Math. J. **26** (1979), 19–32 [70, 196].

[149] D. Khadzhiev, *Some questions in the theory of vector invariants*, Mat. Sb., Nov. Ser. **72 (3)** (1967), 420–435, English Translation: Math. USSR, Sb. 1, 383–396. [157].

[150] Friedrich Knop, *Der kanonische Modul eines Invariantenringes*, J. Algebra **127** (1989), 40–54 [71].

[151] Friedrich Knop, Peter Littelmann, *Der Grad erzeugender Funktionen von Invariantenringen*, Math. Z. **196** (1987), 211–229 [71].

[152] Hanspeter Kraft, *Geometrische Methoden in der Invariantentheorie*, Aspects of Mathematics **D1**, Vieweg, Braunschweig/Wiesbaden 1985 [1, 2, 46, 64, 180].

[153] Hanspeter Kraft, Peter Slodowy, Tonny A. Springer, eds., *Algebraische Transformationsgruppen und Invariantentheorie*, DMV Seminar **13**, Birkhäuser, Basel 1987 [2].

[154] Heinz Kredel, *MAS: Modula-2 algebra system*, in: V. P. Gerdt, V. A. Rostovtsev, D. V. Shirkov, eds., *Fourth International Conference on Computer Algebra in Physical Research*, pp. 31–34, World Scientific Publishing, Singapore 1990 [73, 127].

[155] Martin Kreuzer, Lorenzo Robbiano, *Computational Commutative Algebra 1*, Springer-Verlag, Berlin 2000 [7, 31].

[156] Teresa Krick, Alessandro Logar, *An algorithm for the computation of the radical of an ideal in the ring of polynomials*, in: Harold F. Mattson, Teo Mora, T. R. N. Rao, eds., *Applied Algebra, Algebraic Algorithms and Error-Correcting Codes (AAECC-9)*, Lect. Notes Comput. Sci. **539**, pp. 195–205, Springer-Verlag, Berlin, Heidelberg, New York 1991 [27].

[157] Joseph P. S. Kung, Gian-Carlo Rota, *The invariant theory of binary forms*, Bull. Am. Math. Soc., New Ser. **10** (1984), 27–85 [1].

[158] Serge Lang, *Algebra*, Addison-Wesley Publishing Co., Reading, Mass. 1985 [42, 218].

[159] Ali Lari-Lavassani, William F. Langford, Koncay Huseyin, Karin Gatermann, *Steady-state mode interactions for D_3 and D_4-symmetric systems*, Dynamics of Continuous, Discrete and Impulsive Systems **6** (1999), 169–209 [228].

[160] H. W. Lenstra, *Rational functions invariant under a finite abelian group*, Invent. Math. **25** (1974), 299–325 [216].

[161] Peter Littelmann, *Koreguläre und äquidimensionale Darstellungen*, J. of Algebra **123 (1)** (1989), 193–222 [206].

[162] Peter Littelmann, Claudio Procesi, *On the Poincaré series of the invariants of binary forms*, J. of Algebra **133** (1990), 490–499 [191].

[163] Domingo Luna, *Slices étales*, Bull. Soc. Math. France **33** (1973), 81–105 [151].

[164] F. Jessie MacWilliams, *A theorem on the distribution of weights in a systematic code*, Bell Syst. Tech. J. **42** (1963), 79–84 [225].

[165] Gunter Malle, B. Heinrich Matzat, *Inverse Galois Theory*, Springer-Verlag, Berlin, Heidelberg 1999 [215].

[166] Sabine Meckbach, Gregor Kemper, *Invariants of textile reinforced composites*, preprint, Universität Gh Kassel, Kassel, 1999 [230].

[167] Stephen A. Mitchell, *Finite complexes with $A(n)$-free cohomology*, Topology **24** (1985), 227–246 [79].

[168] H. Michael Möller, Ferdinando Mora, *Upper and lower bounds for the degree of Gröbner bases*, in: John Fitch, ed., *EUROSAM 84, Proc. Int. Symp. on Symbolic and Algebraic Computation*, Lect. Notes Comput. Sci. **174**, pp. 172–183, Springer-Verlag, Berlin, Heidelberg, New York 1984 [13].

[169] David Mumford, John Fogarty, Frances Kirwan, *Geometric Invariant Theory*, Ergebnisse der Math. und ihrer Grenzgebiete **34**, third edn., Springer-Verlag, Berlin, Heidelberg, New York 1994 [1, 52, 54, 60].

[170] Joseph L. Mundy, Andrew Zisserman, *Geometric Invariance in Computer Vision*, MIT Press, Cambridge, Mass. 1992 [231].

[171] Masayoshi Nagata, *On the 14th problem of Hilbert*, Am. J. Math. **81** (1959), 766–772 [40, 43].

[172] Masayoshi Nagata, *Complete reducibility of rational representations of a matrix group*, J. Math. Kyoto Univ. **1** (1961), 87–99 [51].

[173] Masayoshi Nagata, *Invariants of a group in an affine ring*, J. Math. Kyoto Univ. **3** (1963/1964), 369–377 [50].

[174] Masayoshi Nagata, Takehiko Miyata, *Note on semi-reductive groups*, J. Math. Kyoto Univ. **3** (1963/1964), 379–382 [50].

[175] Haruhisa Nakajima, *Invariants of finite groups generated by pseudo-reflections in positive characteristic*, Tsukuba J. Math. **3** (1979), 109–122 [105].

[176] Haruhisa Nakajima, *Modular representations of abelian groups with regular rings of invariants*, Nagoya Math. J **86** (1982), 229–248 [105].

[177] Haruhisa Nakajima, *Regular rings of invariants of unipotent groups*, J. Algebra **85** (1983), 253–286 [105].

[178] Haruhisa Nakajima, *Quotient singularities which are complete intersections*, Manuscr. Math. **48** (1984), 163–187 [97].

[179] Haruhisa Nakajima, *Quotient complete intersections of affine spaces by finite linear groups*, Nagoya Math. J. **98** (1985), 1–36 [97].

[180] Mara D. Neusel, *Inverse invariant theory and Steenrod operations*, Mem. Amer. Math. Soc. **146** (2000) [132].

[181] Mara D. Neusel, Larry Smith, *Invariant Theory of Finite Groups*, Mathematical Surveys and Monographs **94**, Amer. Math. Soc., 2002 [2, 3].

[182] P. E. Newstead, *Intoduction to Moduli Problems and Orbit Spaces*, Springer-Verlag, Berlin, Heidelberg, New York 1978 [52, 54, 57, 59].

[183] Emmy Noether, *Der Endlichkeitssatz der Invarianten endlicher Gruppen*, Math. Ann. **77** (1916), 89–92 [108, 110].

[184] Emmy Noether, *Gleichungen mit vorgeschriebener Gruppe*, Math. Ann. **78** (1918), 221–229 [216].

[185] Emmy Noether, *Der Endlichkeitssatz der Invarianten endlicher linearer Gruppen der Charakteristik p*, Nachr. Ges. Wiss. Göttingen (1926), 28–35 [74].

[186] A. M. Popov, *Finite isotropy subgroups in general position in irreducible semisimple linear Lie groups*, Tr. Mosk. Math. O.-va **50** (1985), 209–248, English transl.: Trans. Mosc. Math. Soc. 1988 (1988), 205–249 [206].

[187] A. M. Popov, *Finite isotropy subgroups in general position in simple linear Lie groups*, Tr. Mosk. Math. O.-va **48** (1985), 7–59, English transl.: Trans. Mosc. Math. Soc. 1986 (1988), 3–63 [206].

[188] Vladimir L. Popov, *Representations with a free module of covariants*, Funkts. Anal. Prilozh. **10** (1976), 91–92, English transl.: Funct. Anal. Appl. **10** (1997), 242–244 [206].

[189] Vladimir L. Popov, *On Hilbert's theorem on invariants*, Dokl. Akad. Nauk SSSR **249** (1979), English translation Soviet Math. Dokl. **20** (1979), 1318–1322 [51].

[190] Vladimir L. Popov, *Constructive Invariant Theory*, Astérique **87–88** (1981), 303–334 [71, 197, 199].

[191] Vladimir L. Popov, *The constructive theory of invariants*, Math. USSR Izvest. **10** (1982), 359–376 [71, 197, 199].

[192] Vladimir L. Popov, *A finiteness theorem for representations with a free algebra of invariants*, Math. USSR Izvest. **20** (1983), 333–354 [70].

[193] Vladimir L. Popov, *Groups, Generators, Syzygies and Orbits in Invariant Theory*, vol. 100, AMS, 1991 [2, 71].

[194] Vladimir L. Popov, Ernest B. Vinberg, *Invariant theory*, in: N. N. Parshin, I. R. Shafarevich, eds., *Algebraic Geometry IV*, Encyclopaedia of Mathematical Sciences **55**, Springer-Verlag, Berlin, Heidelberg 1994 [2, 43, 156, 158, 162, 163, 206, 207].

[195] Maurice Pouzet, *Quelques remarques sur les résultats de Tutte concernant le problème de Ulam*, Publ. Dép. Math. (Lyon) **14** (1977), 1–8 [221].

[196] Maurice Pouzet, Nicolas M. Thiéry, *Invariants algébriques de graphes et reconstruction*, Comptes Rendus de l'Académie des Sciences (2001), to appear [222].

[197] Eric M. Rains, N.J.A. Sloane, *Self-dual codes*, in: Vera S. Pless, ed., *Handbook of Coding Theory*, vol. 1, pp. 177–294, Elsevier, Amsterdam 1998 [226].

[198] Victor Reiner, Larry Smith, *Systems of parameters for rings of invariants*, preprint, Göttingen, 1996 [81].

[199] Thomas H. Reiss, *Recognizing Planar Objects Using Invariant Image Features*, Lecture Notes in Computer Science **676**, Springer-Verlag, Berlin 1993 [231].

[200] David R. Richman, *On vector invariants over finite fields*, Adv. in Math. **81** (1990), 30–65 [112, 120, 121].

[201] David R. Richman, *Explicit generators of the invariants of finite groups*, Adv. in Math. **124** (1996), 49–76 [108].

[202] David R. Richman, *Invariants of finite groups over fields of characteristic p*, Adv. in Math. **124** (1996), 25–48 [112, 113, 115].

[203] Lorenzo Robbiano, Moss Sweedler, *Subalgebra bases*, in: W. Bruns, A. Simis, eds., *Commutative Algebra*, Lecture Notes in Math. **1430**, pp. 61–87, Springer-Verlag, New York 1990 [123, 159].

[204] Paul Roberts, *An infinitely generated symbolic blow-up in a power series ring and a new counterexample to Hilbert's fourteenth problem*, J. Algebra **132** (1990), 461–473 [44].

[205] Charles A. Rothwell, Andrew Zisserman, David A. Forsyth, Joseph L. Mundy, *Fast recognition using algebraic invariants*, in: Joseph L. Mundy, Andrew Zisserman, eds., *Geometric Invariance in Computer Vision*, pp. 398–407, MIT Press, Cambridge, Mass. 1992 [232].

[206] David J. Saltman, *Noether's problem over an algebraically closed field*, Invent. Math. **77** (1984), 71–84 [216].

[207] David J. Saltman, *Groups acting on fields: Noether's problem*, Contemp. Mathematics **43** (1985), 267–277 [216].

[208] D. H. Sattinger, *Group Theoretic Methods in Bifurcation Theory*, Lecture Notes in Math. **762**, Springer-Verlag, Berlin, Heidelberg, New York 1979 [227].

[209] Barbara J. Schmid, *Finite groups and invariant theory*, in: P. Dubreil, M.-P. Malliavin, eds., *Topics in Invariant Theory*, Lect. Notes Math. **1478**, Springer-Verlag, Berlin, Heidelberg, New York 1991 [111].

[210] Frank-Olaf Schreyer, *Die Berechnung von Syzygien mit dem verallgemeinerten Weierstrass'schen Divisionssatz*, Diplomarbeit, Universität Hamburg, 1980 [19].

[211] Issai Schur, *Vorlesungen über Invariantentheorie*, Springer-Verlag, Berlin, Heidelberg, New York 1968 [64].

[212] Gerald Schwarz, *Representations of simple Lie groups with regular rings of invariants*, Invent. Math. **49** (1978), 167–197 [206].

[213] F. Seidelmann, *Die Gesamtheit der kubischen und biquadratischen Gleichungen mit Affekt bei beliebigem Rationalitätsbereich*, Math. Ann. **78** (1918), 230–233 [217].

[214] A. Seidenberg, *Constructions in algebra*, Trans. Amer. Math. Soc. **197** (1974), 273–313 [30].

[215] Jean-Pierre Serre, *Groupes finis d'automorphismes d'anneaux locaux réguliers*, in: *Colloque d'Algèbre*, pp. 8–01 – 8–11, Secrétariat mathématique, Paris 1968 [104, 105].

[216] C. S. Seshadri, *On a theorem of Weitzenböck in invariant theory*, J. Math. Kyoto Univ. **1** (1962), 403–409 [44].

[217] R. James Shank, *S.A.G.B.I. bases for rings of formal modular seminvariants*, Comment. Math. Helvetici **73** (1998), 548–565 [79, 128].

[218] R. James Shank, *SAGBI bases in modular invariant theory*, presented at the Workshop on Symbolic Computation in Geometry and Analysis, MSRI (Berkeley), October 1998 [129].

[219] R. James Shank, David L. Wehlau, *On the depth of the invariants of the symmetric power representations of $SL_2(\mathbf{F}_p)$*, J. of Algebra **218** (1999), 642–653 [100].

[220] R. James Shank, David L. Wehlau, *Computing modular invariants of p-groups*, preprint, Queen's University, Kingston, Ontario, 2001 [120, 129, 131].

[221] G. C. Shephard, J. A. Todd, *Finite unitary reflection groups*, Canad. J. Math. **6** (1954), 274–304 [104, 106, 131, 136, 225].

[222] Tetsuji Shioda, *On the graded ring of invariants of binary octavics*, Am. J. Math. **89** (1967), 1022–1046 [42].

[223] N. J. A. Sloane, *Error-correcting codes and invariant theory: New applications of a nineteenth-century technique*, Amer. Math. Monthly **84** (1977), 82–107 [84, 224, 226].

[224] G.F. Smith, M.M. Smith, R.S. Rivlin, *Integrity bases for a symmetric tensor and a vector. The crystal classes*, Arch. Ration. Mech. Anal. **12** (1963), 93–133 [230].

[225] Larry Smith, *Polynomial Invariants of Finite Groups*, A. K. Peters, Wellesley, Mass. 1995 [2, 73, 83, 111, 129, 132, 223].

[226] Larry Smith, *Noether's bound in the invariant theory of finite groups*, Arch. der Math. **66** (1996), 89–92 [108].

[227] Larry Smith, *Some rings of invariants that are Cohen-Macaulay*, Can. Math. Bull. **39** (1996), 238–240 [87].

[228] Larry Smith, *Polynomial invariants of finite groups: A survey of recent developments*, Bull. Amer. Math. Soc. **34** (1997), 211–250 [79].

[229] Larry Smith, *Putting the squeeze on the Noether gap—the case of the alternating groups A_n*, Math. Ann. **315** (1999), 503–510 [108].

[230] Louis Solomon, *Partition identities and invariants of finite groups*, J. Comb. Theory, Ser. A **23** (1977), 148–175 [223].

[231] Tonny A. Springer, *Invariant Theory*, vol. 585 of *Lect. Notes Math.*, Springer-Verlag, New York 1977 [2, 42, 64].

[232] Tonny A. Springer, *On the Invariant Theory of* SU_2, Nederl. Akad. Wetensch. Indag. Math. **42 (3)** (1980), 339–345 [183, 191].

[233] Tonny A. Springer, *Linear Algebraic Groups*, vol. 9 of *Progress in Mathematics*, Birkhäuser Boston, Inc., Boston 1998 [39, 50].

[234] Richard P. Stanley, *Hilbert functions of graded algebras*, Adv. Math. **28** (1978), 57–83 [103].

[235] Richard P. Stanley, *Combinatorics and invariant theory*, in: *Relations Between Combinatorics and Other Parts of Mathematics (Columbus, Ohio 1978)*, Proc. Symp. Pure Math. **34**, pp. 345–355, Am. Math. Soc., Providence, RI 1979 [222].

[236] Richard P. Stanley, *Invariants of finite groups and their applications to combinatorics*, Bull. Amer. Math. Soc. **1(3)** (1979), 475–511 [81, 103, 115, 222, 223].

[237] Richard P. Stauduhar, *The determination of Galois groups*, Math. Comput. **27** (1973), 981–996 [210].

[238] Michael Stillman, Harrison Tsai, *Using SAGBI bases to compute invariants*, J. Pure Appl. Algebra (1999), 285–302 [129].

[239] Bernd Sturmfels, *Algorithms in Invariant Theory*, Springer-Verlag, Wien, New York 1993 [2, 3, 7, 148, 159, 164, 209, 218].

[240] Bernd Sturmfels, Neil White, *Gröbner bases and Invariant Theory*, Adv. Math. **76** (1989), 245–259 [164].

[241] Richard G. Swan, *Invariant rational functions and a problem of Steenrod*, Invent. Math. **7** (1969), 148–158 [216].

[242] Moss Sweedler, *Using Gröbner bases to determine the algebraic and transcendental nature of field extensions: Return of the killer tag variables*, in: Gérard Cohen, Teo Mora, Oscar Moreno, eds., *Applied Algebra, Algebraic Algorithms and Error-Correcting Codes*, Springer-Verlag, Berlin, Heidelberg, New York 1993 [133, 215].

[243] Gabriel Taubin, David B. Cooper, *Object recognition based on moment (or algerbaic) invariants*, in: Joseph L. Mundy, Andrew Zisserman, eds., *Geometric Invariance in Computer Vision*, pp. 375–397, MIT Press, Cambridge, Mass. 1992 [233].

[244] Jacques Thévenaz, *G-Algebras and Modular Representation Theory*, Clarendon Press, Oxford 1995 [130].

[245] Nicolas M. Thiéry, `PerMuVAR`, *a library for* `mupad` *for computing in invariant rings of permutation groups*, `http://permuvar.sf.net/` [74, 220].

[246] Nicolas M. Thiéry, *Invariants algébriques de graphes et reconstruction; une étude expérimentale*, Dissertation, Université Lyon I, Lyon 1999 [222].

[247] Nicolas M. Thiéry, *Algebraic invariants of graphs; a study based on computer exploration*, SIGSAM Bulletin **34** (2000), 9–20 [220, 221].

[248] Nicolas M. Thiéry, *Computing minimal generating sets of invariant rings of permutation groups with SAGBI-Gröbner basis*, in: *International Conference DM-CCG, Discrete Models - Combinatorics, Computation and Geometry, Paris, July 2-5 2001*, 2001, to appear [221].

[249] S. M. Ulam, *A Collection of Mathematical Problems*, Interscience Publishers, New York, London 1960 [221].

[250] Wolmer V. Vasconcelos, *Computational Methods in Commutative Algebra and Algebraic Geometry*, Algorithms and Computation in Mathematics **2**, Springer-Verlag, Berlin, Heidelberg, New York 1998 [7, 14–16].

[251] Ascher Wagner, *Collineation groups generated by homologies of order greater than 2*, Geom. Dedicata **7** (1978), 387–398 [105].

[252] Ascher Wagner, *Determination of the finite primitive reflection groups over an arbitrary field of characteristic not 2, I*, Geom. Dedicata **9** (1980), 239–253 [105].

[253] Keiichi Watanabe, *Certain invariant subrings are Gorenstein I*, Osaka J. Math. **11** (1974), 1–8 [103].

[254] Keiichi Watanabe, *Certain invariant subrings are Gorenstein II*, Osaka J. Math. **11** (1974), 379–388 [103].

[255] David Wehlau, *Constructive invariant theory for tori*, Ann. Inst. Fourier **43**, 4 (1993) [203, 204].

[256] David Wehlau, *Equidimensional representations of 2-simple groups*, J. Algebra **154** (1993), 437–489 [206].

[257] R. Weitzenböck, *Über die Invarianten von linearen Gruppen*, Acta Math. **58** (1932), 231–293 [44].

[258] Hermann Weyl, *Theorie der Darstellung kontinuierlicher halbeinfacher Gruppen durch lineare Transformationen I*, Math. Z. **23** (1925), 271–309 [181].

[259] Hermann Weyl, *Theorie der Darstellung kontinuierlicher halbeinfacher Gruppen durch lineare Transformationen II,III,IV*, Math. Z. **24** (1926), 328–376, 377–395, 789–791 [181].

[260] Clarence Wilkerson, *A primer on the Dickson invariants*, Amer. Math. Soc. Contemp. Math. Series **19** (1983), 421–434 [215, 217].

[261] R. M. W. Wood, *Differential operators and the Steenrod algebra*, Proc. Lond. Math. Soc. **75** (1997), 194–220 [132].

[262] Patrick A. Worfolk, *Zeros of equivariant vector fields: Algorithms for an invariant approach*, J. Symbolic Comput. **17** (1994), 487–511 [219, 227].

[263] Alfred Young, *On quantitive substitutional analysis (3rd paper)*, Proc. London Math. Soc. **28** (1928), 255–292 [164].

[264] A.E. Zalesskii, V.N. Serezkin, *Linear groups generated by transvections*, Math. USSR, Izv. **10** (1976), 25–46 [105].

[265] A.E. Zalesskii, V.N. Serezkin, *Finite linear groups generated by reflections*, Math. USSR, Izv. **17** (1981), 477–503 [105].

[266] Oscar Zariski, *Interpretations algebrico-geometriques du quartorzieme probleme de Hilbert*, Bull. Sci. Math. **78** (1954), 155–168 [44].

[267] D. P. Zhelobenko, *Compact Lie Groups and Their Representations*, Transl. Math. Monogr. **40**, American Mathematical Society, Providence 1973 [181].

Notation

Index